ANIMAL BIOTECHNOLOGY

ANIMAL BIOTECHNOLOGY

Editor

Gebhard Ohme

scitus
academics

Animal Biotechnology

Edited by **Gebhard Ohme**

This Book is Printed on Acid-Free Paper. ♾

ISBN: 9781681179193

Library of Congress Control Number: 2017948948

SCITUS Academics LLC,

616, Corporate Way, Suite 2, 4766,

Valley Cottage, NY 10989

www.scitusacademics.com

Notice

Printed in United States of America.

Preface

Animal biotechnology is the use of science and engineering to modify living organisms. The goal is to make products, to improve animals and to develop microorganisms for specific agricultural uses. Animals are used in many ways in biotechnology. Biotechnology provides new tools for improving human health and animal health and welfare and increasing livestock productivity. Biotechnology improves the food we eat - meat, milk and eggs. Biotechnology can improve an animal's impact on the environment. The animal biotechnology in use today is built on a long history. Some of the first biotechnology in use includes traditional breeding techniques that date back to 5000 B.C.E. Such techniques include crossing diverse strains of animals to produce greater genetic variety. The offspring from these crosses then are bred selectively to produce the greatest number of desirable traits. For example, female horses have been bred with male donkeys to produce mules, and male horses have been bred with female donkeys to produce hinnies, for use as work animals, for the past 3,000 years. This method continues to be used today. Animal biotechnology in use today is based on the science of genetic engineering. Under the umbrella of genetic engineering exist other technologies, such as transgenics and cloning, that also are used in animal biotechnology. Transgenics is the transferal of a specific gene from one organism to another. Gene splicing is used to introduce one or more genes of an organism into a second organism. A transgenic animal is created once the second organism incorporates the new DNA into its own genetic material.

This Volume 'Animal Biotechnology' encompasses all aspects of animal science and biotechnology, including animal genetics and breeding, animal reproduction and physiology, animal nutrition and biochemistry, feed processing and testing, and animal biotechnology. It is designed to introduce research achievements in animal science and technology to the international community, and to strengthen academic field. The topics including advances in animal assisted reproduction, conservation of animal genetic diversity, biotechnology of gut microbiome, nutrigenomics, fodder and feed biotechnology, immunogenetics, genomics and its applications in animals. We hope that a wide variety of scientists, researchers, and others may benefit from this book.

Table of Contents

CHAPTER 1

An Animal Cell Culture: Advance Technology for Modern Research

Rajeev Nema[1], Sarita Khare[2]

[1]Department of Zoology and Biotechnology, Sarojini Naidu Government Girls Post Graduate (Autonomous) College, Bhopal, India

[2]Center for Microbiology & Bio-Technology Research and Training, Bhopal, India

ABSTRACT

At the present time animal cell culture is more significant and multifarious application tool for current research streams. A lot of field assorted from animal cell culture such: stem cell biology, IVF technology, cancer cell biology, monoclonal antibody production, recombinant protein production, gene therapy, vaccine manufacturing, novel drug selection and improvement. In this review conclude animal cell culture as well as its requirements.

Keywords: Animal Cell Culture

1. INTRODUCTION

Tissue Culture is a general idiom used for the removal of cells, tissues, or organs from an animal and their next placement into an artificial environment conductive to growth. Tissue culture is capable of clear as the growth of tissue or cell separate from the organism. It is also known as techniques of keeping tissues alive and growing in an appropriate culture medium. Growing tissues of living organism outside the body is made possible in an appropriate culture medium, containing mixture of nutrient either in solid or liquid form. At present remarkable association in the field of animal cell culture done by researchers: Human insulin became the earliest recombinant protein to be approved as a therapeutic agent, Human growth hormone produced from recombinant bacteria was established in favor of beneficial use, plasminogen activator (tPA) early recombinant animal cells became commercially accessible, [1]. Historical background in the field of animal cell culture demonstrated in **Table 1**.

Table 1. Historical background of animal cell culture.

Year	Significant work	Scientist
1885	Maintained embryonic chick cells in a saline culture	Roux
1897	Demonstrated the survival of cells isolated from blood and connective tissue in serum and plasma	Loeb
1907	Cultivated frog nerve cells in a lymph clot held by the "hanging drop" method and observed the growth of nerve fibers *in vitro* for several weeks	Harrisone
1911	First liquid media consisted of sea water, serum, embryo extract, salts and peptones	Lewis and Lewwis
1916	Proteolytic enzyme trypsin for the subculture of adherent cells	Rous and Jones
1923	T-flask as the first specifically designed cell culture vessel	Carrel and Baker
1948	Isolated mouse l fibroblasts which formed clones from single cells	Earle
1949	Polio virus could be grown on human embryonic cells in culture	Enders
1952	Continuous cell line from a human cervical carcinoma known as hela (helen lane) cells	Gey
1955	Nutrient requirements	Eagle
1964	Hat medium for cell selection	Littlefield
1975	First hybridoma capable of secreting a monoclonal antibody	Kohlar and Milstein

2. EQUIPMENT REQUIRED FOR CELL CULTURE

2.1. Laminar Flow Hoods

There are two types of laminar flow hoods, vertical and horizontal. The vertical hood, also well-known as a biology safety cabinet, is effective for harmful organisms since Horizontal hoods are designed such that the air flows directly at the operator hence they are not useful for working with hazardous organisms but are the best protection for cultures. Both types of hoods have continuous displacement of air that passes through a HEPA (high efficiency particle) filter use for the purpose of removes particulates from the air. In a vertical hood, the filtered air blows down from the top of the cabinet; in a horizontal hood, the filtered air blows out at the operator in a horizontal fashion. The hoods are equipped with a short-wave UV light that can be turned on for a few minutes to sterilize the surfaces of the hood, but be aware that only exposed surfaces will be accessible to the UV light. Do not put your hands or face near the hood when the UV light is on as the short wave light can cause skin and eye damage. The hoods should be turned on about 10 - 20 minutes before being used (**Figure 1**).

2.2. CO_2 Incubators

Cells are grown-up in an atmosphere of 5% - 10% CO_2 because the medium used is buffered with sodium bicarbonate/carbonic acid and the pH must be firmly maintained. Cells are thought to left out of the incubator for as undersized time as possible and the incubator doors should not be opened for very long. The humidity must also be maintained for individuals cells' growing in tissue culture dishes so a pot of water is kept filled at the entire times (**Figure 2**).

Figure 1. Laminar flow hoods.

2.3. Microscopes

Inverted phase contrast microscopes used for visualizing the cells. Microscopes must be kept enclosed and the lights turned down at the same time as not in use. Because the cells are found on bottom of the tissue culture flask that is by use of an inverted microscope is important to absorb cell culture in vitro. The culture media remains above the growing cells plats. If such plates are put over the stage of an ordinary microscope, the growing cells, at bottom cannot be observed. Therefore, the inverted microscope is used for the intention (**Figure 3**).

2.4. Vessels

Anchorage dependent cells have compulsory of a nontoxic, biologically inert, and optically visible surface that will allow cells to attach and allow improvement for the duration of growth. These consist of petri dishes, multiwell plates, microtiter plates, roller bottles, and screwcap flasks—T-25, T-75, T-150 (cm^2 of surface area) (**Figure 4**).

Figure 2. CO_2 incubators.

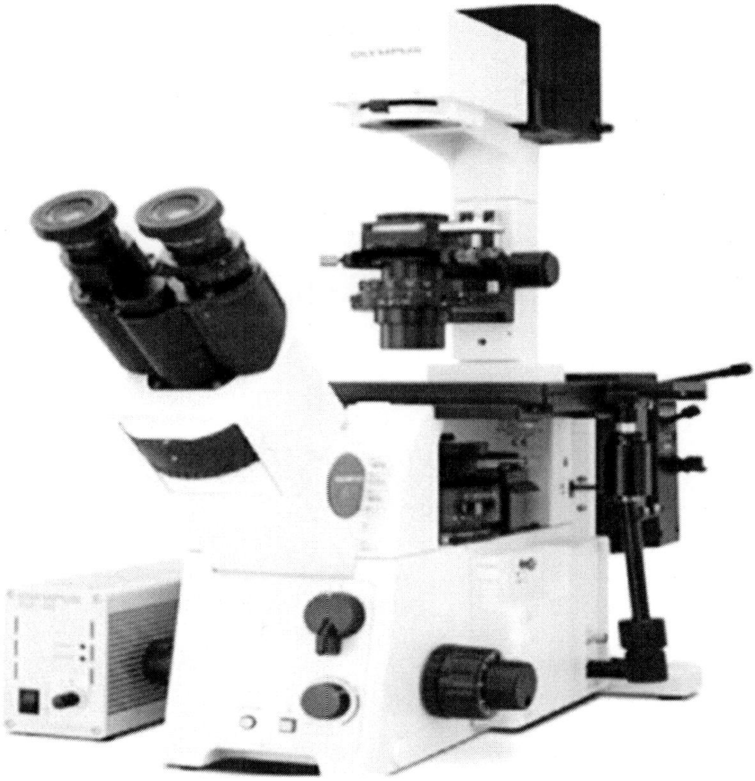

Figure 3. Inverted microscopes.

2.5. Centrifuges

There are different types of centrifuges based on speed. A low speed centrifuge is needed for most of the cell culture. The separated beads of cells are disrupted simply by a gentle breaking action. Frequently cells are centrifuged at 20°C because of motor evolves heat which rises the temperature; therefore make use of low temperature centrifugation is preferred so that the cells should not be exposed to elevated temperature (**Figure 5**).

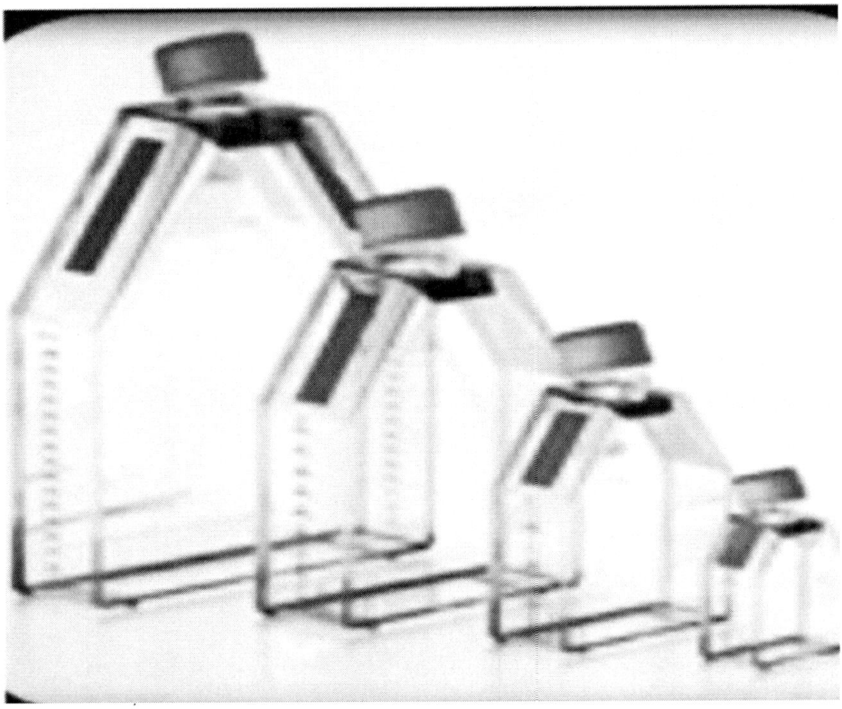

Figure 4. Vessels.

2.6. Freeze

Freezing or solidification is a phase change in which a liquid turns into a solid when its temperature is lowered under its freezing point. The render null and void procedure is melting. Human gametes and embryos can survive freezing and are viable for up to 10 years, a process known as cryopreservation. Investigational attempts to freeze human beings for later revitalization are known as cryonics (**Figure 6**).

3. SUBSTRATES IN FAVOR OF CELL DEVELOPMENT

There are numerous types of vertebrate cell that have need of support for their development in vitro otherwise they will not grow appropriately. Such cell are called anchorage-dependent cell. Used for that reason a large number of substrate which possibly will necessitate for their enlargement (e.g. glass, palladium, metallic surfaces), nonadhesive (e.g. agar agrose, etc.).

4. MEDIA REQUIREMENT FOR CELL CULTURE

When artificial environment formed in the laboratory is in generally known at the same time as media. A media comprises an appropriate source of energy for the cells which they can easily utilize and compounds which regulate the cell cycle. The choice of media is cell type specific and often empirical and there is no "all purpose" medium. It should provide many nutrients, buffering capacity, isotonic, and should be sterile. Characteristics and compositions of the cell culture media vary depending on the particular cellular Requirements. Important parameters include osmolarity, pH, and nutrient formulations.

Figure 5. Centrifuges.

4.1. Basic Components in the Culture Media

Most animal cell culture media are generally having following 10 basic components and they are as follows: Energy sources: Glucose, Fructose, Amino acids, Nitrogen sources: Amino acids.

Figure 6. Freezer.

The various types of media used for tissue culture may be grouped into two broad categories:

1) Natural media;

2) Artificial media.

4.1.1. Natural Media

These media consist solely of naturally occurring biological fluids and are of the following three types:

1) Clots;

2) Biological fluids;

3) Tissue extracts.

4.1.2. Artificial Media

Different artificial media have been devised to serve one of the following purposes:

1) Immediate survival (a balanced salt solution, with specified pH and osmotic pressure is adequate);

2) Prolonged survival (a balanced salt solution supplemented with serum, or with suitable formulation of organic compounds);

3) Indefinite growth;

4) Specialized functions.

4.2. A Variety of Artificial Media Developed for Cell Cultures May Be Grouped into the Subsequent Four Classes

1) Serum containing media;

2) Serum free media;

3) Chemically defined media;

4) Protein free media.

5. CULTURE ENVIRONMENTS

One of the major advantages of cell culture is the capability to manipulate the physicochemical (i.e., temperature, pH, osmotic pressure, O_2 and CO_2 tension) and the physiological environment (i.e., hormone and nutriaent concentrations) in which the cells proliferate. Culture environment is a very responsible for cell growth and their maintenance. (Invitrogen Cell Culture Basics). Some specific part discussed below:

5.1. pH

Most normal mammalian cell lines grow well at pH 7.4, and there is very little variability among different cell strains. However, some transformed cell lines have been shown to grow better at slightly more acidic environments (pH 7.0 - 7.4), and some normal fibroblast cell lines prefer slightly more basic environments (pH 7.4 - 7.7). In laboratory pH control by pH meter (**Figure 7**).

5.2. CO_2

CO_2-bicarbonate based buffer. For the reason that the pH of the medium is dependent on the delicate balance of dissolved carbon dioxide (CO_2) and bicarbonate (HCO_3), changes in the atmospheric CO_2 can alter the pH of the medium. Most researchers usually use 5% - 7% CO_2 in air; 4% - 10% CO_2 is common for most cell culture experiments. However, each medium has a recommended CO_2 tension and bicarbonate concentration to achieve the correct pH and osmolality; refer to the media manufacturer's instructions for more information. Inside laboratory condition CO_2 concentration controlled by CO_2 incubator shaker (**Figure 8**).

5.3. Temperature

The majority human and mammalian cell lines are maintained at 36°C to 37°C for optimal growth while Avian cell lines need 38.5°C in favor of maximum growth. Even though these cells can also be maintained at 37°C, they will grow further slowly but Cell lines derived from cold-blooded animals (e.g., amphibians, cold-water fish) bear an extensive temperature vary between 15°C and 26°C.

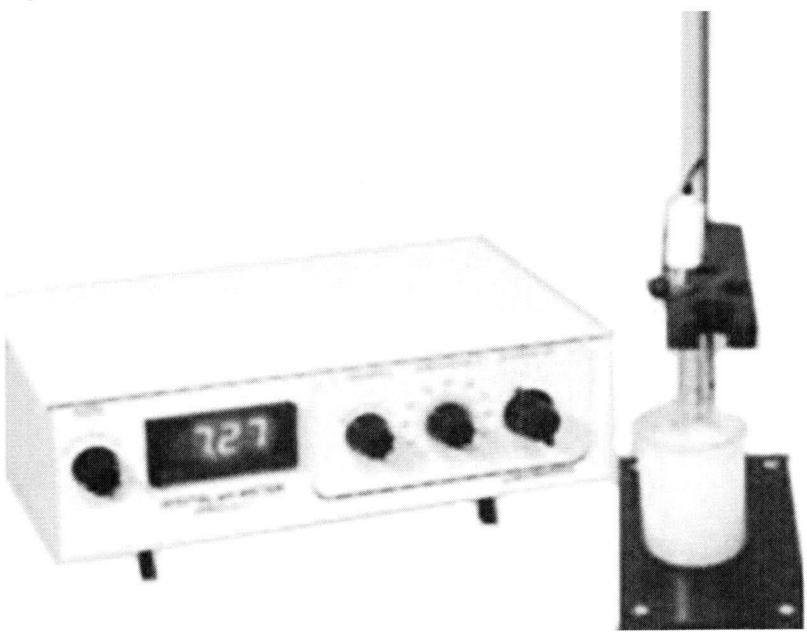

Figure 7. pH meter.

Figure 8. Incubator shaker.

6. TYPES OF CELLS

On the basis of morphology or functional characteristics three type cell considered for cell culture.

6.1. Epithelial Cell

Attached to a substrate and appears flattened and polygonal in shape (**Figure 9**).

6.2. Lymphoblast Cell

Cells do not attach; remain in suspension with a spherical shape (**Figure 10**).

6.3. Fibroblast Cell

Cells attached to a substrate; appears elongated and bipolar (**Figure 11**).

7. PROCEDURE OF CELL CULTURE

7.1. Primary Cell Culture

Primary cell culture is first cultivation of cell in synthetic condition [2]. Primary cultures, which be obtained straight forwardly from an animal furthermore be capable of maintain the Differentiated state for an undersized period (2). Three Basic Steps of primary tissue culture.

Ÿ Isolation of tissue.

Disaggregation of cells—[1] Chemical disaggregation (2) Mechanical disaggregation (**Figure 12**).

Ÿ Incubation in addition to growth.

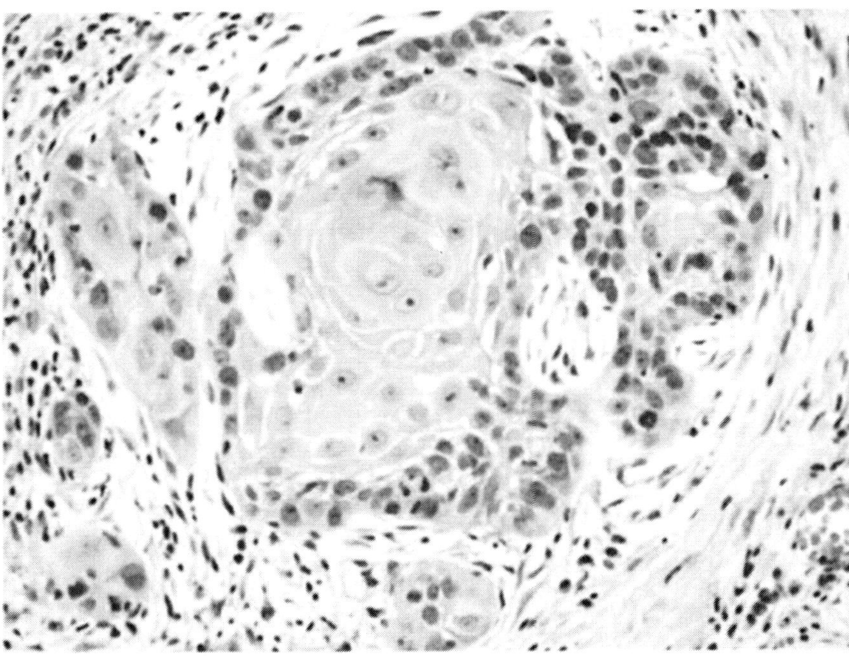

Figure 9. Epithelial cell.

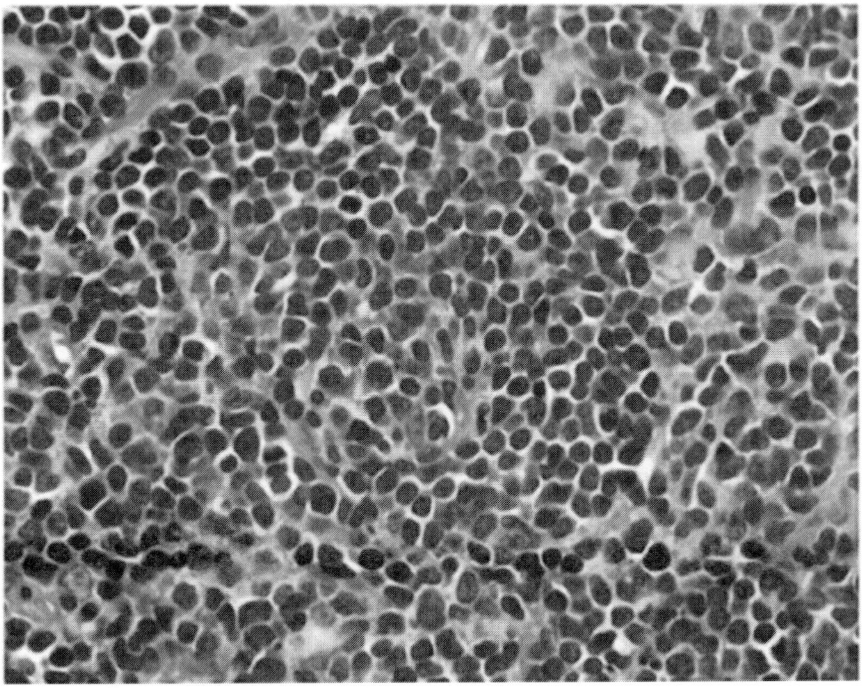

Figure 10. Lymphoblast cell.

7.2. Subculture (Passaging)

In animal cell culture a subculture is a new cell culture made by transferring some or all cells from a previous culture to fresh growth medium. This action is called subculturing or passaging the cells [3].

7.3. Monolayer Culture

At what time the bottom of the culture vessel is covered by means of a continuous layer of cells, frequently one cell in thickness, they are referred to as monolayer cultures [4].

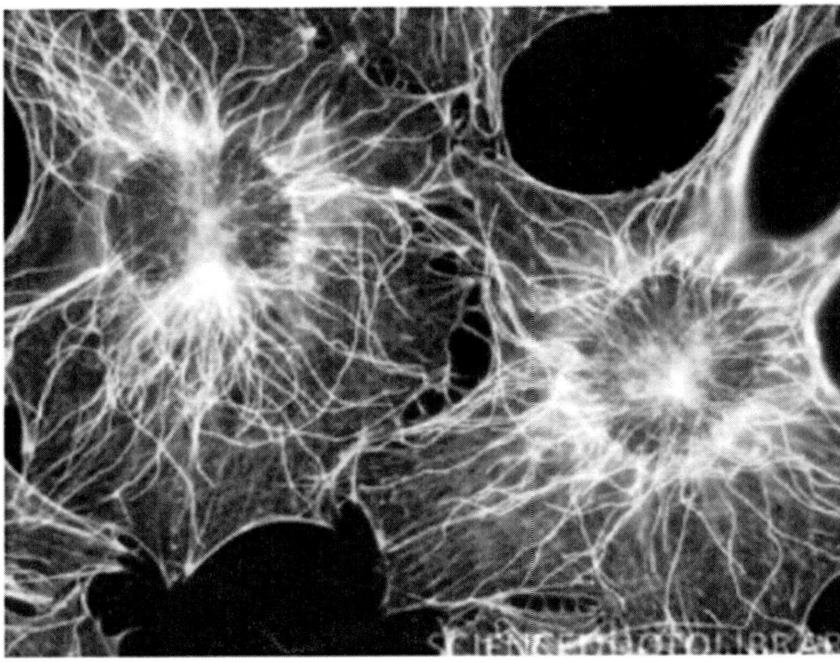

Figure 11. Fibroblast cell.

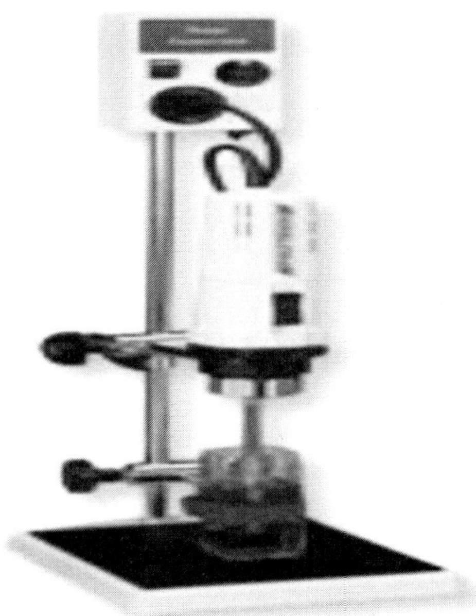

Figure 12. Tissue homogenizing.

7.4. Suspension Cultures

A few of the cells which are non-adhesive e.g. cells of leukemia or convinced cells which can be mechanically kept within suspension, can exist propagated in suspension. There are certain applications in propagation of cells by suspension culture process.

7.5. Type of Cell Culture

7.5.1. Anchorage Dependent Cell Culture

Cells shown to necessitate attachment for growth are set to be anchorage dependent cells. The adherent cells are typically derived from tissues of organs such as kidney where they are immobile in addition to embedded in connective tissue. They cultivate adhering to the cell culture.

7.5.2. Anchorage Independent Cell Culture

Cells which make not required attachment for growth or do not attach to the surface of the culture vessels are anchorage independent cells/suspension cells [5]. Each and every one suspension cultures are derived from cells of the blood system for the reason that these cells are furthermore suspended in plasma in vitro e.g. lymphocytes [6].

8. CELL LINE

A cell line arises from a primary culture at the time of the first successful subculture. The term cell line implies that cultures from it consist of lineages of cells originally present in the primary culture [3].

On the basis of the life span of culture, the cell lines are categorized into two types.

8.1. Finite Cell Lines

Cell lines which encompass a restricted life span and exit from beginning to end a restricted number of cell generations (frequently 20 - 80 population doublings) be wellknown as finite cell lines.

8.2. Continuous Cell Lines

Cell lines transformed under laboratory surroundings or in vitro culture environment give rise in the direction of continuous cell lines. The cell lines demonstrate the property of ploidy (aneupliody or heteroploidy), lack of contact inhibition and anchorage dependence. They produce in monolayer or suspension type. The growth rate is fast and doubling-up time is 12 - 24 hours.

9. PRESERVATION AND STORAGE

Cryopreservation

Liquid N_2 is used to preserve tissue culture cells, either in the liquid phase (–196°C) or in the vapor phase (–156°C). Toward minimize the effects of freezing, several precautions are taken. First, a cryoprotective agent which lowers the freezing point, such as glycerol or DMSO, is added. A typical freezing medium is 90% serum, 10% DMSO. In addition, it is best to use healthy cells that are growing in log phase and to replace the medium 24 hours before freezing. Also, the cells are slowly cooled from room temperature to –80°C to allow the water to move out of the cells before it freezes. The effect of the isopropanol is to allow the tubes to come to the temperature of the freezer slowly, at about 1°C per minute. Cells are stored at liquid nitrogen temperatures because the growth of ice crystals is retarded below –130°C [7,8] (**Figure 13**).

10. CELL CULTURE POSSIBLE PROBLEMS

Protection of aseptic condition is one of the most complex challenges in tissue culture there are quite a lot of rout to contamination which includes malfunction in the sterilization procedures used for glassware & pipettes, particulates cross contamination of air inside the room, weakly maintained incubation, inappropriate handling.

10.1. Cell Culture and Cross-Contamination

Cell line cross-contamination can be a trouble for scientists working through cultured cells. Studies propose anywhere starting 15% - 20% of the instance; cells used in experiments have been misidentified or contaminated with another cell line [9,10] troubles with cell line crosscontamination have even been detected in lines from the NCI-60 panel, which are used regularly for drug screening studies [11,12]. Number of contaminants, from microbiological, most of the fungal and bacterial contamination quickly overwhelms a culture and is usually visible to the naked eye within a short period of time.

Figure 13. Liquid n_2 tank.

10.2. Bacteria and Fungi Contamination

Microorganisms are the most frequent cell culture contaminators for the reason that they thrive in all environments and are effortlessly moveable from any exterior source such as laboratory instruments, gloves or clothing toward the cells. They grow fast and can be easily seen beneath the microscope. Illustration indicators of contamination include media color change because of a shift in pH, turbidity, presence of non-cellular material, cell vacuolization, or constant cell lysis and death [13].

10.3. Mycoplasma Contamination

Mycoplasma is extremely small bacteria-like organisms that are difficult contaminators of cell cultures. Mycoplasma has the capability to alter the host cell culture's morphology, function, metabolism, growth and attachment to the culture vessel. For that reason, the integrity of any experiments performed with mycoplasma-contaminated cells is doubtful because the host cells are not performing normally [14,15].

10.4. Chemical Contamination

Chemical contamination is the occurrence of any lifeless substance in the cell culture that causes unfavorable effects to the cells. This may comprise impure media, serum or even water which may contain unwanted endotoxins or organic compounds if not purified. In addition, toxic levels of even vital nutrients can be harmful, and chemical contaminants may possibly also come from unclean storage vessel.

10.5. Antibiotics

Accurate working perform, antibiotics should not be used for the routine maintenance of cell lines. In the existence of antibiotics, contamination may be suppressed, but could alter the phenotype or genotype of the cells. Antibiotics are toxic and can alter the biochemistry of the cells. If an infection is not understandable, for the reason that it has been suppressed but not eliminated by antibiotics, all other cultures in the laboratory are at hazard [16].

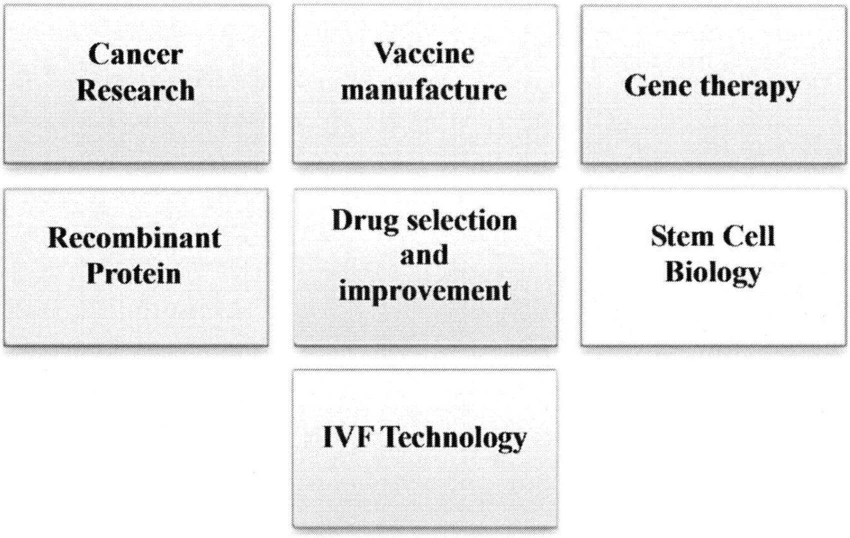

Figure 14. Animal cell culture application.

11. CONCLUSION

At present make use of animal cell culture has undergone a significant spreading out from being a purely investigational procedure to become a conventional technological module of a lot of aspects of biological research. This review summarizes conceptual background and basic techniques of culturing animal cells in a format that is readily easy to get to all researchers in the field. Animal tissue culture functions in therapeutic filed are vast. The evaluation of the cell's response to chemicals, or as a tool to produce cellular-derived protein products that really helps in medical improvement. An animal tissue culture provides an approach to manufacture monoclonal antibody that makes it potential to produce antibody that have specificity controlled toward pathogen. In the field of animal cell culture discovery leads to the opportunity of curing various diseases such as AIDS and cancer. Additional application of animal tissue culture gives a conventional, quick and approachable method for manufacture of well-tolerated and valuable vaccines. Cell culture seems to come with a lot of important towards medical progression. Other than that, there are still differences between in vitro and in vivo system during drug testing and organ transplantation which make it not totally trustworthy.

12. SIGNIFICANCE OF ANIMAL CELL CULTURE

The animal cell cultures are used for a diverse range of research and development. These areas are as **Figure 14**.

13. ACKNOWLEDGEMENTS

Rajeev Nema, Sarita Khare The authors express gratitude to all member of Center for Microbiology & Bio-Technology Research and Training for kind support.

REFERENCES

1. Surachi, U. (1999) Basic techniques in animal cell culture. Drug Delivery System Workshop, Bangkok.
2. Freshney, R.I. (2006) Basic principles of cell culture. John Wiley & Sons, Hoboken.
3. Freshney, R.I. (2005) Culture of animal cells: A manual of basic technique. 5th Edition, Wiley, New York.
4. Phelan, M.C. (1996) Techniques for mammalian cell tissue culture. Current Protocols in Molecular Biology, 3, Appendix 3B.
5. Chen, J.-Y., Penco, S., Ostrowski, J., Balaguer, P., Pons, M., Starrett, J.E., Reczek, P., Chambon, P. and Gronemeyer, H. (1995) RAR-specific agonist/antagonists which dissociate transactivation and AP1 transrepression inhibit anchorage-independent cell proliferation. The EMBO Journal, 14, 1187-1197.
6. Glade, P.R. and Hirschhorn, K. (1970) Products of lymphoid cells in continuous culture. The American Journal of Pathology, 60, 483.
7. Hay, R.J. (1978) Preservation of cell culture stocks in liquid nitrogen. TCA Manual, 4, 787-790.
8. Freshney, R.I. (1994) Culture of animal cells. In: Freshney, R.I., Ed., A Manual of Basic Technique, Wiley-Liss, New York, 387-389.
9. Drexler, H.G. and Uphoff, C.C. (2000) Contamination of cell culture, mycoplasma. Encyclopedia of Cell Technology, 1, 609-627.
10. Cabrera, C.M., et al. (2006) Identity test: Determination of cell line cross-contamination. Cytotechnology, 51, 45-50.
11. Chatterjee, R. (2007) Cell biology: Cases of mistaken identity. Science, 315, 928-931.
12. Liscovitch, M. and Ravid, D. (2006) A case study in misidentification of cancer cell lines: MCF-7/AdrR cells (re-designated NCI/ADR/RES) are derived from OVCAR- 8 human ovarian carcinoma cells. Cancer Letters, 245, 350-352.
13. Cobo, F., Stacey, G.N., Hunt, C., et al. (2005) Microbiological control in stem cell banks: Approaches to standardization. Applied Microbiology and Biotechnology, 68, 456-466.
14. McGarrity, G.J. (1976) Spread and control of mycoplasmal infection of cell cultures. In Vitro Cellular & Developmental Biology—Plant, 12, 643-647.

15. McGarrity, G.J., Vanaman, V. and Sarama, J. (1984) Cytogenetic effects of mycoplasmal infection of cell cultures: A review. In Vitro Cellular & Developmental Biology— Plant, 20, 1-18.

16. Masters, J.R. and Stacey, G.N. (2007) Changing medium and passaging cell lines. Nature Protocols, 2, 2276-2284.

CHAPTER 2

Analysis of the Lifetime and Culling Reasons for AI Boars

Damian Knecht, Anna Jankowska-Mąkosa and Kamil Duziński*

Institute of Animal Breeding, Wroclaw University of Environmental and Life Sciences, Chelmonskiego 38C, 51– 630 Wroclaw, Poland

ABSTRACT

Background

The aim of the study was to analyze the lifetime and culling reasons for boars used in insemination centers (AI centers).

Methods

The data collected from 355 culled boars from 1998 to 2013 included: age at start of semen collection, boar herd life, culling reason, daily gain and lean meat content, and number of ejaculates not meeting sales requirements after dilution. Culling reasons were divided into 7 groups: low semen value (LSV), low or lack of libido (LL), leg problems (LP), infectious diseases (ID), old age (OA), reduced demand for semen from the given boar (RD), and others (OT).

Results

The most common culling reasons for boars were LSV (23.7%) and RD (22.5%). It was observed that the lowest daily gains were noted in boars culled due to OA. Boars culled due to OA and RD were maintained in production for the longest time (over 1000 d), for LSV and ID retention was about 700 d, and due to LL below 400 d. The survival probability was over 0.9 until 1.5 yr, and just over 0.2 until 4 yr. The highest relative frequency was observed in the 36th and 42nd mo of life (over 16%). Hazard risk analysis revealed a more than 10 times higher risk of culling in the case of LL, ID or OT, in comparison to OA.

Conclusions

The results can be used as a direct point of reference for the identification of emerging problems in AI boar exploitation and the development of an appropriate culling policy in AI centers.

Keywords: AI centers Boars Culling Exploitation Management

BACKGROUND

The reasons for culling of boars demonstrate a close relationship with the efficiency and economic profitability of operations in commercial herds and AI centers. With increasing interesting in AI, the most important role started to play AI centers. It was estimated that already 70% of Polish sows are inseminated and 99% using fresh semen, and almost 85% of this demand come from AI centers. An understanding of the reasons for culling is necessary information for the planning and rational management of such units. It should be noted that there are two main forms of boar culling, i.e. unplanned and planned. Unplanned culling (forced), i.e. culling as a result of diseases, sudden falls, behavioral problems, and lameness, clearly adversely affects productivity. In terms of productivity, Safranski [1] noted that the collection and production processes of insemination doses is dependent on a number of factors. These factors hinder the long-term accurate estimation of exploitation predispositions of boar, and thus this creates uncertainty in production. On the other hand, planned culling (decision-making) (old age, poor production results), despite the initial costs incurred, over the long term appears to be the most appropriate and cost-effective decision. Unfortunately, decisions about culling boars from a herd are still taken reluctantly by owners and are also often postponed. Only an appropriate culling schedule and herd replacement program can satisfy the essential prerequisite for the maintenance of production stability and repeatability [2].

Boars are very sensitive to abrupt changes which impair correct semen formation [3]. Preparing detailed records about culling within a specified time (e.g. annual balance) allows for the identification of risks to the boar population, especially in terms of possible diseases, disorders, and behavioral problems [4]. It is important for AI centers to have an active population of boars that produce appropriate ejaculates with high quality and quantity parameters [5], which is also connected with proper preparation of doses [6].

Analysis of culling reasons enables a clear increase in the production efficiency of AI units. Although it might seem that in terms of the profitability of pig production all avenues have already been investigated, research into the control, improvement and acceleration of time required to obtain effective production capacity of boars is still valid. Progress in the production results of modern piggeries may be supported by AI [7]. The development of AI has been made possible due to the constant production of high value semen doses [8]. Smital [9] stated that the economic effectiveness of AI centers is closely dependent on the productivity of boars during exploitation, and is limited in the highest degree by the construction of the testicles, libido, and physical activity (limb, spine defects). However, according to the results of previous studies, the problems of sperm quality are not the only reason for boar culling [3, 4, 10, 11], and this has necessitated a deeper analysis of all possible causes.

Given the above consideration, it can be clearly stated that the choice of boar for AI centers and their accurate monitoring during exploitation are very important issues. Once selected, boars should remain within the active productive population for as long as possible, compensating costs and generating incomes for AI centers [4]. Recouperation of the maintenance and exploitation costs of AI boar stations usually occurs after 2–3 yrs of a boar's life [12]. Only the direct identification of culling reasons can skillfully define the problems faced in herds of boars maintained at AI centers. Unfortunately, this subject is still overlooked in global scientific publications. Most articles focus on an analysis of culling reasons and lifetime of sows and only some concern boars, but these mostly relate to those on breeding farms [3, 4, 10, 11] and not AI boars. Therefore, the aim of our study was to analyze lifetime and culling reasons for boars used in AI center.

METHODS

Experimental location and animals

The study was carried out between 1998–2013 at the Boar Exploitation Station in Częstochowa. The study population included 355 culled boars, whose histories of exploitation were followed from birth until death. The presented population was representative in proportion to the most common breed components used for AI in Poland, such as: Polish Landrace, Polish Large White, Duroc, Pietrain, Hampshire, Duroc × Pietrain, Hampshire × Pietrain, Duroc × Hampshire. The collected data from 1998 to 2013 included: age at start of semen collection, boar herd life, culling reason, daily gain and lean meat content (evaluated before purchase of boar), and number of ejaculates not meeting the requirements of sales after dilution (ejaculates incompatible with requirements). Reasons for culling were divided into 7 groups: low semen value (LSV), low or lack of libido (LL), leg problems (LP), infectious diseases (ID), old age (OA), reduced demand for semen from the given boar (RD), and others (OT). The whole population of boars was divided into experimental groups based on the reason for culling.

Decision process for unplanned and planned culling were different. The structure of unplanned culling were simple and based on the first observation of boars by employees receiving ejaculates. Information were directed to the supervisors and the director, the director usually took the sole decision about unplanned culling. In the case of the planned culling decision-making process was more complex. The long-term replacement plan was developed and approved by the owner and the supervisory board. After all, plan was also constantly being upgraded under the flowing marketing information and the demand for a certain product. On this basis planned culling were made.

Therefore, there were finally 7 groups of culling reasons (LSV, LL, LP, ID, OA, MU, OT). In subsequent years the ratios of culling reasons were similar therefore, the sample period was treated jointly. The overall characteristics of the study population are presented in Table 1.

Table 1. The overall characteristics of the study population of boars ($n = 355$)

Trait	Mean	SD	Min	Max
Age of semen collection entry, d	259.74	33.51	201	458
Boar herd life, d	835.57	503.5	21	2350
Daily gain, g	775.81	94.32	554	1170
Lean meat content, %	61.34	1.94	55	67
Ejaculates incompatible with requirements, n	3.36	5.49	0	37

SD standard deviation, *Min* minimum, *Max* maximum

Daily gain and lean meat content

The assessment of daily gain and lean meat content, before purchase of the boar, was made between d 170 and 210 of life. Daily gains were calculated by dividing the body weight of young boars (during assessment time) by the age on the assessment day. The measurements of the animals for the estimation of daily gains were made using a Mensor WM150P1 electronic scales. Daily gains were standardized to 180 d using a model developed for standardization in accordance with the methodology of Mucha and Różycki [13], in order to reduce age differences during assessment. The percentage of meat content in carcasses was estimated intravitally on the basis of two backfat thickness (points P2 and P4) and lion eye height (point P4) measurements. The measurements were made using a PIGLOG 105 (SFK) ultrasonic device, positioned behind the last rib (between the thoracic and lumbar vertebrae), 3 cm off the midline (point P2) and 8 cm off the midline (point P4). To increase the accuracy of the assessment, standardization of traits was performed to 110 kg body weight. The measurement values were inserted into a specially developed equation [14], which allowed an estimation of the intravital proportion of meat in the carcass on the assessment day. The results for meat content were also standardized to d 180 [13].

Boars performance

Before the start of semen collection, all boars were held in quarantine, the length of which was approximately 37.24 ± 5.02 d. During quarantine semen was collected once a wk just to observation only and not for insemination. Additionally, all boars were exploited in the same manner, developed and adopted according to the methodology of the AI station. Ejaculates were collected by masturbation via the manual method using a container with a filter. The gelatinous fraction was separated. Immediately after collection, the volume of semen was measured using a scalar cylinder. The concentration of spermatozoa was evaluated using a SpermaCue device, Model 12300/0500 (Minitube International, Verona, USA). Based on the semen volume and spermatozoa concentration, the total number of spermatozoa in the ejaculate was calculated. Semen dilution was effected using the same semen extender. Boars until 10 mon gave ejaculates once a week, at the age of 10–14 mon this was three times in two week and from 15 mon twice a week. The average annual replacement rate was 49.8%. Ejaculates were classified as normal for further dilution, when the following requirements were met: color from gray to milky white, flavor specific, lack of foreign admixtures, more than 70% of progressive motile sperm cells, pH 7.0–7.9, morphological abnormality changes to 15% (5% primary, 10%

secondary). Insemination doses of 80 mL contained a constant 2.8×10^9 spermatozoa. Semen was stored at 15 °C for not longer than 48 h.

Housing and feeding

Boars were single-housed and maintained in accordance with the principles of animal welfare [15]. Each individual pen area was 8 m^2/boar. Boars were kept on a solid concrete floor, which was covered with straw. The air temperature in all the boar pens was close to 15 °C (min 12 °C, max 20 °C). Relative humidity was close to 75% (min 65%, max 85%). The air circulation inside the building was equal to 0.15 m/s in Winter and 0.20 m/s in Summer. Preventive care and vaccination was carried out regularly in accordance with the methodology of the unit. The microclimate of the area and ventilation were controlled by computer. Over the whole study period, boars were fed the same all-mash mixture, dosed according to the recommended nutrition standard for boars, with permanent access to water (Table 2).

Statistical analysis

The data was analyzed using the STATISTICA (2014) statistical program. The values in the tables are arithmetical means ($\bar{x}$) and standard deviations (SD). A Chi-squared test was used to examine the significance of differences for the relative frequency (%) of boars removed from the removal groups. Other permanent collected data were checked for normality with the Kolmogorov-Smirnov (K-S) test with the Lilliefors correction. In addition, the Brown-Forsythe test (B-F) determined whether the distributions of the variables had the same variance. An advanced mixed model using the GLM procedure was used and an analysis of variance was conducted for factorial designs to determine the effect of tested parameters by the culling reason. The significance of differences was calculated on the basis of Tukey's multiple range test. The levels of significance of differences were given conventionally: significant $0.01 < P \leq 0.05$ and highly significant $P \leq 0.01$. Pearson's correlation coefficient (r) was calculated between daily gains, lean meat content and age at start of semen collection entry, and boar herd life. Survival analysis was performed using the Cox proportional hazard model. The time variable was defined as age in months from birth. The year effect in herd was included in the analysis. The Cox proportional hazard model expressed risk during t for the tested independent variable system and was expressed by the equation:

Table 2. The nutritional value of 1 kg of compound feed for boars

Item	Value
Dry matter, g	887
Metabolizable energy, MJ	12.5
Crude protein, g	169.63
Fat, g	41.98
Crude fiber, g	59.58
Lysine, g	9.38
Methionine + cystine, g	6.14
Threonine, g	5.9
Tryptophan, g	1.8
Valine, g	5.9
Isoleucine, g	4.5
Ca, g	7.13
P total, g	5.6
Na, g	1.8
Vitamin A, IU	15,000
Vitamin D_3, IU	2000
Vitamin E, mg	100
Vitamin C, mg	1000
Vitamin K, mg	5
Vitamin B_1, mg	2
Vitamin B_2, mg	7
Vitamin B_6, mg	4
Vitamin B_{12}, µg	30
Biotin, µg	400
Folic acid, mg	4
Niacin, mg	30
Pantothenic acid, mg	15
Choline, mg	1185
Mn, mg	50
Fe, mg	145
I, mg	1
Zn, mg	100
Cu, mg	20
Co, µg	500
Se, µg	300

$$h(t) = h_0 e^{\beta i X i}$$

where: h(t) was the risk at a given time interval, h_0 the base risk (the risk obtained if there were no risk factors in the model), and β_i the regression coefficient for the i-th removal reasons. The hazard ratio (HR) for the situation when the risk factor X is present versus the situation when the risk factor X is absent was calculated as:

$$\mathrm{HR} = \frac{h(t, x = 1)}{h(t, x = 0)} = e^{\beta i X i}$$

RESULTS

The results of the assessment of the selected parameters of boar performance, depending on the culling reasons, are presented in Table 3. The most common culling reasons were LSV, followed closely by RD. Differences amounting to only 1.2% ($P > 0.05$) were noted between these reasons. However, of these the two most common culling reasons demonstrated statistically significant differences with other reasons, i.e. LP ($P \leq 0.05$) and ID, OA, LL, OT ($P \leq 0.01$). Boars culled due to LL were characterized by the highest age at the start of semen collection. The difference from the youngest age at the start of semen collection observed for boars culled due to ID was over half mo ($P \leq 0.05$). The longest boar herd life expectancies were noted for boars culled due to OA, and differences with other groups were statistically proven at the level of $P \leq 0.01$. Boar herd life expectancy in excess of 1000 d were achieved by boars culled due to RD. Boars with LSV and ID were held in production for over 700 d, and the shortest period was noted for boars with LL, i.e. less than 400 d. It was observed that the lowest daily gains were noted in boars culled due to OA, and differences with other reasons were at a level from 53.32 g to 68.14 g ($P \leq 0.05$), with the exception of RD ($P > 0.05$). Similar observations have been made for the lean meat content parameter and differences for OA were statistically confirmed with LSV, ID and OT ($P \leq 0.05$). The highest number of ejaculates incompatible with sales requirements after dilution were recorded in boars culled due to ID and the lowest were for LL, RD and OT boars (all below 3; $P \leq 0.01$).

Table 3. Selected parameters of boar performance by culling reasons (mean ± SD)

Item	LSV	LL	LP	ID	OA	RD	OT
Proportionate rate, %	23.7^{Aa}	9.3^{B}	14.9^{b}	9.6^{B}	9.3^{B}	22.5^{Aa}	10.7^{B}
Age of semen collection entry, d	259 ±37	267 $\pm 31^{b}$	261 ±32	251 $\pm 29^{a}$	258 ±48	261 ±27	259 ±31
Boar herd life, d	738 $\pm 463^{B}$	399 $\pm 321^{A}$	695 $\pm 381^{AB}$	759 $\pm 415^{B}$	1605 $\pm 433^{D}$	1105 $\pm 309^{C}$	460 $\pm 276^{AB}$
Daily gain, g	786.13 $\pm 86.89^{a}$	788.55 $\pm 129.72^{a}$	783.91 $\pm 107.62^{a}$	797.29 $\pm 83.4^{a}$	729.15 $\pm 94.03^{b}$	761.3 ±73.41	782.47 $\pm 93.02^{a}$
Lean meat content, %	61.62 $\pm 1.82^{a}$	61.29 ±2.32	61.27 ±1.44	61.65 $\pm 1.77^{a}$	60.43 $\pm 2.2^{b}$	61.25 ±1.95	61.54 $\pm 2.18^{a}$
Ejaculates incompatible with requirements, n	3.42 ±5.72	1.85 $\pm 2.76^{A}$	3.92 ±5.58	5.79 $\pm 8.64^{B}$	3.73 ±5.68	2.54 $\pm 4.66^{A}$	2.97 $\pm 3.74^{A}$

LSV low semen value, *LL* low or lack of libido, *LP* leg problems, *ID* infectious diseases, *OA* old age, *RD* reduced demand for semen of the given boar, *OT* others

^{a,b} .. in the same row signifies statistically significant differences between reasons of culling, with $P \leq 0.05$

^{A,B,C,D} .. in the same row signifies statistically significant differences between reasons of culling, with $P \leq 0.01$

The estimated survival probability of boars in the AI station and the relative frequency of boar removal in each month are presented in Figure 1. A survival probability for boars of over 0.9 was noted until 1.5 yr. A drastic drop in the probability was observed to the age of 4 yr, achieving a value a little over 0.2. From this point, survival probability gently fell (expired) to 7 yr, when there was a complete replacement of the herd. The highest relative frequencies were observed at the 36th and 42nd month (over 16%). Indicators of relative frequency over 10% were also noted in the 18th and 30th mo of life.

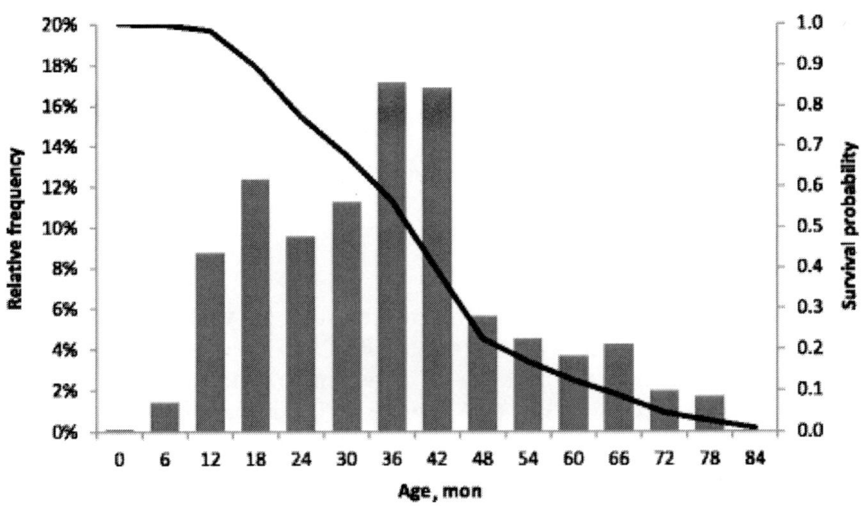

Figure 1: Survival probability of boars in the AI station and relative frequency (column) of boar removal by age in mon

Figure 2 shows the percentage of boars culled both below and above 4 yr, depending on the reason. It was noted that all boars with LL were culled by 4 yr. Equally large culling rates by 4 yr were reported for boars with OT and LP. More than 70% of boars from the ID, LSV or RD groups were culled by 4 yr. The highest levels of culling over 4 yr were obtained for the OA group.

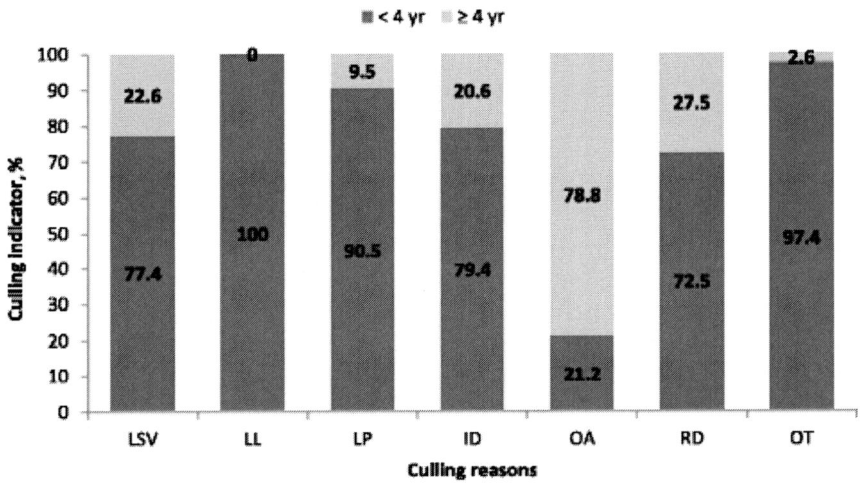

Figure 2 The percentage of boars culled before and after 4 yr, by culling reasons. LSV- low semen value, LL- low or lack of libido, LP- leg problems, ID- infection diseases, OA- old age, RD- reduced demand for semen of the given boar, OT- others

The hazard analysis for culled boars depending on the reason is presented in Table 4. The highest hazard ratios for culling were achieved for OT, LL and ID. The lowest hazard ratios (i.e. 3.08 to 4.38) were noted for LSV, RD and LP. The highest parameter values were observed for LL and OT, and the lowest negative values occurred for RD.

Table 5 shows correlation coefficients of selected parameters related to culling reasons. Statistically confirmed ($P \leq 0.05$) positive correlations between daily gain and age at the start of semen collection were observed for boars from OA, ID and LSV groups. The remaining statistically significant positive correlations were reported for OA between daily gain and boar herd life expectancy, and also lean meat content and age of semen collection entry. Additionally, for LP boars positive correlations were calculated between lean meat content and age at the start of semen collection or boar herd life expectancy ($P \leq 0.05$). Three negative correlations were found ($P < 0.05$). Two of these were noted for boars culled due to LL between daily gains and boar herd life expectancy, and also lean meat content and boar herd life expectancy. The last negative correlations were observed between lean meat content and age at start of semen collection for RD boars.

Table 4. Hazard analysis for culled boars depending on the reason

Item	Parameters	Standard error	Hazard ratio	Confidence interval
LSV	0.031	0.112	3.08	1.86; 5.01
LL	1.063	0.143	10.47	5.45; 20.1
LP	0.191	0.131	4.38	2.49; 7.72
ID	1.002	0.157	10.42	5.79; 18.75
OA	0	0	1	------
RD	−0.575	0.113	3.29	2.01; 5.38
OT	1.055	0.156	16.81	9.51; 29.73

LSV low semen value, *LL* low or lack of libido, *LP* leg problems, *ID* infectious diseases, *OA* old age, *RD* reduced demand for semen of the given boar, *OT* others

Table 5. Correlation coefficients for selected parameters by culling reasons

Item	LSV	LL	LP	ID	OA	RD	OT
Daily gain/age of semen collection entry	0.26*	−0.07	0.09	0.21*	0.32*	−0.16	0.14
Daily gain/boar herd life	0.12	−0.44*	0.03	0.05	0.28*	0.08	−0.01
Lean meat content/age of semen collection entry	0.08	0.17	0.34*	−0.14	0.23*	−0.34*	−0.11
Lean meat content/boar herd life	0.16	−0.49*	0.31*	0.18	0.05	0.01	0.13

LSV low semen value, *LL* low or lack of libido, *LP* leg problems, *ID* infectious diseases, *OA* old age, *RD* reduced demand for semen of the given boar, *OT* others
* – correlation statistically significant, with $P \leq 0.05$

DISCUSSION

The results presented here provide valuable information on the culling reasons for boars. A proportionate rate of culling reasons is a useful indicator of the strength of reasons and helps prioritize the improvement of a herd [10]. Observed culling of boars due to LSV and LL combined amounted to 33%, which was not as high as in the results of previous studies, which amounted to 19–25% [4, 11]. Such discrepancies could be explained by the higher culling ratio in experimental units and farms using

AI, for which the requirements in this regard are much greater [10]. Additionally, current capabilities for the accurate microscopic assessment of ejaculates contribute to changes in the classification of culling reasons especially semen value [11]. The more precise the characterization of culling reasons, the more accurate the analysis of the herd and the problems appearing therein and also the expected herd life of boars.

The average age at the start of semen collection and boar herd life approached the results obtained in piggery farms [4, 11], and was also much longer than in breeding herds from the early 1990s [10].

In contrast to the study of Koketsu and Sasaki [11], age at the start of semen collection had an effect on the reason for culling. Boars culled due to LL were the latest to begin semen collection. This is probably because the jump reflex was poorly noticeable, and this consequently led to a delay in the introduction of boars to ménage and a later analysis of the reason for culling. Libido is determined by many factors, including the level of circulating testosterone in the male body [16]. AI stations do not typically prefer to maintain boars demonstrating low libido, because it impacts upon operational costs and therefore such boars are eliminated from an active herd as soon as possible. Berger and Conley [17] even stated that boars with low libido and demonstrating a low ability of sperm to fertilize are useless for the AI industry, as are diseased animals.

Boars with diagnosed infectious diseases were characterized by the youngest age at start of semen collection. A longer period of quarantine, acclimatization and isolation before starting semen collection is recommended to reduce the occurrence of new pathogens and deaths of animals as a result of infection [18]. In our study, quarantine time was unified; therefore, in the case of the analyzed place, and the age at the start of semen collection could be extended by a few days without any negative effect on the remaining culling reasons. Such preventive action can lead to a reduction in boars culled due to ID.

Herd management in AI stations is significant, especially health status. Health reasons for boar culling include not only infectious diseases but also limb problems or unexpected falls. The health status of boars is a complex factor and results not only from veterinary care but also nutrition, management, exploitation, maintenance conditions or individual characteristics [19]. During culling analysis it is reasonable to separate the major determinants of health, i.e. infectious diseases and limb problems. The high levels of culling due to LP in our study remained close to results from a study conducted by D'Allaire and Leman [10], although other authors have noted both higher [3] and lower values [4, 11]. Regardless of the research, herd life expectancy of boars was similar. Important for the diagnosis of locomotor problems in boars is rapid preventive and curative action, because these disorders are classified as being painful for animals [15].

Due to the use of pigs for high quality pork, particular importance during production parameter analysis should be placed on daily gains and lean meat content [20]. It was observed in our study that boars with higher daily gains and lean meat content had a greater predisposition to infectious diseases. On the other hand, boars culled due to OA were characterized by the lowest daily gains and lean meat content. It can be concluded, therefore, that improvement of AI boars in terms of daily gains and lean meat content is justified only to a certain point. Our earlier (unpublished) results have shown that differences between semen parameters for boars with daily gains 700–750 g and 750–800 g are almost the same, are differences between lean meat content. Such precise analyses help us to conclude, that selection of AI boars with lower daily gains allows for longer boar herd life expectancy and culling herds with greater size due to old age. Wolf [21] stated that production traits (daily gains and lean meat content) have a negligible impact on semen characteristics, but play an important role in boar exploitation, which is especially important for AI centers. Confirmation of this thesis may be presented in our study in the form of the statistically confirmed correlations for exploitation parameters between daily gain and lean meat content. Inheritance of important production traits is essential, particularly in AI.

A significant emphasis is placed on the impact of sires on offspring in all currently existing insemination programs, where an extremely large number of offspring are obtained from each sire [22]. The importance of the proper and efficient functioning of AI stations can be seen in the sale of insemination portions consistent with the expectations of customers [23]. Therefore, great emphasis should be placed on the selection of a boar to achieve profitability of purchase. The identification of the market factor (RD) as one of the main culling reasons clearly shows that artificial insemination has become so popular, and competition in the market so large that in order to maintain the stability of a company adaption to customer requirements and market flexibility is required. As presented in the results, the culling of boars for reasons including market factors (reduced demand) represents an important element in a reasonable culling policy for AI boars and was ranked second. Previous studies on culling boars did not characterize this factor alone and the factor was probably included in the "other reasons" category. Most recent work in this area seems to indicate a higher proportion of culling for "other reasons" (c. 20%) [4, 11] in comparison with our results. Selection of boars is mainly based on the selected parameters of growth and carcass traits, with minimal emphasis on production traits [5]. However, AI stations should not be limited only to selection of boars due to the best production traits for pig producers, but must also take into account factors that affect the individuals' performance, i.e. the construction, temperament, quantity and quality of ejaculate [24]. A company operating on the market and focused on profitability has to meet the needs of customers. The constantly growing popularity of AI contributes to problems with fulfilling needs in terms of the appropriate quality of insemination portions [25]. The best solution is, therefore, to agree upon a compromise choice. If there is excessive pressure to choose traits favoring the AI stations, then market traits may be overlooked. A lack of demand for the offered insemination portions from a specific boar means its maintenance is unjustified. Culling for economic reasons may also take place in the case of new boars of the same genotype which are characterized by much better production parameters from their predecessors. In such a situation a decrease in demand for insemination portions from predecessor may be observed and this raises questions concerning its further use or culling, because the economy affects the company's balance sheet. Production farms are located in a particular economic environment, thus economic culling reasons should be taken into account in research. Hence, it was fully justified for our study to determine the reduced demand for semen of a boar to be a reason which closely affects the survival analysis of the herd.

Survival analysis is the recommended method for the study of boar stayability in herds [26]. Our results are similar to those in Segura-Correa et al. [4] but were only observed in farm A. Other farms in the study by these authors were characterized by much lower survival probability. On the other hand, in research conducted by Koketsu and Sasaki [11] a survival function with a similar shape was shown, although with increasing age the survival probability was higher in our study. However, boar herd life expectancy in high-performing herds and in our AI centre ended after 78 mon (6.5 yr). It is believed that culling boars from a high production herd should take place no later than the age of 6 yr. Nevertheless, it is suggested that culling of boars with low semen values and andrology-fertility indicators significantly below expectations should occur as soon as possible [5]. Culling in the AI center fits the shape of the survival curve. A higher share of culling due to old age proves the probity of the decisions concerning choosing boar and proper management in the unit. The critical point in the economic viability of boars exploitation is the age of 2–3 yrs [12], which in the case of our own research has been confirmed by relative frequency. Additionally, a detailed distribution of culling reasons both before and above 4 yr has been presented to characterize the problems concerning the direct contribution of boars to AI centre profitability.

Analyzing the above-mentioned distribution, the best solution would be quick culling of diseased boars, and those with reduced libido and low sperm quality. However, sometimes individuals suffer from infectious diseases or limb problems in later years, because the disease can appear in all ages and is a time-independent parameter. In the same regard, low semen quality should be considered, because despite the development of boars producing semen with more favorable traits [24, 27], this problem can also still affect boars with a long history of exploitation and at an advanced productive age. The most unpredictable culling reason for boars to be noted from the graph is the market factor (RD), because reduced customer demand for specific semen may result from the appearance of better individuals or arising trends and fashions in pig production.

The hazard function focuses on the occurrence of an event and reflects the instantaneous potential at the time of the event. An important feature of the hazard function is the reference to a specific time [28]. In our study, the hazard risk ratio described the occurrence of boar culling in comparison with the culling of a boar due to its age. A hazard risk higher than 1 indicates an increased risk of culling boars from the active herd in comparison with culling due to age. A hazard risk ratio below 1 indicates a lower risk of culling. Therefore, one can see that there was a more than 10-fold increased risk of boar culling in the case of LL, ID or OT compared to OA. This indicates the significance of these factors in shaping the overall predictability of possible problems in a herd of boars.

CONCLUSIONS

An understanding of culling reasons contributes to the ability to identify problems relating to AI center functioning. The results of the above study may be used as a direct point to develop appropriate strategies for boar culling in AI centers. An effective program of culling affects the economic viability of centers. A too frequent exchange of individuals and a large share of young boars negatively affects the cost of production and the health status of the herd. A large share of culling due to age, and a low rate in the case of diseases, hormonal disorders and quality of sperm provide a good selection of material production and careful management of the unit. An example of good practice is also the separate classification of economic reasons for culling (e.g. reduced demand for semen of the boar), because these reflect the changing preference of customers and the relevancy of products offered on the market. Each decision about boar culling, regardless of the reason, is difficult but necessary. However, if made at the right time and based on relevant observations, such decisions enable the preservation of the liquidity and profitability of AI centers.

ACKNOWLEDGEMENTS

The authors wish to thank Franciszek Wiśniowski, Director of the Boar Exploitation Station in Częstochowa, Małopolska Biotechnology Center for the possibility to conduct research and Karl Bernhardt for revising the English version of the manuscript.

Funding

This experiment and article preparation were financed by Wroclaw University of Environmental and Life Sciences, statutory project no. 2014.

Availability of data and materials

Please contact authors for data requests.

Authors' contributions

DK design of the study and methodology, preparation of manuscript, obtaining results, coordination of research group; AJ-M preparation of manuscript, participating in obtaining results, design of discussion; KD statistical analysis, presentation of results and conclusions, supervision of manuscript submission. All authors read and approved the final manuscript.

Competing interests

The authors declare that they have no competing interests.

Ethics approval

II Local Ethical Committee for Experiments on Animals at the University of Environmental and Life Sciences decided that this project was not subject to Committee action.

REFERENCES

1. Safranski TJ. Genetic selection of boars. Theriogenology. 2008;70:1310–6.
2. de Jong E, Appeltant R, Cools A, Beek J, Boyen F, Chiers K, et al. Slaughterhouse examination of culled sows in commercial pig herds. Livest Sci. 2014;167:362–9.
3. Acosta MJ, Rueda M. A note on causes of boar removal in Cuban pig farms. Livest Res Rural Dev. 2009;21(52):1–5.
4. Segura-Correa JC, Alzina-López A, Santos Ricalde R. Stayability in the herd and culling causes of boars in four pig farms of Yucatan, Mexico. Trop Subtrop Agroecosystems. 2010;12:411–6.
5. Robinson JAB, Buhr MM. Impact of genetic selection on management of boar replacement. Theriogenology. 2005;63:668–78.
6. Goldberg AMG, Argenti LE, Faccin JE, Linck L, Santi M, Lourdes Bernardi M, et al. Risk factors for bacterial contamination during boar semen collection. Res Vet Sci. 2013;95:362–7.
7. Duziński K, Knecht D, Środoń S. The use of oxytocin in liquid semen doses to reduce seasonal fluctuations in the reproductive performance of sows and improve litter parameters – a 2-year study. Theriogenology. 2014;81:780–86.
8. Knecht D, Jasek S, Procak A, Krzyżewski P. Efficiency of inseminating sows with pure breed and crossbreed boars. Med Wet. 2004;60:1208–11.
9. Smital J. Effects influencing boar semen. Anim Reprod Sci. 2009;110:335–46.
10. D'Allaire S, Leman AD. Boar culling patterns in swine breeding herds in Minnesota. Canadian Vet J. 1990;31:581–3.
11. Koketsu Y, Sasaki Y. Boar culling and mortality in commercial swine breeding herds. Theriogenology. 2009;71:1186–91.
12. Vinent-Duany NJ, Parra C, Sagaró-Zamora F, Garzón-Gómez V. Estudio de desecho de verracos en el centro genetic porcino. . 2007. Consulted March 31, 2009 URL: http://www.ilustrados.com/publicaciones/EEZZVEAAVZlANFBfrv.php.
13. Mucha A, Różycki M. Standaryzacja cech określających mięsność tusz w ocenie przyżyciowej świń. Rocz Nauk Zoot. 2005;32:45–50
14. Szyndler-Nędza M, Różycki M. Opracowanie równań regresji do przyżyciowego szacowania procentowej zawartości mięsa w tuszy knurów. Rocz Nauk Zoot. 2005;32:51–60.
15. Ordinance of the Minister of Agriculture and Rural Development. Requirements and how to proceed while maintaining livestock species for which protection standards are provisions of the European Union. 2010. http://isap.sejm.gov.pl/DetailsServlet?id=WDU20100560344. Accessed 15 Feb 2010.
16. Okere C, Joseph A, Ezekwe M. Seasonal and genotype variations in libido, semen production and quality in artificial insemination boars. J Anim Vet Adv. 2005;4:885–8.
17. Berger T, Conley AJ. Reducing endogenous estrogen during prepuberal life does not affect boar libido or sperm fertilizing potential. Theriogenology. 2014;82:627–35.
18. Dee S, Joo H, Pijoan C. Controlling the spread of PRRS virus in the breeding herd through management of the gilt pool. J Swine Healt Prod. 1995;3:64–9.
19. Anil SS, Anil L, Deen J. Evaluation of pattern of removal and associations among culling because of lameness and sow productivity traits in swine breeding herds. J Am Vet Med Assoc. 2005;226:956–61.

20. Knecht D, Jankowska-Mąkosa A, Duziński K. Does the activity of producer group organizations improve the production of pigs? Ann Anim Sci. 2015;15:759–74.
21. Wolf J. Genetic correlations between production and semen traits in pig. Animal. 2009;3:1094–9.
22. Knecht D, Środoń S, Szulc K, Duziński K. The effect of photoperiod on selected parameters of boar semen. Livest Sci. 2013;157:364–71.
23. Knecht D, Środoń S, Duziński K. The influence of boar breed and season on semen parameters. S Afr J Anim Sci. 2014;44:1–9.
24. Wolf J, Smital J. Quantification of factors affecting semen traits in artificial insemination boars from animal model analyses. J Anim Sci. 2009;87:1620–7.
25. Knecht D, Środoń S, Duziński K. Does a boar's season of birth determine semen parameters and reproductive performance? Reprod Domest Anim. 2014;49:183–90.
26. Yazdi MH, Rydhmer L, Ringmar-Cederberg E, Lundeheim N, Johansson K. Genetic study of longetivity in Swedish landrace sows. Livest Prod Sci. 2000;63:255–64.
27. Kondracki S, Banaszewska D, Wysokińska A, Radomyska M. Effect of age on semen traits of Duroc breed used in insemination. Anim Sci Pap Rep. 2004;22:281–8.
28. Hoge MD, Bates RO. Developmental factors that influence sow longevity. J Anim Sci. 2011;89:1238–45.

CHAPTER 3

The Roles of MicroRNAs in Regulation of Mammalian Spermatogenesis

Xiaoxu Chen, Xueliang Li, Jiayin Guo, Pengfei Zhang and Wenxian Zeng*

College of Animal Science and Technology, Northwest A&F University, Yangling, Shaanxi 712100, China

ABSTRACT

Mammalian spermatogenesis contains three continuous and organized processes, by which spermatogonia undergo mitosis and differentiate to spermatocytes, follow on meiosis to form haploid spermatids and ultimately transform into spermatozoa. These processes require an accurately, spatially and temporally regulated gene expression patterns. The microRNAs are a novel class of post-transcriptional regulators. Cumulating evidences have demonstrated that microRNAs are expressed in a cell-specific or stage-specific manner during spermatogenesis. In this review, we focus on the roles of microRNAs in spermatogenesis. We highlight that N6-methyladenosine (m6A) is involved in the biogenesis of microRNAs and miRNA regulates the m6A modification on mRNA, and that specific miRNAs have been exploited as potential biomarkers for the male factor infertility, which will provide insightful understanding of microRNA roles in spermatogenesis.

Keywords: microRNAs RNA methylation Spermatogenesis

BACKGROUND

Male fertility is dependent upon the successful perpetuation of spermatogenesis that is a highly organized process of germ cell differentiation occurring within the seminiferous tubules in the testes. Spermatogonial stem cells (SSCs) are a subset of undifferentiated spermatogonia that are capable of self-renewal to maintain the pool of SSCs or differentiation to give rise to spermatogenic lineage, thus supporting the continuous production of spermatozoa. Spermatogenesis initiates once SSCs enter differentiation process [1]. The spermatogonia go into the meiotic phase and become spermatocytes. After a long-lasting meiosis I, preleptotene spermatocytes transform into second spermatocytes and enter meiosis II to produce haploid round spermatids [2], which undergo spermiogenesis including acrosomal biogenesis, flagellum development, chromatin condensation, cytoplasmic reorganization and exclusion [3]. Ultimately, the round spermatids transform into spermatozoa, which are released into the lumen of seminiferous tubules [4].

This highly organized spermatogenesis requires accurate, spatial and temporal regulation of gene expression governed by transcriptional, post-transcriptional and epigenetic processes [5, 6]. More than a thousand of protein coding genes that are involved in the spermatogenesis have been identified [7, 8]. However, the mechanisms that mediate the expression of these spermatogenesis-related genes have not been fully uncovered. The microRNAs (miRNAs, miR), small (~22 nucleotides) single-strand noncoding RNAs, are linked to cell proliferation, differentiation and apoptosis [9, 10, 11]. Transcriptome data indicate that miRNAs are extensively transcribed during spermatogenesis. The miRNAs are differentially expressed in a cell-specific and step-specific manner ([12, 13], Chen et al. unpublished data). Some miRNAs are specifically expressed in certain type of male germ cells, while the others are universally expressed among different types of cells in the testes. Growing evidences have showed that the miRNAs are essential for male germ cell development and differentiation [14, 15, 16, 17]. A few recent reviews have reported the roles of miRNAs in spermatogenesis and fertility [5, 6, 10, 11]. In this article, we briefly summarize the most recent progress of miRNAs in the regulation of spermatogenesis.

miRNA biogenesis

At present, there are 1881 miRNA loci having been annotated in the human genome in the miRNA database (http://www.mirbase.org). Analysis has revealed that 1% of the human genome is miRNA genes [18, 19, 20], of which about half of miRNA genes located in the introns (intronic miRNAs) of host genes [21]. However, some intronic miRNAs exhibit low correlated expression level with their host genes. It is likely these miRNAs are transcribed from unique transcription units independent of host genes [22, 23, 24].

The biogenesis of miRNAs is modulated at a few levels, including miRNA transcription, processing by Drosha and Dicer, RNA methylation, uridylation and adenylation (Figure 1) [25, 26, 27]. The initial transcripts are termed the primary miRNAs (pri-miRNAs) that are variable in length from several hundreds to thousands of nucleotides [25]. The pri-miRNAs are methylated by the methyltransferase like 3 (METTL3), marking them for recognition and processing by the DiGeorge syndrome critical region 8 (DGCR8) [28]. The pri-miRNAs are thus processed by drosha ribonuclease III (Drosha) and its cofactor DGCR8 into ~ 70 nucleotides (nt) long miRNA precursor (pre-miRNAs) [29, 30]. The pre-miRNAs are then transported into the cytoplasm by exportin 5 (EXP5) in accompanied with Ran-GTP [31, 32] and cleaved by Dicer into ~22 base pair (bp) double-strands RNAs (dsRNAs) [33, 34, 35]. These dsRNAs are loaded onto an Argonaute protein (AGO) so as to form miRNA-induced silencing complex (miRISC), in which one strand of the ~22-nt RNA duplex remains in AGO as a mature miRNA, whereas the other strand is degraded [36]. Interestingly, Alarcon et al. recently reported that RNA-binding protein heterogeneous nuclear ribonucleoprotein A2/B1 (HNRNPA2B1) binds m6A-bearing pri-miRNAs, interacts with DGCR8 and thus facilitates the processing of pri-miRNAs [37]. In consistent with this, loss of HNRNPA2B1 or depletion of METTL3 led to concomitant accumulation of unprocessed pri-miRNAs and decrease of the global mature miRNAs [28, 37]. Therefore, the methylation mark acts as a key post-transcriptional modification that enhances the initiation of miRNA biogenesis.

Mechanisms of miRNA action

Usually, a specific base-pairing between miRNAs and mRNAs induces mRNA degradation or translational repression [38]. In mammals, the overall complementarity between a miRNA and its target is usually imperfect, which allows each miRNA to potentially regulate multiple RNAs [39]. It is estimated that one miRNA may target as many as 400 genes on average [19]. Conversely, the expression of a single gene can also be modulated by multiple miRNAs [40].

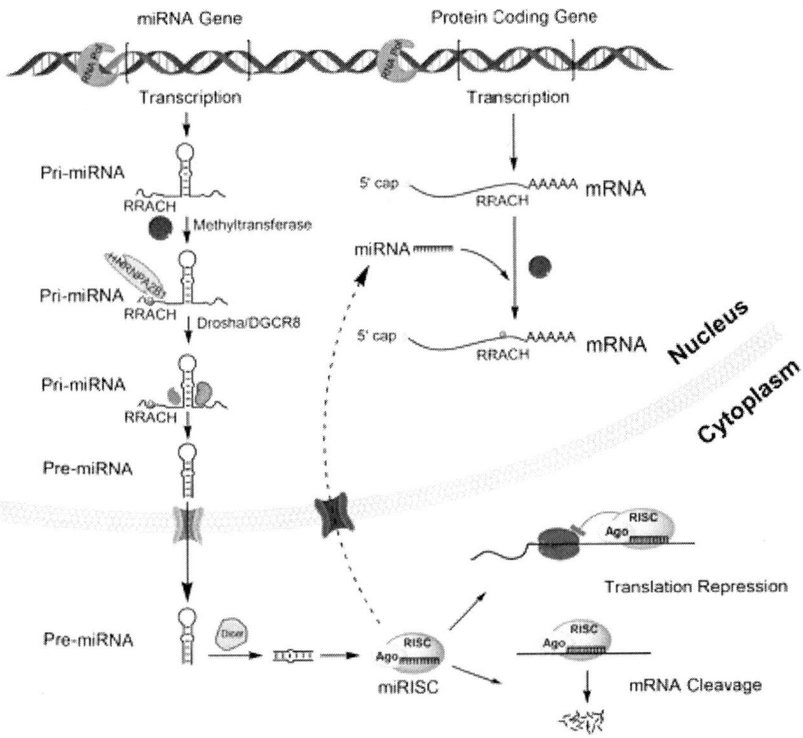

Figure 1. Biogenesis and function of miRNAs

Interestingly, it has been reported recently that miRNAs regulate the m6A modification in mRNAs via a sequence pairing mechanism. As a result, manipulation of miRNA expression leads to change of m6A modification through modulating the binding of METTL3 to mRNAs (Figure 1) [41]. The m6A modification, in turn, modulates mRNA metabolism and thus is another key post-transcriptional control of gene expression [37, 42, 43]. Evidences have indicated that m6A methylation determines stem cell fate by regulating pluripotent transition toward differentiation [41, 44, 45]. Intriguingly, deficiency of ALKBH5, a m6A demethylase, leads to aberrant spermatogenesis and apoptosis in mouse testis through the demethylation of m6A on mRNAs [46].

Functions of miRNAs in spermatogenesis

Conditional Dicer knockout mouse models

The overall importance of miRNA signaling for regulation of spermatogenesis has been demonstrated using conditional knockout of *Dicer* in germ cells. *Dicer1* ablation in prospermatogonia just before birth using *Ddx4* promoter-driven Cre expression led to an alteration in meiotic progression, significant increase of apoptosis in pachytene spermatocytes, a reduced number of round spermatids and morphological defects in spermatozoa [47]. Moreover, *Ngn3* is expressed endogenously in type A spermatogonia starting from postnatal d 5 [48, 49]. In the mouse model of selective deletion of *Dicer1* in type A spermatogonia by *Ngn3* promoter-driven Cre, the first clear defects were displayed in haploid round spermatids. The spermiogenesis was severely compromised [50]. Similarly, conditional depletion of *Dicer1* using the *Stra8Cre* transgene in early spermatogonia resulted in the comparable phenotype to the *Ngn3Cre*-driven *Dicer1* deletion [51, 52]. In addition, deletion of *Dicer1* in haploid spermatids using the *protamine 1 (Prm1)-Cre* transgene led to abnormal morphology in the elongated spermatids and spermatozoa [53]. But, the *Prm1Cre-Dicer1* knockout caused a less severe phenotype compared to those in which Dicer1 was deleted from prospermatogonia and spermatogonia [53].

Collectively, the earlier the ablation of Dicer occurs, the more severe side effects on spermatogenesis are found. Therefore, miRNA-mediated post-transcriptional control is an important regulator for spermatogenesis.

The roles of miRNAs in SSC self-renewal and differentiation

SSCs are the foundation of spermatogenesis that involves a delicate balance between self-renewal and differentiation of SSCs to ensure the lifelong production of spermatozoa. In the testes, the SSCs reside in a unique microenvironment or 'niche'. The niche factor glial cell line-derived neurotropic factor (GDNF) is the first well-defined paracrine factor that promotes SSC self-renewal [54]. GDNF signaling acts via the RET tyrosine kinase [55] and requires a ligand-specific co-receptor GFRα1 [56] in mouse SSCs [57]. Evidences have shown that through the PI3K/AKT-dependent pathway [58] or the SRC family kinase (SFK) pathway [59], GDNF regulates the expression of the transcription factors B cell CLL/lymphoma 6 member B (BCL6B), ETS variant 5 (ETV5), DNA-binding protein 4 (ID4), LIM homeobox 1 (LHX1) and POU class 3 homeobox 1 (POU3F1) to drive SSC self-renewal [59].

miRNAs conduce maintenance of the pool of SSCs. It has been shown that *miR-20* along with *miR-21, −34c, −135a, −146a, −182, −183, −204, −465a-3p, −465b-3p, −465c-3p, −465c-5p* and *−544* were preferentially expressed in the SSC-enriched population (Figure 2) [60, 61]. Importantly, *miR-20, miR-21* and *miR-106a* contribute to maintenance of mouse SSC homeostasis [61]. *miR-135a* mediates the maintenance of rat SSCs by regulating FOXO1 that promotes high levels of Ret protein on the cell surface of SSCs [62]. Moreover, *miR-544* regulates self-renewal of goat SSCs by targeting the promyelocytic leukemia zinc finger gene (PLZF), which is the first transcription factor to be identified as being involved in SSC self-renewal [63]. Similarly, *miR-224* regulates mouse SSC self-renewal via modulating PLZF and GFRα1 [64]. Interestingly, *miR-34c* is expressed in goat SSCs and promotes SSC apoptosis in a p53-depemdent manner [65]. Recently, it was found that *miR-204* was involved in the regulation of dairy goat SSC proliferation via targeting Sirt1 [66]. Collectively, miRNAs are involved in regulating SSCs fate.

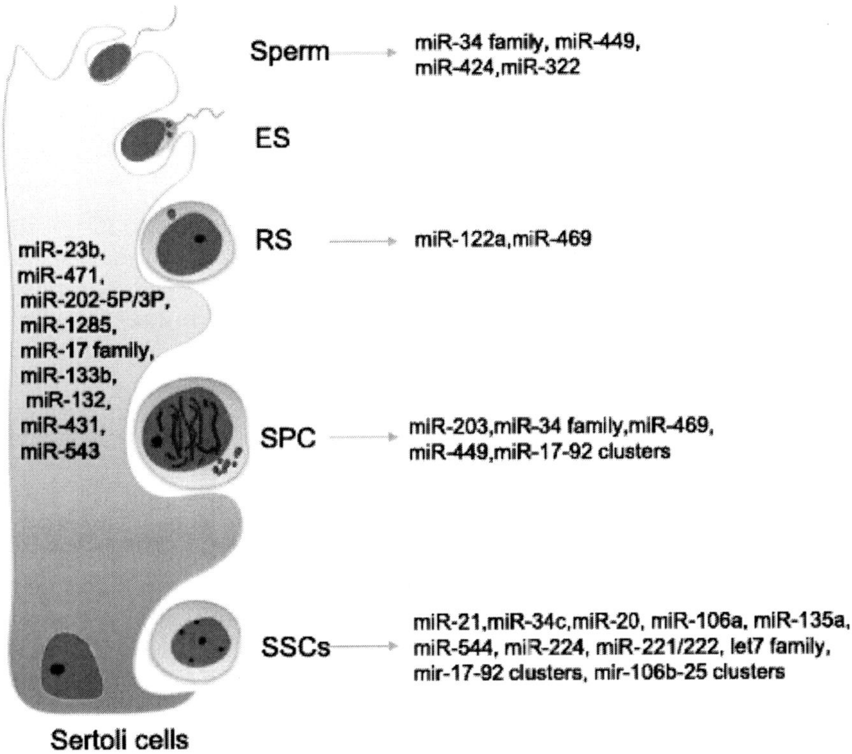

Figure 2. The expression of associate miRNAs in testicular cells

On the other hand, some miRNAs have been identified to mediate spermatogonia differentiation. It is well-known that retinoic acid (RA) directs the sequential programs of spermatogonial differentiation and the entry into meiosis [67, 68]. *miR-146* [69], *miR-let7* family miRNAs [70], *miR-17-92* and *miR-106b-25* clusters [71] are downregulated during RA-induced spermatogonial differentiation. Importantly, male germ cell-specific knockout of *miR-17-92* cluster resulted in the reduced number of SSCs and spermatogonia, and impaired spermatogenesis [71, 72]. Interestingly, exposure to RA downregulates *miR-221/222* expression, while GDNF upregulates *miR-221/222* abundance. Over-expression of *miR-221/222* in undifferentiated spermatogonia made them resisting to RA-induced transition into c-kit-positive differentiated spermatogonia [73]. In addition, *miR-34c* promotes SSC differentiation and meiosis process by targeting NANOS2 and up-regulating meiosis regulated genes *Stra8, Scp3* and *Dazl* [74]. Taken together, miRNAs are related to the post-transcriptional regulation of spermatogonia differentiation.

The roles of miRNAs in meiosis and spermiogenesis

Growing evidences have also demonstrated that specific miRNAs regulate meiosis (Figure 2). The expression of *miR-449* cluster is abundant and is upregulated upon meiotic initiation during testis development and in adult testes. The expression pattern of the *miR-449* cluster is similar to that of *miR-34b/c*. Moreover, *miR-34b/c* and *miR-449* cluster share the same seed region and thus target same sets of mRNAs [75, 76, 77, 78]. Depletion of either *miR-34* cluster or *miR-449* cluster displays no apparent defect in male germ cell development. However, simultaneous knockout of these two clusters led to sexually dimorphic and infertility, suggesting that *miR-34b/c* and the *miR-449* cluster function redundantly in the regulation of spermatogenesis [71]. Furthermore, *miR-18*, one of the *miR-17-92* cluster, is abundantly expressed in spermatocytes. *miR-18* targets heat shock factor2 (Hsf2), which is a critical transcription factor for spermatogenesis [79]. Finally, *miR-34b-5p* regulates meiotic progression by targeting *Cdk6* [80].

A unique chromatin remodelling occurs during spermatogenesis when histones are replaced by DNA packing proteins, such as transition proteins (TPs) and protamines (PRMs), which are exclusive to male germ cells [81, 82]. In the post-mitotic germ cells, the timely expression of TPs and PRMs is prerequisite for compaction and condensation of chromatin during spermiogenesis [83]. To secure this timed expression pattern, *Tp* and *Prm* are subjected to an efficiently post-transcriptional control. It has been demonstrated that *miR-469* suppresses the translation of TP2 and PRM2 by targeting mRNA of *Tp2* and *Prm*2 in pachytene spermatocytes and round spermatids [84]. On the contrary, *miR-122a* that is abundantly expressed in late-stage male germ cells reduces the *Tp2* mRNA expression by RNA cleavage [85].

Although the majority of miRNAs disappear during spermiogenesis, the sperm born miRNAs have also been demonstrated to play important roles. *miR-34* is present in mouse spermatozoa and zygotes but not in the oocytes or in embryos beyond the one-cell stage [86]. Upon fertilization, *miR-34c* is transferred from spermatozoa to zygote where it reduces the expression of Bcl-2 and p27, leading to S-phase entry and the first cleavage. Moreover, injection of *miR-34c* inhibitor into the zygotes inhibits DNA synthesis and suppresses the first cleavage division, suggesting that the sperm-borne *miR-34c* is required for zygote cleavage [86]. In addition, dysregulation of *miR-424/322* induces DNA double-strand breaks in spermatozoa [87]. Importantly, a set of sperm miRNAs are differentially expressed in asthenozoospermic and oligoasthenozoospermic males compared with normozoospermic males [88, 89]. Furthermore, *miR-151a-5p* is abundant in severe asthenozoospermia cases compared with healthy controls and participates in mitochondrial biological functions [53, 90]. Therefore, specific miRNAs have been exploited as potential biomarkers for male factor infertility [91].

miRNAs in testicular somatic cells

Spermatogenesis is supported by the testicular Sertoli cells, peritubular myoid (PTM) cells and Leydig cells [92, 93, 94]. The extrinsic factors derived from these somatic cells trigger specific events in germ cells that dictate or influence spermatogenesis. It has been shown that miRNAs are highly abundant in Sertoli cells (Figure 2) [12, 95, 96]. *MiR-133b* and *miR-202* are involved in pathogenesis of azoospermia or Sertoli-cell-only syndrome [97, 98]. Importantly,conditional depletion of *Dicer1* from Sertoli cells, using the Anti-Müllerian hormone (Amh) promoter-driven Cre in mice, results in disrupted spermatogenesis and progressive testis degeneration, indicating that miRNAs in Sertoli cells play critical roles in spermatogenesis [99, 100]. Specifically, *miR-133b* promotes the proliferation of human Sertoli cells by targeting GLI3 and mediating expression of Cyclin B1 and Cyclin D1 [97]. Moreover, *miR-762* promotes porcine immature Sertoli cell growth via the ring finger protein 4 (*RNF4*) [101].

FSH and androgens are fundamentally important for spermatogenesis. To elucidate the molecular mechanisms by which FSH and androgen act in the Sertoli cells, Nicholls et al. [102] investigated the expression and regulation of micro-RNAs (miRNAs). The authors have found that a subset of miRNAs were up-regulated after hormone suppression in rat model and in vitro culture of primary rat Sertoli cells. Interestingly, Pten, an intracellular phosphatase, and Eps15, a mediator of endocytosis, were down-regulated by the withdrawal of hormones [102]. In consistent with it, overexpression of *miR-23b* in vitro resulted in decreased translation of PTEN and EPS15 protein. Similarly, by using androgen suppression and androgen replacement, Chang et al. [53] identified that androgen regulated the expression of several microRNAs in mouse Sertoli cells [103]. One of the *miRNAs* targets found in this study is desmocollin-1 (*Dsc1*), which plays an essential role in cell-cell adhesion in epithelial cells [104]. On the other hand, elevated estradiol level is associated with male infertility [105]. Evidences indicate that estradiol regulates proliferation of Sertoli cells in a dose-dependent manner, in which *miR-17* family and *miR-1285* are involved in the regulation [106, 107]. Collectively, miRNA transcription is a new paradigm in the hormone dependence of spermatogenesis.

Leydig cells are responsible for androgen production that is essential for sperm production [108]. Basic fibroblast growth factor (bFGF) promotes the development of stem Leydig cells and inhibits LH-stimulated androgen production by regulating miRNAs [109]. Interestingly, *miR-140-5p/140-3p* control mouse Leydig cell numbers in the developing testis. Deletion of *miR-140-5p/miR-140-3p* results in an increase of number of Leydig cells, indicating that the miRNAs are likely to regulate the expression of factors produced by Sertoli cells that regulate differentiation of Leydig cells [110].

Collectively, these findings indicate that miRNAs regulate the development and functions of Sertoli cells and Leydig cells, which create the niche for SSCs and thus provide structural and nutritional support for germ cells. Therefore, miRNAs in somatic cells play important roles in spermatogenesis.

CONCLUSION AND PERSPECTIVES

Extensive and accurate regulation of gene expression is prerequisite for spermatogenesis. miRNAs are expressed in a cell-specific or stage-specific manner during spermatogenesis. However, the roles and underlying mechanisms of many of those miRNAs in spermatogenesis remain largely unknown. Future studies should primarily focus on uncovering the roles of germ-cell specific miRNAs in spermatogenesis. The powerful single-cell small RNA sequencing would help to more accurately profile the miRNAs for certain type of germ cells. Meanwhile, the establishment of long-term culture of SSCs and in vitro induction of differentiation of male germ cells make it possible to elucidate the role of a certain miRNA or miRNA cluster in vitro. The application of CRISPR/Cas9 system and conditional knockout strategies would speed up the understanding of miRNA functions. Secondly, growing evidences have been demonstrated that some specific miRNAs are preferentially expressed in testicular somatic cells. But it is not clear whether these miRNAs act as secreted paracrine factors in the SSC niche, or whether they indirectly mediate the secretion of growth factors, GDNF for instance, which then affect germ cells. More somatic cell expressed miRNAs are needed to be functionally characterized. Thirdly, it has been demonstrated that some transcription factors promote SSC self-renewal (for example, BCL6B, BRACHYURY, ETV5, ID4, LHX1, and POU3F1), while several transcription factors stimulate spermatogonia differentiation

(DMRT1, NGN3, SOHLH1, SOHLH2, SOX3, and STAT3) [111]. However, it is unclear which and how miRNA/miRNA cluster regulates the expression of these transcription factors. Fourthly, it has been discovered recently that RNA methylation is involved in pri-miRNA processing [28, 37], opening the door for exploring RNA methylation in the biogenesis and function of the miRNAs. Future research will pay increasing attention on the understanding of biological functions of epigenetic changes (or marks) during germ cell development. Finally, specific miRNAs in spermatozoa or seminal plasma will be exploited as potential biomarkers for male factor infertility. The annotation of the miRNAs and the elucidation of their regulating mechanisms in pathogenesis will provide insight into the etiology of male sterility and infertility. Together, uncovering these questions will shed new light on the pivotal roles of miRNA in spermatogenesis and fertility.

ACKNOWLEDGEMENTS

We thank Dr. Bin Wu at Arizona Center for Reproductive Endocrinology and Infertility, USA for reviewing and editing the manuscript. This project is funded, in part, by the National Basic Research Program of China (973 program; No.2013CB943103), the National Natural Science Foundation of China (No. 31272439, No. 31230048 and No. 31572401), and Programs Foundation of Ministry of Education of China (No. 20130204110017).

Funding

Not applicable.

Availability of data and materials

Data sharing not applicable to this article as no datasets were generated or analysed during the current study.

Authors' contribution

CXX have searched primary sources, written the manuscript. LXL, GJY, ZPF and ZWX revised and edited the manuscript. All authors read and approved the final manuscript.

Competing interests

The authors declare that they no competing interests.

Consent for publication

Not applicable.

REFERENCES

1. Kanatsu-Shinohara M, Shinohara T. Spermatogonial stem cell self-renewal and development. Annu Rev Cell Dev Biol. 2013;29:163–87.
2. Miller MP, Amon A, Unal E. Meiosis I: when chromosomes undergo extreme makeover. Curr Opin Cell Biol. 2013;25:687–96.
3. Rathke C, Baarends WM, Awe S, Renkawitz-Pohl R. Chromatin dynamics during spermiogenesis. Biochim Biophys Acta. 2014;1839:155–68.
4. Guyonnet B, Dacheux F, Dacheux JL, Gatti JL. The epididymal transcriptome and proteome provide some insights into new epididymal regulations. J Androl. 2011;32:651–64.
5. Kimmins S, Sassone-Corsi P. Chromatin remodelling and epigenetic features of germ cells. Nature. 2005;434:583–9.
6. Papaioannou MD, Nef S. microRNAs in the testis: building up male fertility. J Androl. 2010;31:26–33.
7. da Cruz I, Rodriguez-Casuriaga R, Santinaque FF, Farias J, Curti G, Capoano CA, et al. Transcriptome analysis of highly purified mouse spermatogenic cell populations: gene expression signatures switch from meiotic-to postmeiotic-related processes at pachytene stage. BMC Genomics. 2016;17:294.
8. Zhu Z, Li C, Yang S, Tian R, Wang J, Yuan Q, et al. Dynamics of the transcriptome during human spermatogenesis: predicting the potential key genes regulating male gametes generation. Sci Rep. 2016;6:19069.
9. Brennecke J, Hipfner DR, Stark A, Russell RB, Cohen SM. bantam encodes a developmentally regulated microRNA that controls cell proliferation and regulates the proapoptotic gene hid in Drosophila. Cell. 2003;113:25–36.
10. Chen CZ, Li L, Lodish HF, Bartel DP. MicroRNAs modulate hematopoietic lineage differentiation. Science. 2004;303:83–6.

11. Ambros V. MicroRNA pathways in flies and worms: growth, death, fat, stress, and timing. Cell. 2003;113:673–6.
12. Tan T, Zhang Y, Ji W, Zheng P. miRNA signature in mouse spermatogonial stem cells revealed by high-throughput sequencing. Biomed Res Int. 2014;2014:154251.
13. Liu Y, Niu MH, Yao CC, Hai YN, Yuan QQ, Liu Y, et al. Fractionation of human spermatogenic cells using STA-PUT gravity sedimentation and their miRNA profiling. Scientific Reports. 2015;5:8084.
14. Yao CC, Liu Y, Sun M, Niu MH, Yuan QQ, Hai YA, et al. MicroRNAs and DNA methylation as epigenetic regulators of mitosis, meiosis and spermiogenesis. Reproduction. 2015;150:R25–34.
15. Khazaie Y, Nasr Esfahani MH. MicroRNA and male infertility: a potential for diagnosis. Int J Fertil Steril. 2014;8:113–8.
16. Wang L, Xu C. Role of microRNAs in mammalian spermatogenesis and testicular germ cell tumors. Reproduction. 2015;149:R127–137.
17. Kotaja N. MicroRNAs and spermatogenesis. Fertil Steril. 2014;101:1552–62.
18. Bentwich I, Avniel A, Karov Y, Aharonov R, Gilad S, Barad O, et al. Identification of hundreds of conserved and nonconserved human microRNAs. Nat Genet. 2005;37:766–70.
19. Friedman RC, Farh KKH, Burge CB, Bartel DP. Most mammalian mRNAs are conserved targets of microRNAs. Genome Res. 2009;19:92–105.
20. John B, Enright AJ, Aravin A, Tuschl T, Sander C, Marks DS. Human MicroRNA targets. PLoS Biol. 2004;2:e363.
21. Shomron N, Levy C. MicroRNA-biogenesis and Pre-mRNA splicing crosstalk. J Biomed Biotechnol. 2009;2009:594678.
22. Ramalingam P, Palanichamy JK, Singh A, Das P, Bhagat M, Kassab MA, et al. Biogenesis of intronic miRNAs located in clusters by independent transcription and alternative splicing. RNA. 2014;20:76–87.
23. Wang D, Lu M, Miao J, Li TT, Wang E, Cui QH. Cepred: Predicting the Co-Expression Patterns of the Human Intronic microRNAs with Their Host Genes. PLoS One. 2009;4(2):e4421.
24. Radfar MH, Wong W, Morris Q. Computational Prediction of Intronic microRNA Targets using Host Gene Expression Reveals Novel Regulatory Mechanisms. PLoS One. 2011;6(6):e19312.
25. Ha M, Kim VN. Regulation of microRNA biogenesis. Nat Rev Mol Cell Biol. 2014;15:509–24.
26. Castel SE, Martienssen RA. RNA interference in the nucleus: roles for small RNAs in transcription, epigenetics and beyond. Nat Rev Genet. 2013;14:100–12.
27. Ji L, Chen X. Regulation of small RNA stability: methylation and beyond. Cell Res. 2012;22:624–36.
28. Alarcon CR, Lee H, Goodarzi H, Halberg N, Tavazoie SF. N6-methyladenosine marks primary microRNAs for processing. Nature. 2015;519:482–5.
29. Cai XZ, Hagedorn CH, Cullen BR. Human microRNAs are processed from capped, polyadenylated transcripts that can also function as mRNAs. RNA. 2004;10:1957–66.
30. Lee Y, Kim M, Han J, Yeom KH, Lee S, Baek SH, et al. MicroRNA genes are transcribed by RNA polymerase II. EMBO J. 2004;23:4051–60.
31. Yi R, Qin Y, Macara IG, Cullen BR. Exportin-5 mediates the nuclear export of pre-microRNAs and short hairpin RNAs. Genes Dev. 2003;17:3011–6.
32. Lund E, Guttinger S, Calado A, Dahlberg JE, Kutay U. Nuclear export of microRNA precursors. Science. 2004;303:95–8.
33. Shomron N, Levy C. MicroRNA-biogenesis and pre-mRNA splicing crosstalk. J Biomed Biotechnol. 2009.
34. Truscott M, Islam AB, Frolov MV. Novel regulation and functional interaction of polycistronic miRNAs. RNA. 2016;22:129–38.
35. Kim VN, Han J, Siomi MC. Biogenesis of small RNAs in animals. Nat Rev Mol Cell Biol. 2009;10:126–39.

36. Bartel DP. MicroRNAs: target recognition and regulatory functions. Cell. 2009;136:215–33.
37. Alarcon CR, Goodarzi H, Lee H, Liu XH, Tavazoie S, Tavazoie SF. HNRNPA2B1 is a mediator of m(6)a-dependent nuclear RNA processing events. Cell. 2015;162:1299–308.
38. Thomas M, Lieberman J, Lal A. Desperately seeking microRNA targets. Nat Struct Mol Biol. 2010;17:1169–74.
39. Pasquinelli AE. MicroRNAs and their targets: recognition, regulation and an emerging reciprocal relationship. Nat Rev Genet. 2012;13:271–82.
40. Sood P, Krek A, Zavolan M, Macino G, Rajewsky N. Cell-type-specific signatures of microRNAs on target mRNA expression. Proc Natl Acad Sci U S A. 2006;103:2746–51.
41. Wang Y, Li Y, Toth JI, Petroski MD, Zhang ZL, Zhao JC. N-6-methyladenosine modification destabilizes developmental regulators in embryonic stem cells. Nat Cell Biol. 2014;16:191–8.
42. Berulava T, Rahmann S, Rademacher K, Klein-Hitpass L, Horsthemke B. N6-adenosine methylation in MiRNAs. PLoS One. 2015;10:e0118438.
43. Chen T, Hao YJ, Zhang Y, Li MM, Wang M, Han WF, et al. m(6)a RNA methylation is regulated by MicroRNAs and promotes reprogramming to pluripotency. Cell Stem Cell. 2015;16:289–301.
44. Batista PJ, Molinie B, Wang JK, Qu K, Zhang JJ, Li LJ, et al. m(6)a RNA modification controls cell fate transition in mammalian embryonic stem cells. Cell Stem Cell. 2014;15:707–19.
45. Geula S, Moshitch-Moshkovitz S, Dominissini D, Mansour AA, Kol N, Salmon-Divon M, et al. Stem cells. m6A mRNA methylation facilitates resolution of naive pluripotency toward differentiation. Science. 2015;347:1002–6.
46. Zheng GQ, Dahl JA, Niu YM, Fedorcsak P, Huang CM, Li CJ, et al. ALKBH5 is a mammalian RNA demethylase that impacts RNA metabolism and mouse fertility. Mol Cell. 2013;49:18–29.
47. Romero Y, Meikar O, Papaioannou MD, Conne B, Grey C, Weier M, et al. Dicer1 Depletion in Male Germ Cells Leads to Infertility Due to Cumulative Meiotic and Spermiogenic Defects. PLoS One. 2011;6(10):e25241.
48. Yoshida S, Takakura A, Ohbo K, Abe K, Wakabayashi J, Yamamoto M, et al. Neurogenin3 delineates the earliest stages of spermatogenesis in the mouse testis. Dev Biol. 2004;269:447–58.
49. Suzuki H, Ahn HW, Chu T, Bowden W, Gassei K, Orwig K, et al. SOHLH1 and SOHLH2 coordinate spermatogonial differentiation. Dev Biol. 2012;361:301–12.
50. Korhonen HM, Meikar O, Yadav RP, Papaioannou MD, Romero Y, Da Ros M, et al. Dicer is required for haploid male germ cell differentiation in mice. PLoS One. 2011;6:e24821.
51. Wu QX, Song R, Ortogero N, Zheng HL, Evanoff R, Small CL, et al. The RNase III. Enzyme DROSHA is essential for MicroRNA production and spermatogenesis. J Biol Chem. 2012;287:25173–90.
52. Zimmermann C, Romero Y, Warnefors M, Bilican A, Borel C, Smith LB, et al. Germ Cell-Specific Targeting of DICER or DGCR8 Reveals a Novel Role for Endo-siRNAs in the Progression of Mammalian Spermatogenesis and Male Fertility. PLoS One. 2014; 9(9):e107023.
53. Chang YF, Lee-Chang JS, Imam JS, Buddavarapu KC, Subaran SS, Sinha-Hikim AP, et al. Interaction between microRNAs and actin-associated protein Arpc5 regulates translational suppression during male germ cell differentiation. Proc Natl Acad Sci U S A. 2012;109:5750–5.
54. Yang QE, Oatley JM. Spermatogonial stem cell functions in physiological and pathological conditions. Curr Top Dev Biol. 2014;107:235–67.
55. Naughton CK, Jain S, Strickland AM, Gupta A, Milbrandt J. Glial cell-line derived neurotrophic factor-mediated RET signaling regulates spermatogonial stem cell fate. Biol Reprod. 2006;74:314–21.

56. Jing SQ, Wen DZ, Yu YB, Holst PL, Luo Y, Fang M, et al. GDNF-induced activation of the Ret protein tyrosine kinase is mediated by GDNFR-alpha, a novel receptor for GDNF. Cell. 1996;85:1113–24.
57. Oatley JA, Brinster RL. Regulation of spermatogonial stem cell self-renewal in mammals. Annu Rev Cell Dev Biol. 2008;24:263–86.
58. Lee J, Kanatsu-Shinohara M, Inoue K, Ogonuki N, Miki H, Toyokuni S, et al. Akt mediates self-renewal division of mouse spermatogonial stem cells. Development. 2007;134:1853–9.
59. Oatley JM, Avarbock MR, Brinster RL. Glial cell line-derived neurotrophic factor regulation of genes essential for self-renewal of mouse spermatogonial stem cells is dependent on src family kinase signaling. J Biol Chem. 2007;282:25842–51.
60. Niu ZY, Goodyear SM, Rao S, Wu X, Tobias JW, Avarbock MR, et al. MicroRNA-21 regulates the self-renewal of mouse spermatogonial stem cells. Proc Natl Acad Sci U S A. 2011;108:12740–5.
61. He ZP, Jiang JJ, Kokkinaki M, Tang L, Zeng WX, Gallicano I, et al. MiRNA-20 and mirna-106a regulate spermatogonial stem cell renewal at the post-transcriptional level via targeting STAT3 and Ccnd1. Stem Cells. 2013;31:2205–17.
62. Moritoki Y, Hayashi Y, Mizuno K, Kamisawa H, Nishio H, Kurokawa S, et al. Expression profiling of microRNA in cryptorchid testes: miR-135a contributes to the maintenance of spermatogonial stem cells by regulating FoxO1. J Urol. 2014;191:1174–80.
63. Song W, Mu H, Wu J, Liao M, Zhu H, Zheng L, et al. miR-544 regulates dairy goat male germline stem cell self-renewal via targeting PLZF. J Cell Biochem. 2015;116:2155–65.
64. Cui N, Hao G, Zhao Z, Wang F, Cao J, Yang A. MicroRNA-224 regulates self-renewal of mouse spermatogonial stem cells via targeting DMRT1. J Cell Mol Med. 2016;20(8):1503–12.
65. Li M, Yu M, Liu C, Zhu H, He X, Peng S, et al. miR-34c works downstream of p53 leading to dairy goat male germline stem-cell (mGSCs) apoptosis. Cell Prolif. 2013;46:223–31.
66. Niu BW, Wu J, Mu HL, Li B, Wu CY, He X, et al. miR-204 regulates the proliferation of dairy goat spermatogonial stem cells via targeting to Sirt1. Rejuvenation Res. 2016;19:120–30.
67. Busada JT, Geyer CB. The role of retinoic acid (RA) in spermatogonial differentiation. Biol Reprod. 2016;94:10.
68. Zhou Q, Nie R, Li Y, Friel P, Mitchell D, Hess RA, et al. Expression of stimulated by retinoic acid gene 8 (Stra8) in spermatogenic cells induced by retinoic acid: an in vivo study in vitamin A-sufficient postnatal murine testes. Biol Reprod. 2008;79:35–42.
69. Huszar JM, Payne CJ. MicroRNA 146 (Mir146) modulates spermatogonial differentiation by retinoic acid in mice. Biol Reprod. 2013;88:15.
70. Tong MH, Mitchell D, Evanoff R, Griswold MD. Expression of Mirlet7 family microRNAs in response to retinoic acid-induced spermatogonial differentiation in mice. Biol Reprod. 2011;85:189–97.
71. Tong MH, Mitchell DA, McGowan SD, Evanoff R, Griswold MD. Two miRNA Clusters, Mir-17-92 (Mirc1) and Mir-106b-25 (Mirc3), Are Involved in the Regulation of Spermatogonial Differentiation in Mice. Biol Reprod. 2012;86(3):72.
72. Xie RY, Lin XL, Du T, Xu K, Shen HF, Wei F, et al. Targeted Disruption of miR-17-92 ImpairsMouse Spermatogenesis by Activating mTOR Signaling Pathway. Medicine. 2016;95(7):e2713.
73. Yang QE, Racicot KE, Kaucher AV, Oatley MJ, Oatley JM. MicroRNAs 221 and 222 regulate the undifferentiated state in mammalian male germ cells. Development. 2013;140:280–90.
74. Yu M, Mu H, Niu Z, Chu Z, Zhu H, Hua J. miR-34c enhances mouse spermatogonial stem cells differentiation by targeting Nanos2. J Cell Biochem. 2014;115:232–42.

75. Choi YJ, Lin CP, Ho JJ, He XY, Okada N, Bu PC, et al. miR-34 miRNAs provide a barrier for somatic cell reprogramming. Nat Cell Biol. 2011;13:1353–U1154.
76. Yuan S, Tang C, Zhang Y, Wu J, Bao J, Zheng H, et al. mir-34b/c and mir-449a/b/c are required for spermatogenesis, but not for the first cleavage division in mice. Biol Open. 2015;4:212–23.
77. Bao J, Li D, Wang L, Wu J, Hu Y, Wang Z, et al. MicroRNA-449 and microRNA-34b/c function redundantly in murine testes by targeting E2F transcription factor-retinoblastoma protein (E2F-pRb) pathway. J Biol Chem. 2012;287:21686–98.
78. Marcet B, Chevalier B, Luxardi G, Coraux C, Zaragosi LE, Cibois M, et al. Control of vertebrate multiciliogenesis by miR-449 through direct repression of the Delta/Notch pathway. Nat Cell Biol. 2011;13:693–U157.
79. Bjork JK, Sandqvist A, Elsing AN, Kotaja N, Sistonen L. miR-18, a member of Oncomir-1, targets heat shock transcription factor 2 in spermatogenesis. Development. 2010;137:3177–84.
80. Smorag L, Zheng Y, Nolte J, Zechner U, Engel W, Pantakani DVK. MicroRNA signature in various cell types of mouse spermatogenesis: Evidence for stage-specifically expressed miRNA-221, –203 and -34b-5p mediated spermatogenesis regulation. Biol Cell. 2012;104:677–92.
81. Kleene KC. Patterns, mechanisms, and functions of translation regulation in mammalian spermatogenic cells. Cytogenet Genome Res. 2003;103:217–24.
82. Pradeepa MM, Manjunatha S, Sathish V, Agrawal S, Rao MRS. Involvement of importin-4 in the transport of transition protein 2 into the spermatid nucleus. Mol Cell Biol. 2008;28:4331–41.
83. Meikar O, Da Ros M, Kotaja N. Epigenetic regulation of male germ cell differentiation. Subcell Biochem. 2013;61:119–38.
84. Dai LS, Tsai-Morris CH, Sato H, Villar J, Kang JH, Zhang JB, et al. Testis-specific miRNA-469 Up-regulated in gonadotropin-regulated testicular RNA helicase (GRTH/DDX25)-null mice silences transition protein 2 and protamine 2 messages at sites within coding region: implications of its role in germ cell development. J Biol Chem. 2011;286:44306–18.
85. Yu ZR, Raabe T, Hecht NB. MicroRNA Mirn122a reduces expression of the posttranscriptionally regulated germ cell transition protein 2 (Tnp2) messenger RNA (mRNA) by mRNA cleavage. Biol Reprod. 2005;73:427–33.
86. Liu WM, Pang RTK, Chiu PCN, Wong BPC, Lao KQ, Lee KF, et al. Sperm-borne microRNA-34c is required for the first cleavage division in mouse. Proc Natl Acad Sci U S A. 2012;109:490–4.
87. Zhao K, Chen YP, Yang RF, Bai Y, Li CL, Li HG, et al. miR-424/322 is downregulated in the semen of patients with severe DNA damage and may regulate sperm DNA damage. Reprod Fertil Dev. 2016;28:1598–607.
88. Liu T, Cheng WW, Gao YT, Wang H, Liu ZX. Microarray analysis of microRNA expression patterns in the semen of infertile men with semen abnormalities. Mol Med Rep. 2012;6:535–42.PubMedGoogle Scholar
89. Abu-Halima M, Hammadeh M, Schmitt J, Leidinger P, Keller A, Meese E, Backes C. Altered microRNA expression profiles of human spermatozoa in patients with different spermatogenic impairments. Fertil Steril. 2013;99:1249.
90. Zhou R, Wang R, Qin YF, Ji J, Xu MF, Wu W, Chen MJ, Wu D, Song L, Shen HB, Sha JH, Miao DS, et al. Mitochondria-related miR-151a-5p reduces cellular ATP production by targeting CYTB in asthenozoospermia. Sci Rep. 2015;5:17743.
91. Yadav RP, Kotaja N. Small RNAs in spermatogenesis. Mol Cell Endocrinol. 2014;382:498–508.
92. Yoshida S, Sukeno M, Nabeshima YI. A vasculature-associated niche for undifferentiated spermatogonia in the mouse testis. Science. 2007;317:1722–6.
93. Li LH, Xie T. Stem cell niche: structure and function. Annu Rev Cell Dev Biol. 2005;21:605–31.
94. Oatley JM, Brinster RL. The germline stem cell niche unit in mammalian testes. Physiol Rev. 2012;92:577–95.

95. Ortogero N, Hennig GW, Langille C, Ro S, McCarrey JR, Yan W. Computer-assisted annotation of murine Sertoli cell small RNA transcriptome. Biol Reprod. 2013;88:3.
96. Wainwright EN, Jorgensen JS, Kim Y, Truong V, Bagheri-Fam S, Davidson T, et al. SOX9 Regulates MicroRNA miR-202-5p/3p Expression During Mouse Testis Differentiation. Biol Reprod. 2013;89(2):34.
97. Yao C, Sun M, Yuan Q, Niu M, Chen Z, Hou J, et al. MiRNA-133b promotes the proliferation of human Sertoli cells through targeting GLI3. Oncotarget. 2016;7:2201–19.
98. Dabaja AA, Mielnik A, Robinson BD, Wosnitzer MS, Schlegel PN, Paduch DA. Possible germ cell-Sertoli cell interactions are critical for establishing appropriate expression levels for the Sertoli cell-specific MicroRNA, miR-202-5p, in human testis. Basic Clin Androl. 2015;25:2.
99. Papaioannou MD, Pitetti JL, Ro S, Park C, Aubry F, Schaad O, et al. Sertoli cell Dicer is essential for spermatogenesis in mice. Dev Biol. 2009;326:250–9.
100. Kim GJ, Georg I, Scherthan H, Merkenschlager M, Guillou F, Scherer G, et al. Dicer is required for Sertoli cell function and survival. Int J Dev Biol. 2010;54:867–75.
101. Ma CP, Song HB, Yu L, Guan KF, Hu PD, Li Y, et al. miR-762 promotes porcine immature Sertoli cell growth via the ring finger protein 4 (RNF4) gene. Sci Rep. 2016;6:32783.
102. Nicholls PK, Harrison CA, Walton KL, McLachlan RI, O'Donnell L, Stanton PG. Hormonal regulation of sertoli cell micro-RNAs at spermiation. Endocrinology. 2011;152:1670–83.
103. Panneerdoss S, Chang YF, Buddavarapu KC, Chen HIH, Shetty G, Wang HZ, et al. Androgen-Responsive MicroRNAs in Mouse Sertoli Cells. Plos One. 2012;7(7):e41146.
104. King IA, OBrien TJ, Buxton RS. Expression of the "skin-type" desmosomal cadherin DSC1 is closely linked to the keratinization of epithelial tissues during mouse development. J Investig Dermatol. 1996;107:531–8.
105. Leavy M, Trottmann M, Liedl B, Reese S, Stief C, Freitag B, et al. Effects of elevated beta-estradiol levels on the functional morphology of the testis - new insights. Sci Rep. 2017;7:39931.
106. Kumar N, Srivastava S, Burek M, Forster CY, Roy P. Assessment of estradiol-induced gene regulation and proliferation in an immortalized mouse immature Sertoli cell line. Life Sci. 2016;148:268–78.
107. Jiao ZJ, Yi W, Rong YW, Kee JD, Zhong WX. MicroRNA-1285 regulates 17beta-estradiol-inhibited immature boar sertoli cell proliferation via adenosine monophosphate-activated protein kinase activation. Endocrinology. 2015;156:4059–70.
108. Ge RS, Hardy MP. Variation in the end products of androgen biosynthesis and metabolism during postnatal differentiation of rat Leydig cells. Endocrinology. 1998;139:3787–95.
109. Liu H, Yang Y, Zhang L, Liang R, Ge RS, Zhang YF, et al. Basic fibroblast growth factor promotes stem Leydig cell development and inhibits LH-stimulated androgen production by regulating microRNA expression. J Steroid Biochem Mol Biol. 2014;144:483–91.
110. Rakoczy J, Fernandez-Valverde SL, Glazov EA, Wainwright EN, Sato T, Takada S, et al. MicroRNAs-140-5p/140-3p Modulate Leydig Cell Numbers in the Developing Mouse Testis. Biol Reprod. 2013;88(6):143.
111. Song HW, Wilkinson MF. Transcriptional control of spermatogonial maintenance and differentiation. Semin Cell Dev Biol. 2014;30:14–26.

CHAPTER 4

Fatty Acid Binding Protein 3 and Transzonal Projections Are Involved in Lipid Accumulation During *in Vitro*Maturation of Bovine Oocytes

Maite del Collado , Juliano Coelho da Silveira, Juliano Rodrigues Sangalli, Gabriella Mamede Andrade, Letícia Rabello da Silva Sousa, Luciano Andrade Silva, Flavio Vieira Meirelles & Felipe Perecin

Veterinary Medicine Department, Faculty of Animal Sciences and Food Engineering, University of Sao Paulo, Av. Duque de Caxias Norte 225, 13635-900, Pirassununga, SP, Brazil

ABSTRACT

Oocytes that undergo *in vitro* maturation (IVM) are metabolically abnormal and accumulate excess lipid content. However, the mechanism of lipid accumulation and the role of cumulus cells in this process are unclear. Recently, it was shown that fatty acid binding proteins (FABPs) performed intra- and extracellular fatty acid transport. We postulated that FABP3 might be responsible for fatty acid transport from cumulus cells to the oocytes *via* transzonal projections (TZPs) during IVM. Transcript and protein levels of FABP3 were analyzed in both *in vivo*- and *in vitro*-matured cumulus-oocyte-complexes and were increased in IVM samples. Further analysis showed increased lipid content in oocytes and cumulus cells in IVM samples compared to *in vivo*-derived. We therefore speculated that altered traffic of fatty acids *via* FABP3 during IVM was the mechanism leading to the excess of lipids accumulated within IVM oocytes. Furthermore, we demonstrated an increase in FABP3 levels and lipid content during the first 9 h of IVM, further strengthening the possibility of fatty acid transport *via* FABP3 and TZPs. Additionally, disruptions of TZPs during IVM decreased lipid accumulation in oocytes. Our results shed light on a possible mechanism involving FABP3 and TZPs that causes excess lipid accumulation in oocytes during IVM.

INTRODUCTION

Assisted reproductive technologies (ARTs) expose gametes and embryos to non-physiological conditions that may cause abnormal development. Increased lipid accumulation during *in vitro* production (IVP) of bovine embryos is one of the most well recognized metabolic abnormalities that accompany the outcome of successful ARTs. The negative effects of increased lipid content in blastomeres include reduced potential for cryopreservation and poor post-thawing embryo survival rates, resulting in low rates of pregnancy and embryo losses[1,2]. Studies on bovine embryos produced using different IVP systems have provided compelling evidence for correlation between lipid-rich culture media supplemented with fetal bovine serum (FBS) and high lipid content as well as low cryotolerance of blastocysts[3,4,5]. It was demonstrated that the lipid accumulation in species such as bovine and mouse occurs during the first step of IVP, the *in vitro* maturation (IVM) of oocytes[3,6], and that it does not occur when the maturation takes place *in vivo*[3]. Furthermore, the lipid accumulation was accentuated when the oocytes were matured in the presence of FBS[3]. Additionally, similar events were observed *in vivo*, when cumulus-oocyte complexes (COCs) were exposed to lipid-rich environments such as follicular fluids of obese women; this exposure may be responsible for excessive lipid accumulation in oocytes, leading to impaired fertility[7,8,9,10,11].

Communication between cumulus cells and oocytes is important for optimal oocyte maturation. This communication is mediated by paracrine signals, transzonal projections (TZPs), gap junctions and possibly extracellular vesicles[12,13]. Despite the paracrine signals, molecules could be transported within TZPs and be transferred to the oocyte by GAP junctions or by the recently proposed extracellular vesicle mediated intercellular transport[13,14]. Gap junctions can transport low-weight molecules (<1 kDa) important for oocyte development, such as ions, nucleotides, amino acids, and metabolites, and are functional during the first 6 h of IVM[15,16]. TZPs are large channels, ~2 μm in diameter, that allow the transport of large molecules such as mRNAs to oocytes until ~9 h after the resumption of meiosis[13]. Exchange of lipids between cumulus cells and oocytes or lipid carriers for fatty acid transport have not been observed in this system so far.

Fatty acid molecules are transported intra- and extracellularly by a family of lipid-binding proteins called Fatty Acid Binding Proteins (FABP); 9 isoforms of FABPs have been reported[17]. These proteins bind mostly to long chain fatty acids (C16-C20) and transport them to the peroxisome, mitochondria or endoplasmic reticulum[18,19,20]. FABPs also transport lipids extracellularly, and have been observed free or enclosed in extracellular vesicles in several tissues and body fluids; they thus serve pleiotropic functions in systemic metabolism[17,21]. Isoform 3 of these lipid-binding proteins, called FABP3 or heart-FABP, seems to have altered functionality within COCs because its transcript levels were reported to be increased during IVM[22].

Even though the dynamic interaction between cumulus cells and oocytes as well as the transport of several molecules into COCs are well known, there is still a gap in the knowledge of the mechanism of lipid metabolism in these complexes during *in vivo* or IVM. In this study, we used an animal model to investigate lipid transport in COCs during oocyte maturation. This model allowed us to study the roles of FABP3 and TZPs in lipid accumulation as well as the influence of the *in vitro* environment imposed by ARTs on the lipid accumulation in COCs.

We demonstrated the transport of FABP3 between cumulus cells and oocytes through TZPs during IVM. This traffic is maintained until around 9 h after the beginning of IVM, and is consistent with the period at which higher FABP3 expression and lipid accumulation was observed in the oocytes. Additionally, blocking TZP formation resulted in reduced lipid droplets content in oocytes undergoing IVM. Our data also suggested that IVM altered the FABP3-mediated accumulation of lipids in COCs. To the best of our knowledge, this is the first manuscript to describe a mechanism of lipid transport within COCs. The data presented here will provide new insights that help elucidate the mechanism behind lipid accumulation during IVM and potentially lead to the development of new approaches to improve ARTs.

RESULTS

IVM leads to increased lipid content in COCs

To investigate the dynamics of lipid accumulation *in vivo* and IVM, we estimated the lipid content in cumulus cells and oocytes. Higher amounts of lipid droplets were observed in both cumulus cells (Figure 1a and Supplementary Figure S2a) and oocytes obtained after IVM (Figure 1b and Supplementary Figure S1). The results demonstrated that IVM caused an increase in lipid content, whereas *in vivo* cumulus cells and oocytes had lipid content similar to that in the immature group.

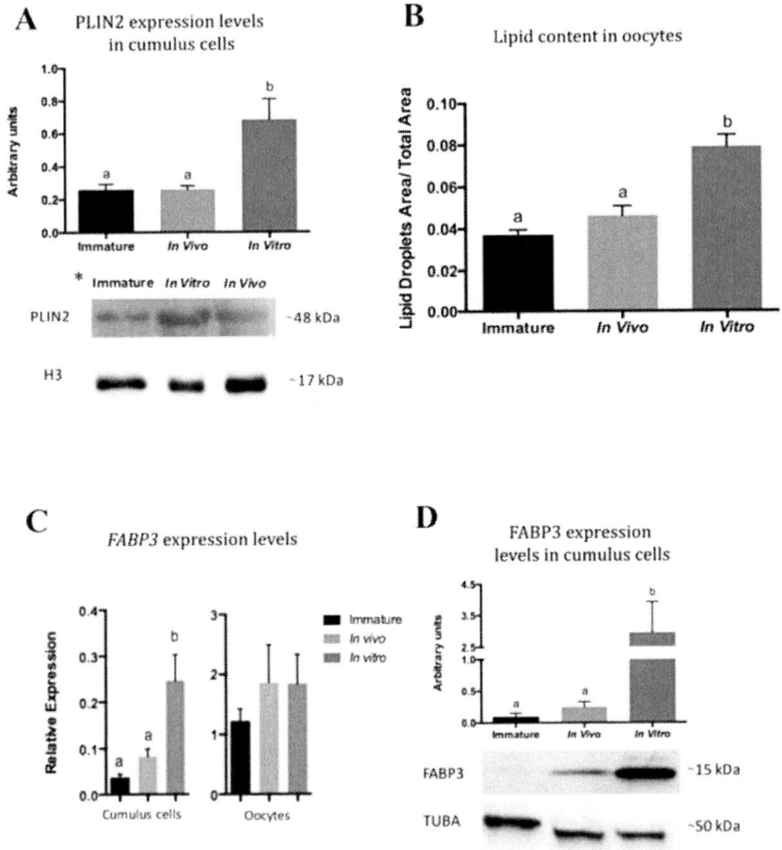

Figure 1. Quantification of lipid content in cumulus cells and oocytes derived from immature, *in vivo*-matured and *in vitro*-matured cumulus-oocyte complexes (COCs). (**A**) Lipid content in cumulus cells was determined by western blot analysis of perilipin 2 (PLIN2) using histone 3 (H3) as a normalizer. (**B**) Lipid quantification in oocytes was performed by fluorescence confocal microscopy of lipid droplets. The values for lipid content represent the ratio of area of lipid droplets to total oocyte area. Representative immature, *in vivo*-matured and *in vitro*-matured oocytes stained with BODIPY 493/503 are shown in Supplementary Figure S1. (**C**) Relative amounts of fatty acid binding protein 3 (*FABP3)* transcripts in immature, *in vivo*-, and *in vitro*- matured cumulus cells and oocytes. (**D**) FABP3 protein levels in cumulus cells normalized by x-Tubulin (TUBA). Lowercase letters above bars in the same graph indicate significant differences ($P < 0.05$). Values are presented as mean ± standard error of the mean. The full western blot images in (**A**) and (**D**) are shown in Supplementary Figure S2a and S2b, respectively. *The order of the groups is different between the graph and the representative immunoblot images.

IVM increases FABP3 transcript and protein levels in cumulus cells

In order to investigate the effect of IVM on FABP3 function, we determined the transcript and protein levels of FABP3 in cumulus cells derived from immature, *in vivo*- and *in vitro*-matured COCs. We observed that the relative amount of FABP3 transcripts increased in cumulus cells after IVM of COCs (Figure 1c). Interestingly, no difference was observed in the relative levels of FABP3 transcripts in oocytes before and after IVM (Figure 1c). Similarly, FABP3 protein levels in cumulus cells obtained from COCs that underwent IVM were also found to be higher than those in immature and *in vivo*-matured cells (Figure 1d and Supplementary Figure S2b). Our data demonstrated that *in vitro*-matured COCs had higher transcript and protein levels of FABP3 in cumulus cells.

FABP3 is localized within TZPs during IVM

Next, we aimed to investigate whether FABP3 was localized along with the TZPs during IVM. We obtained a series of immunostained images of immature COCs as well as COCs matured *in vitro* for 9 or 18 h. We observed that FABP3 was present within TZPs in immature COCs (Figure 2 and Supplementary Figure S3). After 9 h of IVM, the amount of FABP3 present in TZPs increased (Figure 3 and Supplementary Figure S4). To verify the presence of FABP3 within TZPs across the zona pellucida (ZP), we subjected oocytes that were either denuded and partially denuded to maturation for 9 h, and then again carried out immunostaining for FABP3 (Figure 4 and Supplementary Figs S5 and S6). We were unable to detect FABP3 in the ZP of denuded oocytes, suggesting that the presence of TZPs is necessary for the transport of FABP3 between the cumulus and oocytes (Figure 4a and b). Furthermore, in partially denuded oocytes, we only detected FABP3 along with the TZPs in the ZP (Figure 4c and d), ruling out the possibility that FABP3 moves between the cumulus and oocytes using other mechanisms. Additionally, because the TZP-mediated transport ceased when oocytes became mature, we investigated the location of FABP3 and TZPs after 18 h of IVM. As expected, we observed that the TZPs were disconnected from the ooplasm and the FABP3 was localized at the terminal portion of the TZPs (Figure 5 and Supplementary Figure S7), suggesting that it was derived from cumulus cells. In summary, our results demonstrated that FABP3 was present within TZPs in immature and 9-h matured COCs (Figure 6a and b). In addition, after 18 h of maturation, when the first polar body was extruded, the TZPs were disconnected from the ooplasm and the FABP3 molecules accumulated at the terminal portion of these projections (Figure 6c). We also performed a negative control for each immunoreaction: we exposed the samples to the maximum laser potency during confocal microscopy and did not detect any signals in the zona pellucida, ruling out the possibility that our signals are artifacts (Supplemental Figure S8).

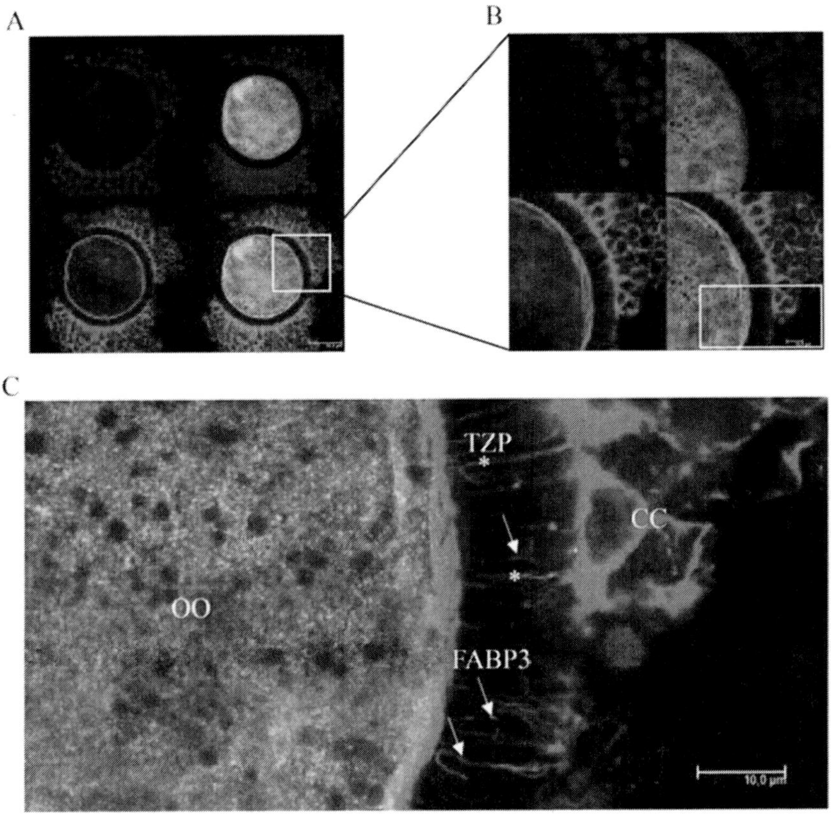

Figure 2. Confocal microscopic analysis demonstrating the presence of fatty acid binding protein 3 (FABP3) within the transzonal projections (TZPs) of immature cumulus-oocyte-complexes (COCs). FABP3 was immunostained using 488 Alexa Fluor (green); TZPs (actin) were labeled with Alexa Fluor 647 phalloidin (red) and nuclei were stained with DAPI (blue). (**A**) Photomicrographs of immature COCs were obtained in 63 x objective and (**B**) in 63 x objective zoomed 3.5 x. (**C**) Digital zoom showing FABP3 (arrow) within TZPs (asterisk) in zona pellucida (OO = oocyte; CC = cumulus cells). A total of 12 immature COCs were analysed and showed similar pattern; 6 additional COCs are shown in Supplementary Figure S3.

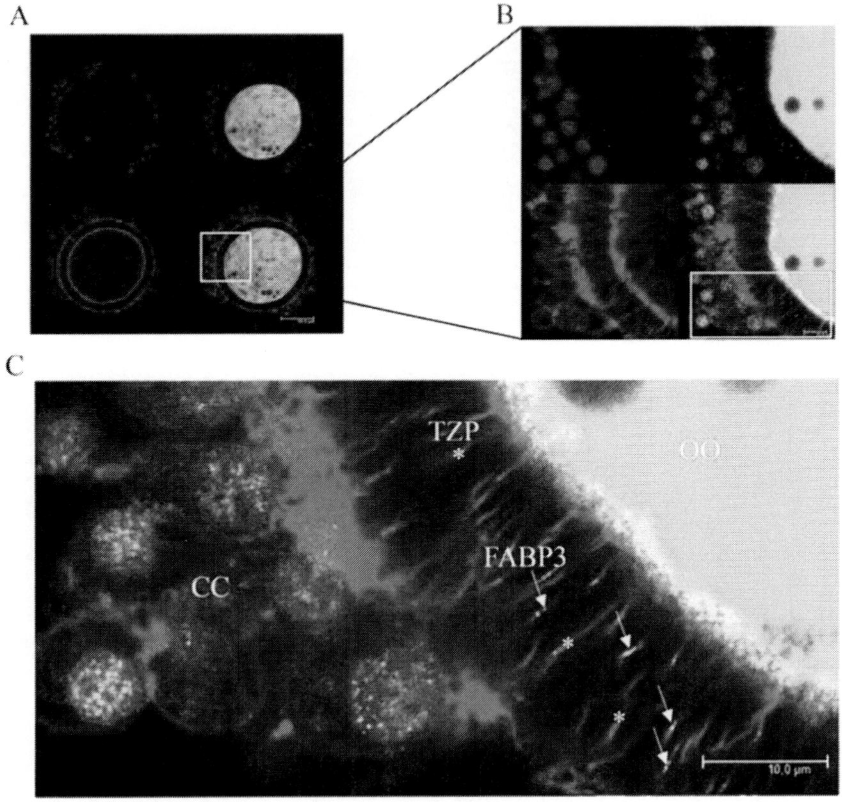

Figure 3. Confocal photomicrographs of fatty acid binding protein 3 (FABP3) immunodetection in transzonal projections (TZPs) after 9 h of *in vitro* maturation (IVM). FABP3 is immunostained using 488 Alexa Fluor (green); TZPs (actin) are stained with Alexa Fluor 647 phalloidin (red); nuclei are stained with DAPI (blue). (**A**) Photomicrographs of COCs after 9 h of IVM were obtained in 63x objective and (**B**) in 63x objective zoomed 3.5x. (**C**) Digital zoom showing FABP3 (arrow) along TZPs (asterisk) in the zona pellucida (OO = oocyte; CC = cumulus cells). A total of 12 COCs *in vitro*-matured for 9 hours were analysed and showed similar pattern; 6 additional COCs are shown in Supplementary Figure S4.

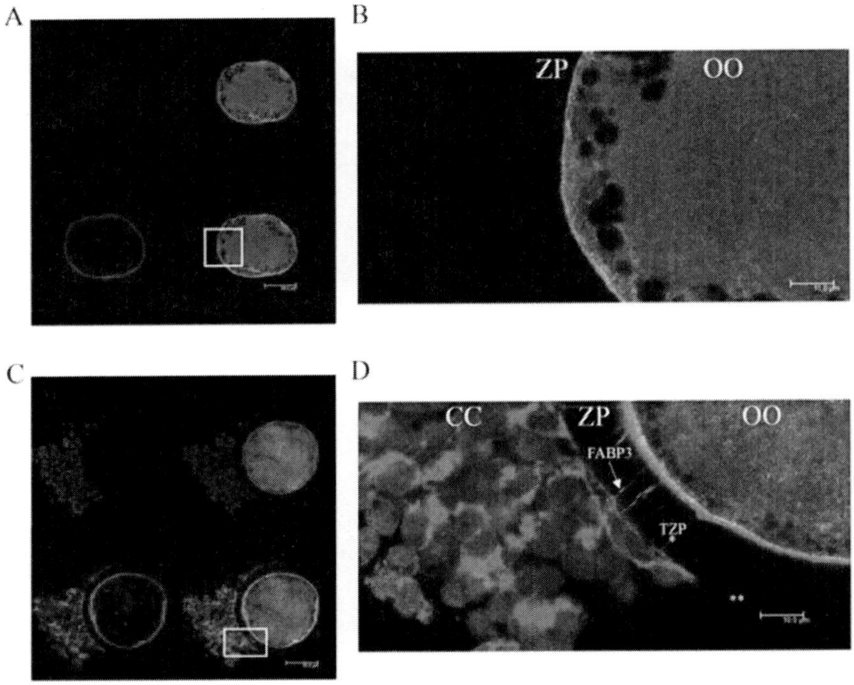

Figure 4. Confocal microscopic analysis of fatty acid binding protein 3 (FABP3) in tranzonal projections (TZPs) of denuded and partially denuded oocytes cultured in the maturation media for 9 h. (**A**) Images acquired from denuded oocytes show no FABP3 and TZPs within the zona pellucida. (**B**) The image of a portion of the denuded oocytes demonstrating the complete absence of TZPs and FABP3 within the zona pellucida. (**C**) Images acquired from a partially denuded oocyte demonstrating the presence of FABP3 at the same location as TZPs within the zona pellucida. (**D**) The portion of the partially denuded COCs demonstrating the complete absence of TZPs and FABP3 within the zona pellucida in the denuded area, and the presence of FABP3 and TZPs in the cumulus-enclosed area. FABP3 (arrow) is immunostained using 488 Alexa Fluor (green); TZPs (actin; asterisk) are stained with Alexa Fluor 647 phalloidin (red); nuclei are stained with DAPI (blue). Photomicrographs were obtained in 63x objective (**A** and **C**), and merged with z-stack captures with digital zoom (**B** and **D**) to illustrate the absence (**B**) or presence (**D**) of FABP3 along the TZPs. (OO = oocyte; CC = cumulus cells). ** indicates the lack of cumulus cells and TZPs. A total of 10 denuded oocytes and 11 partially denuded oocytes *in vitro*-matured for 9 hours were analysed and showed similar pattern; 6 additional denuded oocytes and 6 additional partially denuded oocytes are shown in Supplementary Figs S5 and S6, respectively.

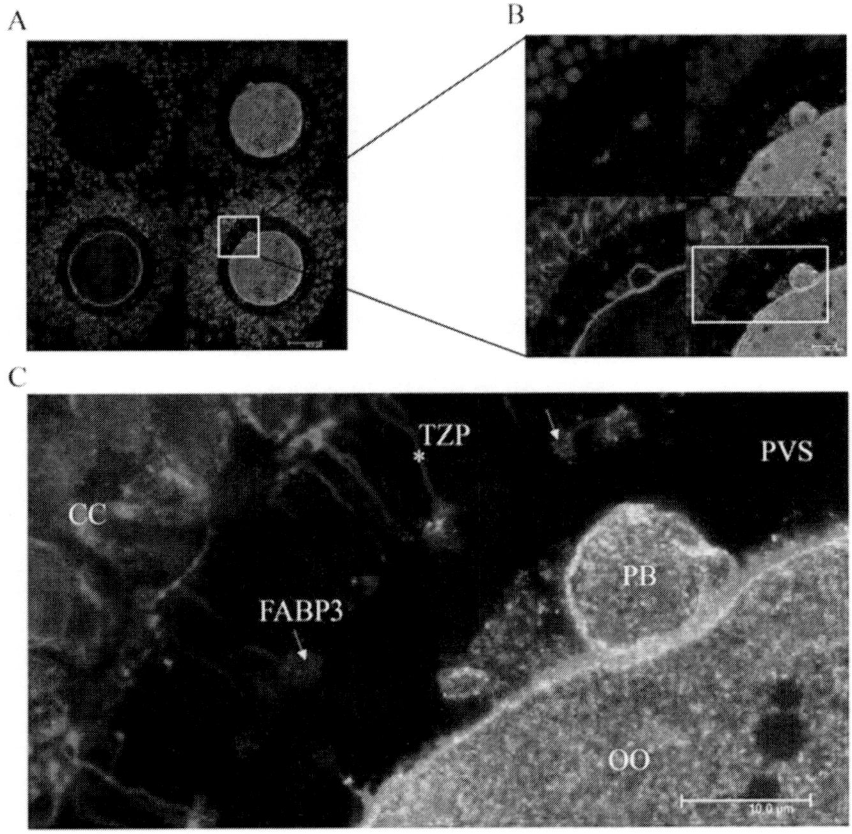

Figure 5. Confocal microscopic analysis of cumulus-oocytes complexes (COCs) after 18 h of IVM. (**A**) Confocal photomicrographs of immunodetection of fatty acid binding protein 3 (FABP3) in tranzonal projections (TZPs) in 18-h matured COCs. (**B**) Photomicrograph showing a detailed view of the COC using a 63x objective and zoomed 3.5x. (**C**) The digital zoom showing the details of the terminal portions of the TZPs disconnected from the ooplasm; FABP3 are immunolocalized at the terminal portion of the TZPs. FABP3 is immunostained using 488 Alexa Fluor (green); TZPs (actin) are stained with Alexa Fluor 647 phalloidin (red); nuclei are stained with DAPI (blue). Photomicrographs were obtained in 63 x objective (**A**) and in 63 x objective zoomed 3.5 x (**B**). Digital zoom (**C**) showing a few FABP3 molecules (arrow) in the terminal portion of TZPs (asterisk) and the perivetellin space (PVS) between oocyte (OO) and zona pellucida. (CC = cumulus cells; PB = polar body). A total of 15 COCs *in vitro*-matured for 18 hours were analysed and showed similar pattern; 6 additional COCs are shown in Supplementary Figure S7.

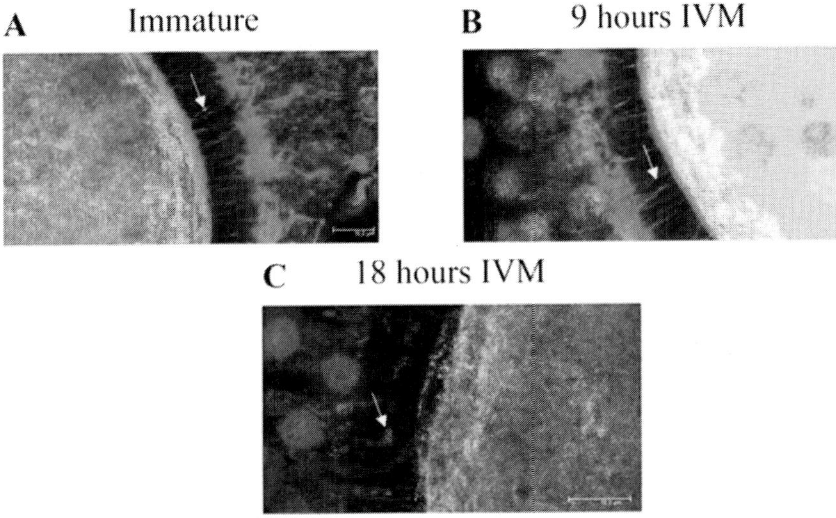

Figure 6. Confocal microscopic analysis of cumulus-oocytes complexes (COCs) at different stages of oocyte maturation (merge of z-stack images) in 63 x objective with digital zoom. (**A**) Photomicrograph showing the localization of fatty acid binding protein 3 (FABP3) and tranzonal projections (TZPs) in immature COCs. (**B**) Photomicrograph showing the localization of FABP3 and TZPs in COCs after 9 h of maturation. (**C**) Photomicrograph showing the localization of FABP3 and TZPs in COCs after 18 h of maturation. Arrows indicate FABP3 immunolocalized in TZPs. FABP3 is immunostained using 488 Alexa Fluor (green); TZPs (actin) are stained with Alexa Fluor 647 phalloidin (red); nuclei are stained with DAPI (blue).

FABP3 protein levels and lipid droplets increase until up to 9 h in oocytes during IVM

Since our data suggested the movement of FABP3 from cumulus to oocytes, we expected an increase in the FABP3 protein levels as well as the lipid content in oocytes during IVM. To verify this, we conducted western blot analysis and lipid droplets staining in immature oocytes and in oocytes that underwent IVM for 9 and 18 h. The protein analysis revealed a ~1.55-fold increase in FABP3 levels between immature oocytes and 9-h matured ones, while no difference was observed between 9-h and 18-h matured oocytes (Figure 7a and Supplementary Figure S2c). The lipid droplet analysis showed a ~1.44-fold increase in lipid accumulation in 9-h matured oocytes compared to immature ones (Figure 7b). Interestingly, no differences were observed between 9- and 18-h matured oocytes, suggesting that there was no lipid accumulation during this time. Thus, our results suggested that there was simultaneous accumulation of FABP3 as well as lipids in the oocytes during the first 9 h of IVM.

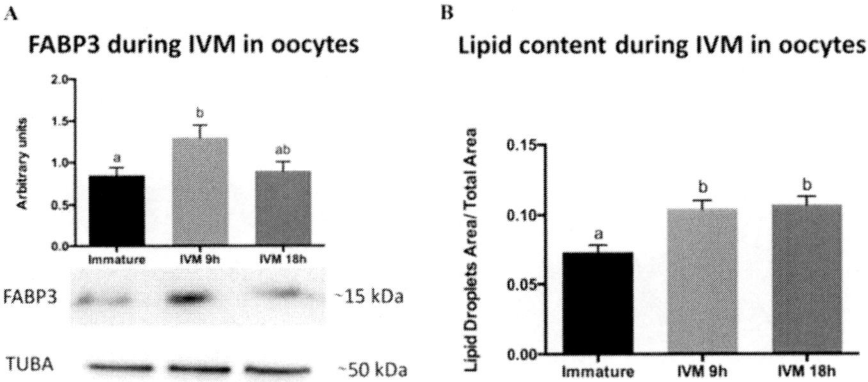

Figure 7. Fatty acid binding protein 3 (FABP3) protein levels and lipid content in oocytes from immature cumulus-oocytes complexes (COCs) and COCs *in vitro*-matured for 9 and 18 h. (**A**) FABP3 protein quantification in oocytes was performed by western blot analysis using x-Tubulin (TUBA) as the normalizer. The full western blot image is shown in Supplementary Figure S2c. (**B**) Lipid quantification in oocytes. The values for lipid content represent the ratio of the area of lipid droplets to the total oocyte area. Letters above bars in the same graph indicate significant difference ($P < 0.05$). Values are represented as mean ± standard error of the mean.

The disruption of TZPs in oocytes causes decrease in lipid content

Because our results suggested that lipids are transported from cumulus cells to oocytes in a TZP-dependent manner, we decided to disrupt the TZPs to further clarify its role in lipid accumulation. To this end, we utilized cytochalasin B, an agent that blocks the polymerization of actin filaments and thus disrupts the TZPs. We exposed the COCs to cytochalasin B during the first 9 h of IVM to verify whether the lipid content was affected. Our results demonstrated that most of the TZPs were disrupted by cytochalasin B during IVM (Figs 8a–d and Supplementary Figure S10). The total lipid content in oocytes that matured in the presence of cytochalasin B was lower than that in the control oocytes (Figure 8e), indicating that functional TZPs are necessary to drive lipid accumulation.

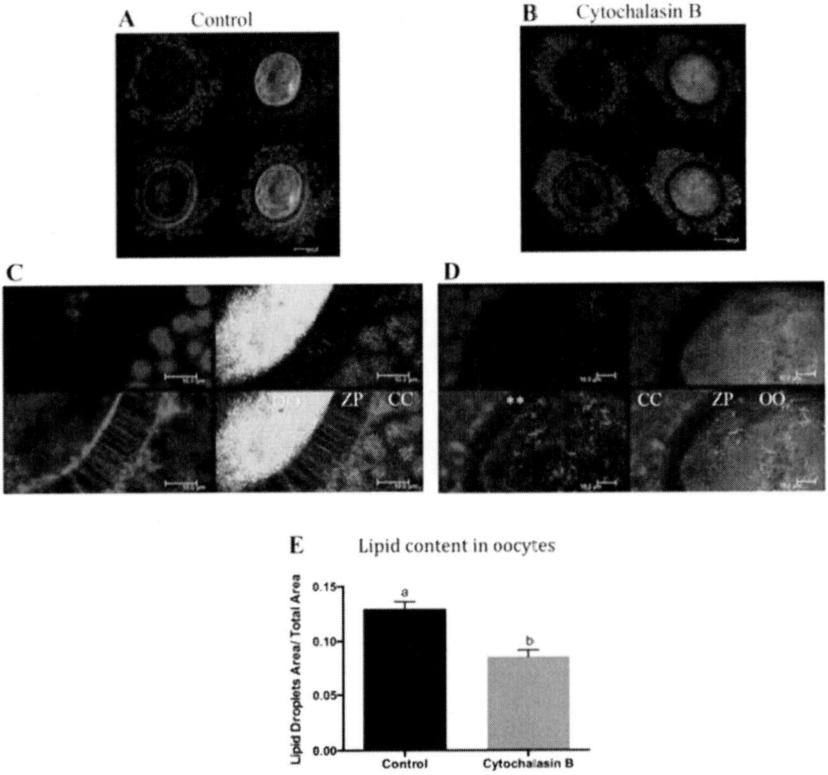

Figure 8. Confocal microscopic analysis of fatty acid binding protein 3 (FABP3) in cumulus-oocytes complexes (COCs) after *in vitro* maturation (IVM) in the absence (control) and presence of cytochalasin B. (**A**) and (**C**) The photomicrographs showing the presence of tranzonal projections (TZPs) and FABP3 in the control group. (**B**) and (**D**) Photomicrographs showing the lack of most parts of TZPs and FABP3 within the zona pellucida. FABP3 was immunostained using 488 Alexa Fluor (green); TZPs (actin) are stained with Alexa Fluor 647 phalloidin (red); nuclei are stained with DAPI (blue). Photomicrographs were obtained in 63x objective (**A** and **C**); and in 63 x objective zoomed 3.5 x (**B** and **D**). CC = cumulus cells; ZP = zona pellucida; OO = oocyte. **Indicate the disassembly and blocking of TZP formation by cytochalasin B. A total of 10 control COCs and 10 cytochalasin treated COCs *in vitro*-matured for 9 hours were analysed and showed similar pattern; 6 additional control COCs and 6 additional cytochalasin B treated COCs are shown in Supplementary Figure S10. (**E**) Lipid content in oocytes from COCs after IVM for 9 h in the absence (control) or presence of cytochalasin B. Lipid quantification was performed by fluorescence confocal microscopy to detect lipid droplets. Representative control and cytochalasin B-treated oocytes stained with BODIPY 493/503 are shown in Supplementary Figure S11. The values for lipid content represent the ratio of the area of lipid droplets to the total oocyte area. Letters above bars indicate significant differences ($P < 0.05$). Values are represented as mean ± standard error of the mean.

DISCUSSION

The increased lipid accumulation in *in vitro*-produced embryos is one of the major disadvantages of IVP, causing impaired cryotolerance and low pregnancy rates[1,2]. It was demonstrated that this accumulation is one of the consequences of IVM in species such as bovine and mouse, and in bovine is caused by exposure of the gametes to a non-physiological environment[3,6]. In the present study, we proposed that the excessive lipid accumulation in *in vitro*-matured oocytes is mediated by the dysregulated transport of a fatty acid lipid-carrier protein named FABP3 through TZPs. We demonstrated that the *in vitro* system caused an increase in FABP3 levels. Additionally, an increase in lipid content during IVM was associated with the increase in FABP3 protein levels in *in vitro*-matured cumulus cells. By comparing immature oocytes with 9- and 18-h matured ones, we observed the presence of FABP3 within TZPs during IVM as well as dynamic changes in FABP3 protein levels and lipid levels in oocytes during the maturation process. In addition, we showed that the disruption of TZPs in oocytes during the first 9 h of maturation decreased the lipid accumulation. Thus, our results demonstrated, for the first time, a probable mechanism for the transport of fatty acids from cumulus cells to the oocytes, which may be responsible for the lipid accumulation in oocytes.

Lipid transportation from the medium to the cells or among the cells in *in vitro* systems has been extensively investigated[3,4,23] as a possible cause of lipids accumulation within *in vitro*-produced embryos. Many studies have reported that embryos cultured in media containing fetal bovine serum acquired more lipids and were less cryotolerant than those grown without serum[4,5,23]. Based on that, a relationship between the lipid content of the maturation medium and the elevated lipid levels observed in oocytes after IVM has been established. However, the mechanisms mediating lipid transport within COCs during IVM have never been identified and remains unknown.

We observed that FABP3 was localized along the TZPs during IVM, suggesting that FABP3 molecules that originated from the cumulus may transport fatty acids from these cells to the oocytes, and eventually end up in the oocytes. TZPs, also known as cumulus cells process endings (CCPEs)[24,25], are able to transport molecules <1 kDa to the oocyte by GAP junction located at the final portion of TZPs[26,27] or by the recently proposed traffic of extracellular vesicles into the cleft between the end of TZPs and the oolema[13,28]. Based on the molecular weight of FABP3 (>1 kDa), the transport to oocytes mediated by vesicles is the probable mechanism. This supposition finds a parallel in the reports of transport of large molecules such as mRNA from cumulus cells to the oocytes mediated by TZPs within COCs[13,29]. Additionally, FABP5 as well as other members of this protein family such as FABP4 and FABP3 were described in extracellular vesicles from different body fluids[17,30,31,32,33,34]. Thus, based on the presence of extracellular vesicles in the cleft between TZPs and the ooplasm[13], a reasonable hypothesis is that FABP3s are transported to the oocyte by extracellular vesicles secreted by cumulus cells.

The family of FABP proteins was for a long time associated with intracellular fatty acid transport, mostly to the mitochondria, where β-oxidation takes place. However, recently, FABP proteins were discovered to possess extracellular functions as well, carrying lipids between tissues[35]. Additionally, studies have reported that extracellular FABP proteins were altered in metabolism-associated diseases[17,30,36,37,38,39,40,41]. Cardiac FABP, also known as H-FABP or FABP3, was first observed in reproductive tissues such as ovary and placenta, where it performed both intra- and extracellular functions[17,20]. Based on the current literature, FABP3 and FABP5 are the only FABPs expressed in bovine COCs, but only FABP3 increased during IVM in bovine oocytes, suggesting a possible role in lipid accumulation[22,42].

Exposure of muscle cells to high levels of fatty acids both *in vivo* and *in vitro* induced the expression of FABP3[43,44]. Because we observed an increased lipid accumulation in COCs, we decided to investigate whether FABP3 in cumulus cells were altered after IVM. We demonstrated an increase in lipid content as well as FABP3 transcript and protein levels in cumulus cells undergoing IVM. We also observed an increase in lipid content in oocytes after exposure to an *in vitro* environment, suggesting that FABP3 might be involved in extracellular fatty acid transport as well, instead of only in the canonical intracellular function. Additionally, *in vivo*-matured cumulus cells did not show any increase in the levels of FABP3 compared to immature cumulus cells, indicating that the *in vitro* system caused the dysregulation of lipid transport. Altogether, our data suggested that the mechanism leading to the increased lipid content observed in oocytes after IVM was mediated by FABP3.

From the results of our immunolocalization experiments we verified the relationship between FABP3 and lipid content in oocytes during the initial stage of IVM, by observing the co-localization of FABP3 and TZPs during the first 9 h of IVM. Therefore, the increased FABP3 levels and lipid content observed at this stage might have been caused by the transport of FABP3 through TZPs. We also demonstrated that after 18 h of IVM, the number of TZPs near the ooplasm decreased drastically, and that FABP3 molecules were localized at the terminal portion of the projections. The fact that FABP3 migrated only until the terminal portion of the TZPs and accumulated there, suggested that traffic occurs only from cumulus cells to oocytes and not in the opposite direction. The FABP3 levels and lipid content in the oocytes remained stable from 9 h to 18 h of IVM. Thus, the increase in FABP3 and lipids seemed to be dependent on the presence of active TZPs.

The necessity of functional TZPs to allow for lipid accumulation in oocytes is supported by the literature. The exposure of COCs to high levels of free fatty acid during the last 6 hours of *in vitro* maturation (considering a 23 h IVM) resulted in lipid accumulation in cumulus cells, but not into oocytes45. Therefore, in moment where TZPs are no longer functional, the traffic of lipid from cumulus to oocytes seems to be impaired. Thus, previous work demonstrated increased accumulation of lipids within the cumulus cells, following the exposure of the COCs to rich-lipids environment45; however, here we demonstrated a possible fatty acid transport mechanism from cumulus cells to oocyte during IVM.

To verify whether TZPs are involved in fatty acid transport from cumulus cells to oocytes, we disrupted TZPs using cytochalasin B during the first 9 h of IVM. Initially, we observed that cytochalasin B-treated oocytes lacked most of the TZPs because of the disruption and that FABP3 was not present within the ZP. Therefore, our treatment disrupted the TZPs and blocked the transport of FABP3 from cumulus cells through the ZP. Additionally, we demonstrated a decrease in lipid content in the cytochalasin B-treated oocytes compared to the non-treated controls. Hence, disruption of TZPs decreased lipid accumulation in oocytes, impairing fatty acid transport from cumulus cells to oocytes, thus supporting our initial speculation.

The results of this work shed light on the effects of exposing gametes to *in vitro* culture conditions during ARTs. The reasons why *in vitro* maturation triggers lipid accumulation and FABP3 upregulation are still elusive. Lipid-rich *in vitro* environment do not seen be responsible, as previously discussed3. Metabolic dysregulation (probably of cumulus cells), including abnormal energy metabolism and abnormal triacylglycerol synthesis, are the candidate pathways to trigger lipid accumulation. However, these hypotheses are yet to be confirmed. The results of this work also shed light on fertility impairment in women suffering from metabolic disorders. It is well known that women who are overweight and/or suffering from metabolic syndromes have several fertility problems associated with high lipid content in the blood, follicular fluids, and oocytes. This elevated level of lipid may trigger endoplasmic reticulum stress and consequently apoptosis in COCs due to a lipotoxicity response, leading to the decreased fertilization rates observed in these women9,10,11. However, the mechanism behind this lipid accumulation remains unknown. Recent studies have identified high levels of FABP4 in the blood and tissues of patients with obesity17. Likewise, FABP3 levels were found to be increased in patients with metabolic syndrome46. Even though these types of pathologies are increasingly becoming frequent among women and are amongst the major causes of modern fertility-related problems, we still do not understand the molecular mechanism underlying these issues.

Our results demonstrated that lipid accumulation during IVM was dependent upon the functional communication between cumulus cells and oocytes. We also suggested a possible mechanism for fatty acid transport from cumulus cells to oocytes *via* FABP3 and TZPs. We demonstrated the dysregulation of FABP3 synthesis and its accumulation in oocytes during IVM, concomitant with the lipid accumulation. We suggested that an increase of lipid accumulation in cumulus cells during IVM could lead to an increase of FABP3 expression levels to transport those fatty acids. Thus, this dysregulation in cumulus cells could lead to increased lipid accumulation in oocytes during IVM. Further experiments are necessary to verify this new mechanism of lipid accumulation in oocytes during IVM. Our data strongly suggested that fatty acid transport within COCs is mediated by FABP3 and that FABP3 may play a role in fertility issues. Finally, a better understanding of this mechanism of transport will help in the development of tools to modulate lipid accumulation in oocytes during IVM; this could have great implications for ARTs in animals and humans and for fertility treatments for women suffering from metabolic disorders.

METHODS

All reagents were purchased from Sigma Chemical Co. (St. Louis, MO, USA) unless otherwise indicated. The Bioethical Committee of the FZEA - University of Sao Paulo, Pirassununga, SP, Brazil has approved this study under the protocol number 14.1.675.74.7. We adopted the International Guiding Principles for Biomedical Research Involving Animals (Society for the Study of Reproduction) as well.

The lipid content and FABP3 levels during *in vivo* and IVM

Recovery of immature, *in vitro-*, and *in vivo-*matured oocytes

In order to compare immature, *in vitro-*, and *in vivo-*matured oocytes and cumulus cells, Nellore (*Bos indicus*) COCs were obtained from slaughterhouses or from live cows by the ovum pick-up (OPU) method. To obtain oocytes for IVM, we performed follicular aspiration of the Nellore cow-ovaries obtained post-mortem. To obtain *in vivo-*matured oocytes, we synchronized the ovarian follicular wave of cyclic Nellore cows and then subjected them to follicular superstimulation and OPU. The detailed protocols are presented in the supplementary materials and methods (Supplementary File).

Confocal microscopic analysis of lipid content in oocytes

Lipid quantifications analyses were performed by staining the lipid droplets with a dye specific for neutral lipids, – BODIPY 493/503 (Molecular Probes, Eugene, OR, USA), followed by confocal microscopy imaging. We utilized denuded oocytes from the three groups (49 immature, 46 *in vivo*-matured, and 45 *in vitro*-matured oocytes, obtained from 12 biological replicates) for confocal analysis. The denuded oocytes were fixed, permeabilized, and stained following the manufacturer's instructions. The oocytes were then analyzed using the LSM 710 confocal microscope (Carl Zeiss, Oberkochen, Germany) at 63X magnitude in oil with an Argon 488 laser. One image per oocyte was captured choosing the largest diameter section and later analyzed with the ImageJ program (NIH; http://rsb.info.nih.gov/ij/). The ratio of the lipid droplet area to the total oocyte area was determined and analyzed with ImageJ, as previously described3. Briefly, lipid content in oocytes was visualized using fluorescence confocal microscopy and quantified based on the fraction of area occupied by lipid droplets within the ooplasm. To obtain area of lipid droplets, we transformed the original image to 8 bits image; and then we determined the lipid droplets area with a plugin named "nucleus counter" (detailed in Supplemental Figure S9).

Quantification of PLIN2 and FABP3 in cumulus cells by western blot

Lipid content in cumulus cells was determined based on the amount of PLIN2, a protein present in lipid droplets and commonly used as a marker for lipid accumulation studies6, 11, 47, 48. To evaluate the levels of PLIN2 and FABP3 in cumulus cells, we performed western blot analysis for immature, *in vivo*-, and *in vitro*-matured cumulus cells. The proteins utilized in these analyses were from the same pool of samples that we used for the gene expression studies as we had stored the organic phase left over after RNA extraction using TRIzol reagent. The protein was isolated from the organic phase of the TRIzol reagent according to the manufacturer's instruction. To achieve the protein quantity required for western blot analysis, samples from the same biological group were pooled and analyzed in three technical replicates. A total of 50 μg of protein was loaded and resolved using a 10%-SDS- PAGE commercial kit, Mini-Protean TGX Gels (456–1033, Bio-RAD, Hercules, CA, USA). Electrophoresis was performed at 100 V for 70 min, and the proteins were transferred onto a polyvinylidene difluoride PVDF membrane using a Trans-Blot Turbo Transfer Pack (170–4156, Bio-RAD) in a semidry transfer apparatus, Trans-Blot Turbo, using a "mini TGX program" (3 min at 25 V constant). The membranes were then placed in a blocking buffer that had 5% bovine serum albumin (BSA) in tris buffered saline with tween (TBST) for 1 h at room temperature. Next, the membranes were incubated overnight with the primary antibody for PLIN2, anti-ADRP rabbit antibody (1:2000, SC32888, Santa Cruz Biotechnology, Dallas, Texas, USA), at 4 °C. Following overnight incubation, the membranes were washed thrice for 5 min with TBST and then incubated with the secondary antibody, anti-rabbit conjugated with HRP antibody (1:2000, A0545, Sigma), for 1 h at room temperature. To detect FABP3, we used an anti-cardiac FABP rabbit antibody (1:2000, AB45966, Abcam) and the secondary anti-rabbit antibody conjugated with HRP (1:4000; A0545, Sigma). The membranes were again washed thrice in 1x TBST for 5 min. For detection, the membranes were subjected to a chemiluminescent reaction using the reagent ECL Plus Prime Western Blotting Detection System solution (Amersham™, Buckinghamshire, UK). Imaging and band density analyses were performed using ChemiDoc MP Image System (Bio-Rad,) and the software Image Lab 5.1, respectively. The relative amounts of PLIN2 were normalized using the anti-histone H3 rabbit antibody (1:2000, H0164, Sigma) blotted in the same membrane. The relative amounts of FABP3 were normalized using the anti-x-Tubulin mouse antibody (1:2000, T9026, Sigma) and an anti-mouse antibody (1:2000, #70765, Cell Signaling) as endogenous controls.

Relative mRNA analysis

To investigate the levels of FABP3 transcript, we extracted mRNA from five pools, each of then containing 20 immatures and *in vivo*- and *in vitro*-matured denuded oocytes (in MII phase) that were collected from 12 biological replicates of *in vivo* and *in vitro* maturation. For the analysis of cumulus cells, we used 8 different pools of cumulus cells retrieved from 20 immature, 10 *in vivo*-, and 10 *in vitro*-matured COCs, representing the same biological replicates. The total RNA was extracted using the TRIzol reagent (ThermoFisher, Waltham, Massachusetts, USA) and the miRNeasy Kit (Qiagen, Valencia, CA, USA). Total RNA was quantified using the NanoDrop 2000 (Thermo Scientific) and the quality was estimated using the 260/280 ratio. Reverse transcription of mRNA was performed using the High-Capacity cDNA Reverse Transcription kit (Applied Biosystems, Foster City, CA, USA) according to manufacturer's instructions. Approximately 100 ng of the total RNA was incubated with 10x RT Buffer, 25x dNTP mix (100 mM), 10x RT Random Primers, and nuclease-free water at 25 °C for 10 min and at 37 °C for 120 min, followed by 5 min at 85 °C to stop the reaction. For real-time quantitative reverse transcription-PCR (RT-qPCR), 10 ng of RNA at the initial concentration was used for each gene. RT-qPCRs were performed in duplicate using 1 µl of cDNA of each gene and 75 nM of primers using the Power SYBR® Green PCR Master Mix kit (Applied Biosystems) in the QuantStudio 6 Flex PCR System (Applied Biosystems). The PCR cycle conditions were as follows: 95 °C for 10 min, 40 cycles of 95 °C for 15 s, and 60 °C for 60 s. The normalized Ct levels for *FABP3* were obtained by the subtracting the Ct of the target gene using the geometric means of three reference genes (*PPIA*, *SDHA*, and *YWAHZ)*. These endogenous genes have been demonstrated to be stably expressed and suitable for use as normalizers in gene expression studies of bovine oocytes49. The primers were designed based on GenBank sequences (Supplementary Table S1) and tested for efficiency by running a standard curve; the specificity was confirmed by sequencing the amplicon.

Immunolocalization of FABP3 within TZPs and quantification of FABP3 levels and lipid content in oocytes during IVM

Through the *in vivo* and IVM studies, we aimed to understand FABP3 and lipid behavior during IVM. First, we performed immunodetection of FABP3 in TZPs in immature and 9- and 18-h matured COCs in order to evaluate changes in the location of this protein. Subsequently, we studied FABP3 levels and the lipid content during IVM to understand the role of FABP3 in mediating fatty acid transport to oocytes. For these experiments, we collected grade I and II COCs from 3–6-mm follicles by postmortem follicular aspiration of bovine ovaries. The COCs had undergone IVM as described previously, and were then removed from the maturation media for analysis at 9 and 18 h after the beginning of IVM.

Immunodetection of FABP3 in TZPs within COC

FABP3 were detected in TZPs by immunofluorescence analysis after staining the actin filaments in order to visualize the TZPs. We stained immature COCs (n = 12), as well as COCs *in vitro*-matured for 9 (n = 12) and 18 h (n = 15). We also analyzed 10 denuded and 11 partial denuded oocytes submitted to IVM for 9 hours. The COCs assigned for immunodetection of FABP3 were obtained from three biological replicates. The COCs were fixed with 4% paraformaldehyde diluted in phosphate buffered saline (PBS) with 0.1% polyvinylpyrrolidone (PBS-PVP) for 12 minutes. COCs were then permeabilized for 20 min in PBS-PVP with 1% Triton X-100. After blocking of non-specific binding sites with 5% BSA in PBS for 1 h, COCs were incubated overnight at 4 °C with the primary rabbit antibody anti-FABP3 (1:200, SC15974R, Santa Cruz Biotechnology) and 1% BSA in PBS for

FABP3 detection. Following incubation, the samples were washed in PBS-PVP and incubated with the secondary antibody, anti-rabbit conjugated with Alexa 488 (A11008, Life Technologies, Thermo Fisher) diluted at 1:200. After incubation, the samples were washed ten times in PBS-PVP to remove all non-specifically bound secondary antibodies. To label the TZPs, COCs were stained with 165 nM of Alexa Fluor 647 phalloidin (A22287, Life Technologies, Thermo Fisher), an actin filament stain, for 30 minutes. Oocytes were mounted on glass slides with coverslips using Prolong Antifade with DAPI reagent (Life Technologies, Thermo Fisher). For capturing images, a SP5 confocal microscope (Leica, Wetzlar, Germany) was used with Diode 405 nm, Argon 488 nm, and HeNe 633 nm adjusted to each probe. All images were captured by sequential acquisition, in 63x magnification in oil, and in 3.5x magnification for detecting TZPs. For the negative control, we omitted the incubation with primary antibody and acquired the images utilizing maximum laser potency to rule out the possibility of artifacts.

Confocal microscopic analysis of lipid content in oocytes

To evaluate a possible relationship between FABP3 transport and lipid accumulation in oocytes, we carried out a time course experiment on immature oocytes and oocytes *in vitro*-matured for 9 and 18 h. We collected a total of 27 immature oocytes, 31 oocytes after 9 h of IVM, and 28 oocytes after 18 h of IVM. We then quantified the lipid content in these oocytes as described in the previous section.

Quantification of FABP3 in oocytes by western blot

Western blot analyses were performed in 5 pools with 50 denuded oocytes from each group. The pools of oocytes were directly lysed in 7.5 µL of RIPA buffer, mixed with 2.5 µL Laemmli buffer (4x), boiled at 98 °C for 5 min, and then loaded onto a SDS-PAGE gel. All the other steps were performed as described in the previous sections.

Disruption of TZPs during IVM using cytochalasin B

To investigate the effects of blocking TZPs on lipid accumulation, we disrupted the TZPs during oocyte maturation using cytochalasin B, a potent fungal toxin that hampers actin polymerization. We treated the oocytes with cytochalasin B (15.7 µM) in the maturation medium for 9 h in order to disrupt the TZPs. To observe the effects of TZP-disruption, we stained the actin filaments and FABP3 in 10 treated COCs and 10 untreated (control) COCs as described previously. This experiment was conducted in three biological replicates. The reduction in TZPs after cytochalasin B treatment was confirmed using an immunofluorescence analysis. Additionally, we quantified the lipid content using fluorimetric analysis of oocytes after 9 h of IVM in the presence (n = 31) and absence (n = 31) of cytochalasin B.

Statistical Analysis

The mRNA expression data were analyzed using the one-way ANOVA and the averages were compared using Tukey's test. The data from the western blot were subjected to logarithmic transformation before ANOVA and Tukey's test. These analyses were performed in SAS software. The lipid accumulation data were compared using the non-parametric Kruskal Wallis test followed by Dunn's multiple comparison test, and were performed using PRISM from Graphpad software. A 5% level of significance was adopted for all tests. Data were presented as mean ± standard error of the mean.

ADDITIONAL INFORMATION

Publisher's note: Springer Nature remains neutral with regard to jurisdictional claims in published maps and institutional affiliations.

ACKNOWLEDGEMENTS

The authors would like to thank the LMMD staff and students for the dedicated work during samples collections and laboratory procedures. The authors are also thankful to Elizabete Rosa Milani from FMRP-USP for technical assistance during acquisition of confocal microscopy images. This work was supported by grants from the Sao Paulo Research Foundation (FAPESP grant 2014/21034-3; grant 2014/03281-3; grant 2014/22887-0; grant 2015/26056-8 and grant 2013/08135-2) and the National Counsel of Technological and Scientific Development (CNPq). The funders had no role in study design, data collection and analysis, decision to publish, or preparation of the manuscript.

REFERENCES

1. Rizos, D., Ward, F., Duffy, P., Boland, M. P. & Lonergan, P. Consequences of bovine oocyte maturation, fertilization or early embryo development in vitro versus in vivo: implications for blastocyst yield and blastocyst quality. Molecular reproduction and development 61, 234–248,
2. Seidel, G. E. Modifying oocytes and embryos to improve their cryopreservation. Theriogenology 65, 228–235,
3. del Collado, M. et al. Influence of bovine serum albumin and fetal bovine serum supplementation during in vitro maturation on lipid and mitochondrial behaviour in oocytes and lipid accumulation in bovine embryos. Reproduction, fertility, and development 28, 1721,
4. Abe, H., Yamashita, S., Satoh, T. & Hoshi, H. Accumulation of cytoplasmic lipid droplets in bovine embryos and cryotolerance of embryos developed in different culture systems using serum-free or serum-containing media. Molecular reproduction and development 61, 57–66,).
5. Rizos, D. Bovine Embryo Culture in the Presence or Absence of Serum: Implications for Blastocyst Development, Cryotolerance, and Messenger RNA Expression. Biology of Reproduction 68, 236–243,
6. Yang, X. et al. Identification of perilipin-2 as a lipid droplet protein regulated in oocytes during maturation. Reproduction, fertility, and development 22, 1262–1271,
7. Jungheim, E. S. et al. Associations between free fatty acids, cumulus oocyte complex morphology and ovarian function during in vitro fertilization. Fertility and sterility 95, 1970–1974,
8. Cardozo, E., Pavone, M. E. & Hirshfeld-Cytron, J. E. Metabolic syndrome and oocyte quality. Trends Endocrin Met 22, 103–109,
9. Robker, R. L. et al. Obese Women Exhibit Differences in Ovarian Metabolites, Hormones, and Gene Expression Compared with Moderate-Weight Women. J Clin Endocr Metab 94, 1533–1540,
10. Robker, R. L., Wu, L. L. Y. & Yang, X. Inflammatory pathways linking obesity and ovarian dysfunction. J Reprod Immunol 88, 142–148,
11. Yang, X. et al. Exposure to lipid-rich follicular fluid is associated with endoplasmic reticulum stress and impaired oocyte maturation in cumulus-oocyte complexes. Fertility and sterility97, 1438–1443,
12. Sutton, M. L. et al. Influence of oocyte-secreted factors and culture duration on the metabolic activity of bovine cumulus cell complexes. Reproduction 126, 27–34,
13. Macaulay, A. D. et al. The gametic synapse: RNA transfer to the bovine oocyte. Biol Reprod 91, 90,
14. Sutton, M. L., Gilchrist, R. B. & Thompson, J. G. Effects of in-vivo and in-vitro environments on the metabolism of the cumulus-oocyte complex and its influence on oocyte developmental capacity. Hum Reprod Update 9, 35–48,
15. Lodde, V., Franciosi, F., Tessaro, I., Modina, S. C. & Luciano, A. M. Role of gap junction-mediated communications in regulating large-scale chromatin configuration remodeling and embryonic developmental competence acquisition in fully grown bovine oocyte. J Assist Reprod Gen 30, 1219–1226,
16. Thomas, R. E., Armstrong, D. T. & Gilchrist, R. B. Bovine cumulus cell-oocyte gap junctional communication during in vitro maturation in response to manipulation of cell-specific cyclic adenosine 3′,5′-monophosophate levels. Biology of Reproduction70, 548–556,
17. Hotamisligil, G. S. & Bernlohr, D. A. Metabolic functions of FABPs-mechanisms and therapeutic implications. Nat Rev Endocrinol 11, 592–605, Ockner, R. K. Fatty-Acid Binding-Proteins - Considerable Progress - More Ground to Cover - a Citation-Classic Commentary on a Binding-Protein for Fatty-Acids in Cytosol of Intestinal-Mucosa,

Liver, Myocardium, and Other Tissues by Ockner, R. K., Manning, J. A., Poppenhausen, R. B. & Ho, W. K. L. Cc/Life Sci, 9–9 (1991).

18. Ockner, R. K. & Manning, J. A. Fatty Acid-Binding Protein in Small-Intestine - Identification, Isolation, and Evidence for Its Role in Cellular Fatty-Acid Transport. J Clin Invest 54, 326–338,
19. Furuhashi, M. & Hotamisligil, G. S. Fatty acid-binding proteins: role in metabolic diseases and potential as drug targets. Nat Rev Drug Discov 7, 489–503,
20. Hertzel, A. V. & Bernlohr, D. A. The mammalian fatty acid-binding protein multigene family: molecular and genetic insights into function. Trends Endocrinol Metab 11, 175–180,
21. Sanchez-Lazo, L. et al. Fatty acid synthesis and oxidation in cumulus cells support oocyte maturation in bovine. Molecular endocrinology 28, 1502–1521,
22. Ferguson, E. M. & Leese, H. J. Triglyceride content of bovine oocytes and early embryos. Journal of reproduction and fertility116, 373–378,
23. de Loos, F., van Vliet, C., van Maurik, P. & Kruip, T. A. Morphology of immature bovine oocytes. Gamete Res 24, 197–204,
24. de Loos, F., Kastrop, P., Van Maurik, P., Van Beneden, T. H. & Kruip, T. A. Heterologous cell contacts and metabolic coupling in bovine cumulus oocyte complexes. Molecular reproduction and development 28, 255–259,
25. Kruip, T. A. M., Cran, D. G., Vanbeneden, T. H. & Dieleman, S. J. Structural-Changes in Bovine Oocytes during Final Maturation Invivo. Gamete Res 8, 29–47,
26. Simon, A. M. & Goodenough, D. A. Diverse functions of vertebrate gap junctions. Trends Cell Biol 8, 477–483,
27. Albertini, D. F., Combelles, C. M., Benecchi, E. & Carabatsos, M. J. Cellular basis for paracrine regulation of ovarian follicle development. Reproduction 121, 647–653,
28. Macaulay, A. D. et al. Cumulus Cell Transcripts Transit to the Bovine Oocyte in Preparation for Maturation. Biol Reprod 94, 16,
29. Subra, C. et al. Exosomes account for vesicle-mediated transcellular transport of activatable phospholipases and prostaglandins. Journal of Lipid Research 51, 2105–2120,
30. Hurwitz, S. N. et al. Proteomic profiling of NCI-60 extracellular vesicles uncovers common protein cargo and cancer type-specific biomarkers. Oncotarget 7, 86999–87015,
31. Fujita, K. et al. Proteomic analysis of urinary extracellular vesicles from high Gleason score prostate cancer. Sci Rep 7, 42961,
32. Kralisch, S. et al. Adipocyte fatty acid-binding protein is released from adipocytes by a non-conventional mechanism. Int J Obesity38, 1251–1254,
33. Kalra, H. et al. Vesiclepedia: a compendium for extracellular vesicles with continuous community annotation. PLoS Biol 10, e1001450,
34. Iso, T. et al. Capillary endothelial fatty acid binding proteins 4 and 5 play a critical role in fatty acid uptake in heart and skeletal muscle. Arterioscler Thromb Vasc Biol 33, 2549–2557,
35. Diaz, P., Harris, J., Rosario, F. J., Powell, T. L. & Jansson, T. Increased placental fatty acid transporter 6 and binding protein 3 expression and fetal liver lipid accumulation in a mouse model of obesity in pregnancy. Am J Physiol-Reg I 309, R1569–R1577,
36. Zhang, Y. et al. Fatty acid binding protein 3 (fabp3) is associated with insulin, lipids and cardiovascular phenotypes of the metabolic syndrome through epigenetic modifications in a northern european family population. Bmc Med Genomics 6,
37. Karakas, S. E., Almario, R. U. & Kim, K. Serum fatty acid binding protein 4, free fatty acids, and metabolic risk markers. Metabolism58, 1002–1007,
38. Haider, D. G. et al. Plasma adipocyte and epidermal fatty acid binding protein is reduced after weight loss in obesity. Diabetes Obes Metab 9, 761a–763,
39. Haluzik, M. M. et al. Serum adipocyte fatty acid binding protein levels in patients with type 2 diabetes mellitus and obesity: the influence of fenofibrate treatment. Physiol Res 58, 93–99 (2009).

40. Krzystek-Korpacka, M. et al. Circulating adipocyte fatty acid-binding protein, juvenile obesity, and metabolic syndrome. J Pediatr Endocrinol Metab 24, 921–928,
41. Charlier, C. et al. Oocyte-somatic cells interactions, lessons from evolution. Bmc Genomics 13, 560,
42. Zanotti, G. Muscle fatty acid-binding protein. Bba-Mol Cell Biol L1441, 94–105,
43. Veerkamp, J. H. & Vanmoerkerk, H. T. B. Fatty-Acid-Binding Protein and Its Relation to Fatty-Acid Oxidation. Mol Cell Biochem123, 101–106,
44. Aardema, H. et al. Bovine cumulus cells protect maturing oocytes from increased fatty acid levels by massive intracellular lipid storage. Biol Reprod 88, 164–164,
45. Akbal, E. et al. Serum heart type fatty acid binding protein levels in metabolic syndrome. Endocrine 36, 433–437,
46. Bickel, P. E., Tansey, J. T. & Welte, M. A. PAT proteins, an ancient family of lipid droplet proteins that regulate cellular lipid stores. Biochimica et biophysica acta 1791, 419–440,
47. Sastre, D. et al. Expression of PLIN2 and PLIN3 during oocyte maturation and early embryo development in cattle. Theriogenology 81, 326–331,
48. Macabelli, C. H. et al. Reference gene selection for gene expression analysis of oocytes collected from dairy cattle and buffaloes during winter and summer. PloS one 9, e93287,

CHAPTER 5

Using Engineered microRNAs as Vectors for Animal RNA Interference: Promises and Challenges

Jerry S. Chen[1], Robert W. Zeller[2]

[1]*Computational Science Research Center, San Diego State University, San Diego, USA*

[2]*Department of Biology, San Diego State University, San Diego, USA*

ABSTRACT

microRNAs are post-transcriptional regulators of gene expression that recruit RNA silencing complexes to target transcripts to prevent translation and promote their degradation. Experimental studies suggest that microRNA binding to target transcripts can result in as much as a 90% decrease in gene expression. Because of this feature, the microRNA pathway has been utilized as a vehicle for potent RNA interference (RNAi). In recent years, significant advances have been made in engineering artificial microRNA vectors for RNAi in a number of biological systems, with the most progress in plants but also some success in mouse and human cell lines. In this mini-review, we provide a brief discussion of the potential of this technology in comparison with other RNAi strategies, and the current challenges in the design of microRNA-based RNAi vectors, particularly for animal systems.

Keywords: microRNAs; RNA Interference; RNAi

1. INTRODUCTION

microRNAs (miRNAs) are small RNA molecules that play important developmental roles in both animals and plants, known clinically for their proposed role in preventing cancer formation [1] -[3] . They were first identified in 1993 by the laboratories of Victor Ambros and Gary Ruvkun [4] [5] . Using the nematode worm C. elegans, Ambros and Ruvkun showed that the gene product of a regulatory molecule called lin-4 was necessary for suppressing the activity of a transcription factor (TF) protein called LIN-14 during early larval development. Interestingly, the product of lin-4 was not a protein as would typically be expected for a molecule with a repressive function, but rather was a tiny RNA molecule less than 70 nucleotides (nt) long. It was then discovered that the sequence of the lin-4 RNA product was highly complementary to a region within the 3' untranslated region (3'UTR) of the LIN-14 messenger RNA (mRNA) transcript. They then suggested that perhaps this RNA-RNA interaction controls the expression of the LIN-14 protein. A few other studies soon followed [6] -[9] but it was not until the complete sequencing of several genomes (including the human genome) at the turn of the century [10] -[13] that microRNA biology blossomed. A flurry of research between 2000 and 2003 demonstrated that these microRNAs were not just chance occurrences, but were in fact abundant in the genome and pervasive throughout the plant and animal kingdom [14] -[36] . This sparked an international scientific endeavor to identify and discover the function of these curious runts of the genome, which continues to this day [3] [37] .

2. USING MICRORNAS AS VECTORS FOR RNA INTERFERENCE

2.1. How microRNAs Are Processed

microRNAs, like protein-coding genes, are encoded in the genome and transcribed by RNA polymerase II. They can reside as stand-alone genes in independent genetic loci or reside within the introns of host genes. In rare cases, they may even bury themselves within the exons of protein-coding genes [38] [39] . Their processed hairpin transcripts are small, only ~70 base pairs (bp) in length, explaining why they evaded detection in the early days of genome research.

What facilitated the widespread discovery of microRNAs was the resolution of their RNA structure. After the primary miRNA is transcribed from the genome, it folds into a stem-loop structure, which is trimmed into a distinguishable ~70 bp hairpin-shaped pre-miRNA molecule by the RNAse III enzyme Drosha [38] [39] (Figure 1). The stems of the pre-miRNA are highly complementary, enabling scientists to find predicted microRNAs by searching for genomic regions with adjacent inverse complementary sequences separated by a small gap that corresponds to the pre-miRNA loop. Since regions of the genome may have consecutive inverse complementary regions by chance, these computational studies generally predict more microRNAs than are found experimentally [18] [40] . Nonetheless, these initial computational studies guided subsequent experimental efforts, which used small-scale methods such as in situ hybridization [41] -[43] and Northern blot [16] [19] [35] as well as largescale methods such as RNA deep sequencing [32] [34] [44] [45] and miRNA microarrays [35] [46] . To date, the current numbers of identified microRNAs range from 200 in C. elegans [47] [48] to over 1000 in humans [48] [49] . Interestingly, whereas the number of protein-coding genes increases at only a modest rate, the number of microRNAs in the genome increases quite significantly in proportion with the complexity of the organism, indicating that the increased complexity of higher organisms may in large part be the result of microRNA expansion [50] .

After trimming by Drosha, the 70 nucleotide hairpin pre-miRNA is then shuttled from the nucleus out into the cytoplasm by the nuclear transport receptor Exportin 5, where the loop of the hairpin is then cut by the RNAse III enzyme Dicer [38] [39] . From the resultant ~20 - 24 bp RNA duplex, one strand becomes the mature miRNA product. The other strand, known as the miRNA*, is generally thought to be degraded, although recent reports suggest that in a few cases they may also have regulatory capability [51] [52] . The mature miRNA is loaded into a RNA silencing complex containing the RNA-binding protein Argonaute and serves as a guide strand to target and bind partially complementary sites in 3'UTR regions of target mRNAs to regulate their expression [3] [38] [39] . Evidence also exists that Argonaute-complexes frequently bind within coding regions, suggesting that miRNAs may also target these regions as well [44] [45] .

2.2. Engineered microRNA Vectors for RNAi: The Promises

RNA interference was first achieved by introduction of synthetic long double-stranded RNA into cells or animals, which transiently produce siRNA strands complementary to the desired target gene transcript [53] [54] . RNA interference was first demonstrated in C. elegans [53] and soon after in mammalian cell lines [54] [55] . In more recent years, RNAi has also been used as a therapeutic tool to knock-down the expression of genes

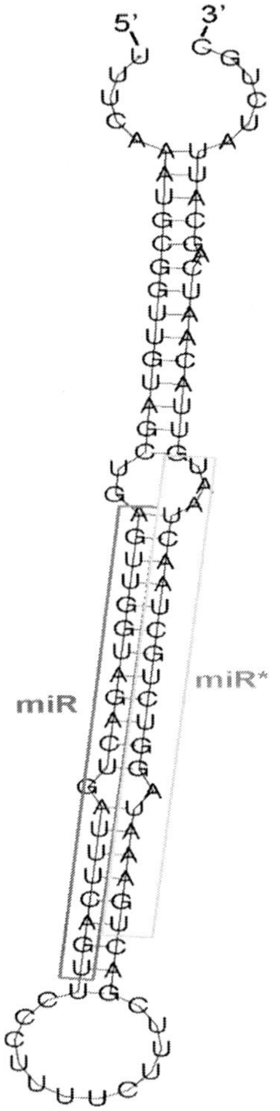

Figure 1. microRNA hairpin structure. Shown here is the pre-miRNA hairpin structure for Ciona miR-155, with the identified mature miR (red) and miRstar (blue) indicated. pre-miRNA hairpins can vary in length from 60 to ~100 nt, and sometimes have tail flanking regions that are also complementary [36] . The ends of the miR-miR* duplex are cleaved by the RNAse III enzyme Dicer, which leaves overhangs on both ends.

implicated in the pathophysiology of many cancers, viral infections, and other diseases [56] -[61] .

For more stable and controlled expression of siRNAs in model organisms, vectors have been developed for transcribing short-hairpin RNAs (shRNAs) under the control of RNA polymerase III promoters such as U6 or H1 [62]

[63] , which can either be expressed via extra-chromosomal arrays or directly integrated into the genome using retroviral transduction [62] [64] . Many labs have utilized this technology to create shRNA libraries and stable transgenic lines [62] [65] . However, this strategy does not work in all organisms and in some systems suffers from low efficiency and high off-target effects [66] . Furthermore, since it appears that stable shRNA expression is only achievable using RNA polymerase III promoters, which are constitutively expressed at a certain level in all cell types, the ability to control the timing and level of shRNA expression is currently not possible [62] .

An alternative approach is to use microRNA hairpin transcripts as a vector for generating silencing RNAs [62] [67] -[69] (Figure 2). Using this technique, the microRNA hairpin undergoes endogenous processing by the microRNA biogenesis machinery, producing an siRNA-like strand which then targets a desired gene transcript. The hairpin is artificially designed such that the mature and miRNA* strands are replaced by a double-stranded miRNA/siRNA-like duplex, and the entire hairpin cassette is cloned into an appropriate RNA-polymerase II promoter-driven vector (Figure 2). In order to more accurately mimic miRNA duplexes and distinguish themselves from double stranded siRNAs, a bulge is sometimes introduced in the strand replacing the miRNA* so that the opposing strands of the duplex are not perfectly complementary [54] . The result is a vehicle for RNAi that uses the miRNA pathway for processing and can produce tissue-specific miRNA-like small RNAs. Using transcripts containing miRNA clusters, one can engineer a multi-cassette vector producing multiple miRNA-like small RNAs to target multiple transcripts. Furthermore, since this strategy essentially mimics the processing of endogenous miRNA transcripts, introduction of a miRNA-based vector into cells may circumvent unwanted biological side effects such as type 1 interferon responses that sometimes occur when introducing double-stranded siRNA [68] [70] . RNAi vectors using engineered microRNAs have been successfully developed in a number of systems, most notably in plants [71] -[74] , algae [75] [76] and mouse and human cell lines [55] [62] . Collectively, these studies have reported high-specificity of targeting and better knockdown efficiencies than shRNAs. In agreement with previous miRNA studies, miRNA-based RNAi vectors used in human cell lines were shown to result in as much as ~80% knockdown of gene expression, and the degree of RNAi knockdown is dose-dependent [62] [68] . Notably, it has been shown that efficient RNAi knockdown can result in observable loss-offunction phenotypes [64] . Importantly though, because miRNA target binding in plants requires near-perfect complementarity while in animals requires only partial complementarity, the potential for off-target effects is an important factor to consider when designing and using artificial miRNA RNAi vectors in animal systems. We discuss these challenges below.

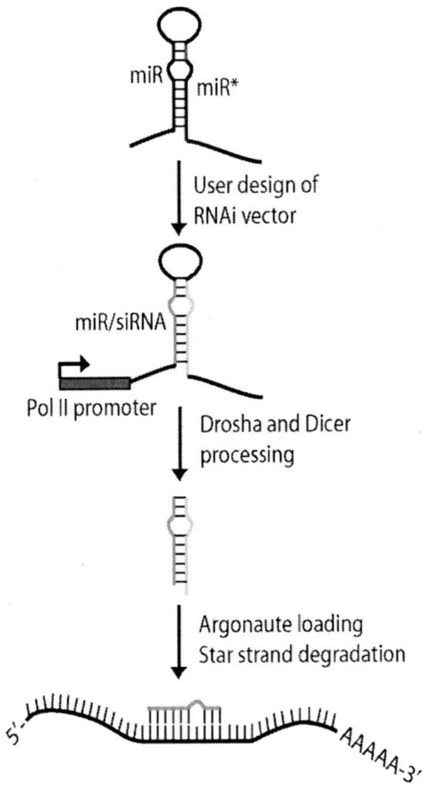

Figure 2. RNA interference using the microRNA pathway. User-designed artifical microRNA hairpins are designed such that the endogenous miR/miR* strands are replaced by siRNA-like strands for which the guide miR/siRNA strand (orange) is processed by the endogenous miRNA pathway and targets a desired gene transcript. The RNAi vector is under the control of a selected Pol II promoter, allowing for tissue-specific expression.

2.3. Engineered microRNA Vectors for RNAi: The Challenges

One of the primary challenges in using microRNAs for RNA interference in animals is in determining the precise rules for how microRNAs bind to their targets. In plants, microRNAs bind to target sites with near-perfect or perfect complementarity, and therefore design of corresponding RNAi vectors is more straightforward. However, in animals, microRNAs can bind to targets with only partial complementarity, and we still do not know exactly how a microRNA binds to an mRNA target. Through experimental

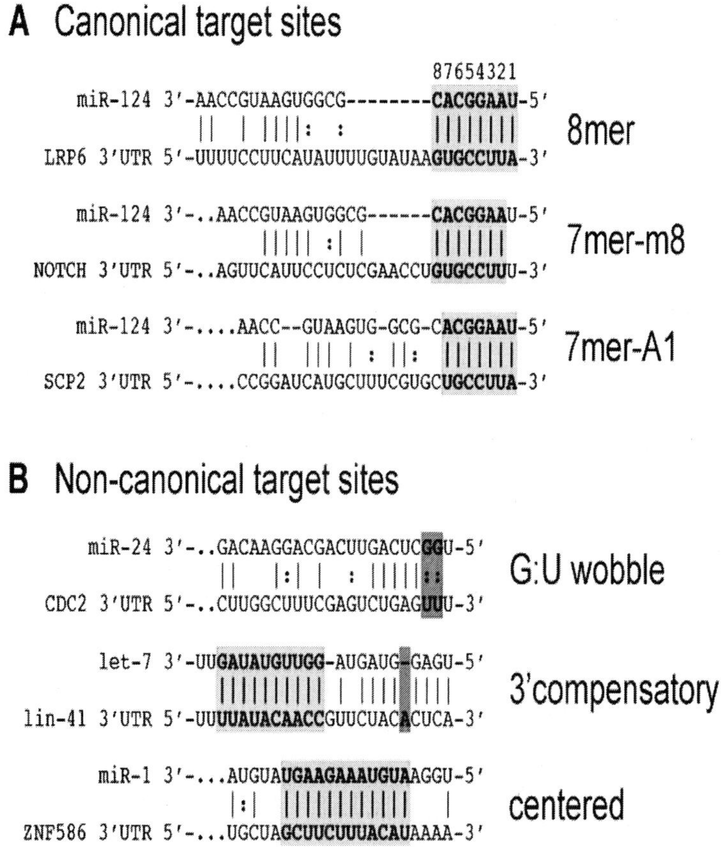

Figure 3. Types of microRNA target sites. (A) There are three types of canonical miRNA target sites: 1) sites perfectly complementary to the entire miRNA seed (8 mer), 2) sites complementary to miRNA nt 2 - 8 (7 mer-m8), and 3) sites complementary to miRNA nt 2 - 7 with an A at position 1 (7 mer-A1) [3] . Shown here are examples of the three canonical target sites complementary to the conserved neuronal miRNA miR-124. (B) There are other types of functional miRNA target sites such as those with a G:U wobble pairing in the seed region, extensive miRNA 3'end compensatory binding, and pairing of ~10 consecutive nucleotides in the central region of the miRNA.

studies over the past decade, it is becoming increasingly clear, however, that there are three large classes of targets based on mRNA binding within the first eight nucleotides of the miRNA, which we refer to as the "seed" region [3] . These are called 8 mer, 7 mer-m8 and 7 mer-A1 seed targets, respectively (Figure 3(A)) [3] [40] [43] [77] [78] . If we merely searched for these kinds of sites within 3'UTR of mRNA transcripts, we would predict that a single microRNA would target on average about 2000 transcripts in human, around 800 per miRNA in simple chordates, and about 300 - 400 in worms and fruit flies [42] [43] [78] -[80] . Many computational target prediction programs are based on finding seed sites [43] [77] [79] . However, it is generally thought that these are overestimates, and that there are many other biological factors involved such as 3'UTR structure and local thermodynamics which may nullify many of these predicted targets [3] [81] . Although it seems very likely that any mRNA containing a seed site in its 3'UTR is a bona fide target, the lack of precise large-scale protein-based methods hinders our ability to solve this problem definitively. However, from many target expression studies, we do know that the level of downregulation among verified seed targets can vary widely, ranging from as little as 10% to over 90% downregulation [43] [77] . To complicate matters, there are several other types of target binding that have been reported to be functional in some, but not all cases. These include mRNA binding with the presence of a single G:U wobble pair within the seed, binding of only miRNA nucleotides 2 - 7, binding of the center of the miRNA, and compensatory binding where complementarity of only 4 - 5 base pairs of the miRNA seed is compensated by extensive binding of the rest of the miRNA [3] [4] [40] [82] -[84] (Figure 3(B)). More recently, an alternative approach to finding miRNA targets utilizes a biochemical assay to isolate and sequence Argonaute-bound messenger transcripts, which has expanded our understanding of base-pairing rules and interestingly has shown that a large percentage of targets are bound in exon regions [44] [45] [84] [85] . In particular, a ligation-based method has been developed for directly associating miRNAs with their RNA targets [84] . Although the efficiency of this protocol is currently very low (only 2% of sequenced reads are ligated miRNA-RNA hybrid reads), this method holds exciting promise for uncovering miRNA target binding rules and improving microRNA-based RNAi technology.

A second major challenge is in considering the regulation of microRNA biogenesis. Once miRNAs are transcribed, the hairpin forms within a larger primary transcript. It has been demonstrated that sequences within the flanking 5' and 3' tails are required for proper Drosha processing [86] . Sequences in the 3' tail have been shown to contain miRNA binding sites, suggesting that the 3' tails may serve as a regulatory region akin to 3'UTRs for messenger RNAs [87] [88] . Furthermore, both the tail and loop regions contain binding sites for regulatory RNA binding proteins [89] [90] . Therefore, when expressing miRNA-based RNAi vectors outside of their endogenous context, potential regulation in these loop and tail regions must be considered. Even for the miRNA duplex, one must consider the potential production of miRNA isoforms (isomiRs) that differ in either the 5' or 3' end [91] [92] , which may be a source of unexpected off-target effects. Since many facets of the microRNA biogenesis pathway are still poorly understood [38] , there is currently still an element of trial-and-error in designing proper miRNA RNAi vectors.

3. CONCLUSION

Despite intensive research over the past decade, the field of microRNA biology is still in its infancy and many fundamental questions remain unanswered. There are still many gaps in our understanding of the molecular machinery involved in microRNA biogenesis, processing, targeting, and degradation. Nonetheless, microRNAs are naturally produced in almost all model systems, have the demonstrated ability to induce near-complete knockdown of target transcript and/or protein levels, and offer the versatility necessary to be used as vehicles for RNA interference in animals.

ACKNOWLEDGEMENTS

Jerry S. Chen, Robert W. Zeller J.S.C. thanks V. Hurless for helpful discussions. This work was supported by National Science Foundation Grants IOS-0347937 and IOS-0951347 to R.W.Z.

REFERENCES

1. Lu, J., Getz, G., Miska, E.A., Alvarez-Saavedra, E., Lamb, J., et al. (2005) MicroRNA Expression Profiles Classify Human Cancers. Nature, 435, 834-838.
2. Calin, G.A. and Croce, C.M. (2006) MicroRNA-Cancer Connection: The Beginning of a New Tale. Cancer Research, 66, 7390-7394.
3. Bartel, D.P. (2009) MicroRNAs: Target Recognition and Regulatory Functions. Cell, 136, 215-233.
4. Lee, R.C., Feinbaum, R.L. and Ambros, V. (1993) The C. elegans Heterochronic Gene Lin-4 Encodes Small RNAs with Antisense Complementarity to Lin-14. Cell, 75, 843-854.
5. Wightman, B., Ha, I. and Ruvkun, G. (1993) Posttranscriptional Regulation of the Heterochronic Gene Lin-14 by Lin-4 Mediates Temporal Pattern Formation in C. elegans. Cell, 75, 855-862.
6. Ha, I., Wightman, B. and Ruvkun, G. (1996) A Bulged Lin-4/Lin-14 RNA Duplex Is Sufficient for Caenorhabditis elegans Lin-14 Temporal Gradient Formation. Genes & Development, 10, 3041-3050.
7. Moss, E.G., Lee, R.C. and Ambros, V. (1997) The Cold Shock Domain Protein LIN-28 Controls Developmental Timing in C. elegans and Is Regulated by the Lin-4 RNA. Cell, 88, 637-646.

8. Olsen, P.H. and Ambros, V. (1999) The Lin-4 Regulatory RNA Controls Developmental Timing in Caenorhabditis elegans by Blocking LIN-14 Protein Synthesis after the Initiation of Translation. Developmental Biology, 216, 671-680.
9. Reinhart, B.J., Slack, F.J., Basson, M., Pasquinelli, A.E., Bettinger, J.C., et al. (2000) The 21-Nucleotide Let-7 RNA Regulates Developmental Timing in Caenorhabditis elegans. Nature, 403, 901-906.
10. Waterston, R. and Sulston, J. (1998) The Human Genome Project: Reaching the Finish Line. Science, 282, 53-54.
11. Adams, M.D., Celniker, S.E., Holt, R.A., Evans, C.A., Gocayne, J.D., et al. (2000) The Genome Sequence of Drosophila Melanogaster. Science, 287, 2185-2195.
12. Venter, J.C., Adams, M.D., Myers, E.W., Li, P.W., Mural, R.J., et al. (2001) The Sequence of the Human Genome. Science, 291, 1304-1351.
13. Chinwalla, A.T., Cook, L.L., Delehaunty, K.D., Fewell, G.A., Fulton, L.A., et al. (2002) Initial Sequencing and Comparative Analysis of the Mouse Genome. Nature, 420, 520-562.
14. Pasquinelli, A.E., Reinhart, B.J., Slack, F., Martindale, M.Q., Kuroda, M.I., et al. (2000) Conservation of the Sequence and Temporal Expression of Let-7 Heterochronic Regulatory RNA. Nature, 408, 86-89.
15. Lau, N.C., Lim, L.P., Weinstein, E.G. and Bartel, D.P. (2001) An Abundant Class of Tiny RNAs with Probable Regulatory Roles in Caenorhabditis elegans. Science, 294, 858-862.
16. Lagos-Quintana, M., Rauhut, R., Yalcin, A., Meyer, J., Lendeckel, W., et al. (2002) Identification of Tissue-Specific microRNAs from Mouse. Current Biology, 12, 735-739.
17. Aravin, A.A., Lagos-Quintana, M., Yalcin, A., Zavolan, M., Marks, D., et al. (2003) The Small RNA Profile during Drosophila melanogaster Development. Developmental Cell, 5, 337-350.
18. Grad, Y., Aach, J., Hayes, G.D., Reinhart, B.J., Church, G.M., et al. (2003) Computational and Experimental Identification of C. elegans microRNAs. Molecular Cell, 11, 1253-1263.
19. Lagos-Quintana, M., Rauhut, R., Meyer, J., Borkhardt, A. and Tuschl, T. (2003) New microRNAs from Mouse and Human. RNA, 9, 175-179.
20. Lim, L.P., Lau, N.C., Weinstein, E.G., Abdelhakim, A., Yekta, S., et al. (2003) The microRNAs of Caenorhabditis elegans. Genes & Development, 17, 991-1008.
21. Ambros, V. (2004) The Functions of Animal microRNAs. Nature, 431, 350-355.
22. Bartel, D.P. (2004) MicroRNAs: Genomics, Biogenesis, Mechanism, and Function. Cell, 116, 281-297.
23. Bentwich, I., Avniel, A., Karov, Y., Aharonov, R., Gilad, S., et al. (2005) Identification of Hundreds of Conserved and Nonconserved Human microRNAs. Nature Genetics, 37, 766-770.
24. Chen, P.Y., Manninga, H., Slanchev, K., Chien, M., Russo, J.J., et al. (2005) The Developmental miRNA Profiles of Zebrafish as Determined by Small RNA Cloning. Genes & Development, 19, 1288-1293.
25. Watanabe, T., Takeda, A., Mise, K., Okuno, T., Suzuki, T., et al. (2005) Stage-Specific Expression of microRNAs during Xenopus Development. FEBS Letters, 579, 318-324.
26. Ruby, J.G., Stark, A., Johnston, W.K., Kellis, M., Bartel, D.P., et al. (2007) Evolution, Biogenesis, Expression, and Target Predictions of a Substantially Expanded Set of Drosophila microRNAs. Genome Research, 17, 1850-1864.
27. Thatcher, E.J., Paydar, I., Anderson, K.K. and Patton, J.G. (2008) Regulation of Zebrafish Fin Regeneration by microRNAs. PNAS, 105, 18384-18389.
28. Glazov, E.A., Cottee, P.A., Barris, W.C., Moore, R.J., Dalrymple, B.P., et al. (2008) A microRNA Catalog of the Developing Chicken Embryo Identified by a Deep Sequencing Approach. Genome Research, 18, 957-964.
29. He, P.-A., Nie, Z., Chen, J., Chen, J., Lv, Z., et al. (2008) Identification and Characteristics of microRNAs from Bombyx mori. BMC Genomics, 9, 248.

30. Yu, X., Zhou, Q., Li, S.C., Luo, Q., Cai, Y., et al. (2008) The Silkworm (Bombyx mori) microRNAs and Their Expressions in Multiple Developmental Stages. PLoS One, 3, e2997.
31. Chen, X., Li, Q., Wang, J., Guo, X., Jiang, X., et al. (2009) Identification and Characterization of Novel Amphioxus microRNAs by Solexa Sequencing. Genome Biology, 10, R78.
32. Shi, W., Hendrix, D., Levine, M. and Haley, B. (2009) A Distinct Class of Small RNAs Arises from Pre-miRNAProximal Regions in a Simple Chordate. Nature Structural & Molecular Biology, 16, 183-189.
33. Wheeler, B.M., Heimberg, A.M., Moy, V.N., Sperling, E.A., Holstein, T.W., et al. (2009) The Deep Evolution of Metazoan microRNAs. Evolution & Development, 11, 50-68.
34. Hendrix, D., Levine, M. and Shi, W. (2010) Method miRTRAP, a Computational Method for the Systematic Identification of miRNAs from High Throughput Sequencing Data. Genome Biology, 11, R39.
35. Keshavan, R., Virata, M., Keshavan, A. and Zeller, R.W. (2010) Computational Identification of Ciona intestinalis microRNAs. Zoological Science, 27, 162-170.
36. Berezikov, E., Robine, N., Samsonova, A., Westholm, J.O., Naqvi, A., Hung, J.H., Okamura, K., Dai, Q., BortolamiolBecet, D., Martin, R., Zhao, Y.J., Zamore, P.D., Hannon, G.J., Marra, M.A., Weng, Z.P., Perrimon, N. and Lai, E.C. (2011) Deep Annotation of Drosophila Melanogaster microRNAs Yields Insights into Their Processing, Modification, and Emergence. Genome Research, 21, 203-215.
37. Lai, E.C. (2003) microRNAs: Runts of the Genome Assert Themselves. Current Biology, 13, R925-R936.
38. Du, T. and Zamore, P.D. (2005) microPrimer: The Biogenesis and Function of microRNA. Development, 132, 4645- 4652.
39. Kim, V.N., Han, J. and Siomi, M.C. (2009) Biogenesis of Small RNAs in Animals. Nature Reviews Molecular Cell Biology, 10, 126-139.
40. Lai, E.C. (2004) Predicting and Validating microRNA Targets. Genome Biology, 5, 115.
41. Aboobaker, A.A., Tomancak, P., Patel, N., Rubin, G.M. and Lai, E.C. (2005) Drosophila microRNAs Exhibit Diverse Spatial Expression Patterns during Embryonic Development. Proceedings of the National Academy of Sciences of the United States of America, 102, 18017-18022.
42. Chen, J. and Zeller, R. (2009) Regulation of Gene Expression by the microRNA miR-124 in the Developing Nervous System of C. Intestinalis. ACSESS Proceedings AP0904, San Diego, 27 March 2009, 1-6.
43. Chen, J.S., San Pedro, M. and Zeller, R.W. (2011) miR-124 Function during Ciona intestinalis Neuronal Development Includes Extensive Interaction with the Notch Signaling Pathway. Development, 138, 4943-4953.
44. Chi, S.W., Zang, J.B., Mele, A. and Darnell, R.B. (2009) Argonaute HITS-CLIP Decodes microRNA-mRNA Interaction Maps. Nature, 460, 479-486.
45. Zisoulis, D.G., Lovci, M.T., Wilbert, M.L., Hutt, K.R., Liang, T.Y., Pasquinelli, A.E. and Yeo, G.W. (2010) Comprehensive Discovery of Endogenous Argonaute Binding Sites in Caenorhabditis elegans. Nature Structural & Molecular Biology, 17, 173-179.
46. Krichevsky, A.M., King, K.S., Donahue, C.P., Khrapko, K. and Kosik, K.S. (2003) A microRNA Array Reveals Extensive Regulation of microRNAs during Brain Development. RNA, 9, 1274-1281.
47. Ruby, J.G., Jan, C., Player, C., Axtell, M.J., Lee, W., Nusbaum, C., Ge, H. and Bartel, D.P. (2006) Large-Scale Sequencing Reveals 21U-RNAs and Additional microRNAs and Endogenous siRNAs in C. elegans. Cell, 127, 1193- 1207.
48. Kozomara, A. and Griffiths-Jones, S. (2011) miRbase: Integrating microRNA Annotation and Deep-Sequencing Data. Nucleic Acids Research, 39, D152-D157.
49. Landgraf, P., Rusu, M., Sheridan, R., Sewer, A., Iovino, N., et al. (2007) A Mammalian microRNA Expression Atlas Based on Small RNA Library Sequencing. Cell, 129, 1401-1414.

50. Hertel, J., Lindemeyer, M., Missal, K., Fried, C., Tanzer, A., Flamm, C., Hofacker, I.L., Stadler, P.F. and The Students of Bioinformatics Computer Labs 2004 and 2005 (2006) The Expansion of the Metazoan microRNA Repertoire. BMC Genomics, 7, 25.
51. Okamura, K., Phillips, M.D., Tyler, D.M., Duan, H , Chou, Y.T. and La, E.C. (2008) The Regulatory Activity of microRNA* Species Has Substantial Influence on microRNA and 3'UTR Evolution. Nature Structural & Molecular Biology, 15, 354-363.
52. Yang, J.S., Phillips, M.D., Betel, D., Mu, P., Ventura, A., Siepel, A.C., Chen, K.C. and Lai, E.C. (2011) Widespread Regulatory Activity of Vertebrate microRNA* Species. RNA, 17, 312-326.
53. Fire, A., Xu, S., Montgomery, M.K., Kostas, S.A., Driver, S.E. and Mello, C.C. (1998) Potent and Specific Genetic Interference by Double-Stranded RNA in Caenorhabditis elegans. Nature, 391, 806-811.
54. Kim, D.H. and Rossi, J.J. (2007) Strategies for Silencing Human Disease Using RNA Interference. Nature Reviews Genetics, 8, 173-184.
55. Elbashir, S.M., Harborth, J., Lendeckel, W., Yalcin, A., Weber, K. and Tuschl, T. (2001) Duplexes of 21-Nucleotide RNAs Mediate RNA Interference in Cultured Mammalian Cells. Nature, 411, 494-498.
56. McFarland, T.J., Zhang, Y., Appukuttan, B. and Stout, J.T. (2004) Gene Therapy for Proliferative Ocular Diseases. Expert Opinion on Biological Therapy, 4, 1053-1058.
57. Bitko, V., Musiyenko, A., Shulyayeva, O. and Barik, S. (2004) Inhibition of Respiratory Viruses by Nasally Administered siRNA. Nature Medicine, 11, 50-55.
58. Rossi, J.J. (2006) RNAi as a Treatment for HIV-1 Infection. Biotechniques, 40, S25-S29.
59. Dykxhoorn, D.M. and Lieberman, J. (2006) Silencing Viral Infection. PLoS Medicine, 3, e242.
60. Raoul, C., Barker, S. and Aebischer, P. (2005) Viral-Based Modelling and Correction of Neurodegenerative Diseases by RNA Interference. Gene Therapy, 13, 487-495.
61. Pai, S., Lin, Y., Macaes, B., Meneshian, A., Hung, C. and Wu, T.C. (2005) Prospects of RNA Interference Therapy for Cancer. Gene Therapy, 13, 464-477.
62. Stegmeier, F., Hu, G., Rickles, R.J., Hannon, G.J. and Elledge, S.J. (2005) A Lentiviral microRNA-Based System for Single-Copy Polymerase II-Regulated RNA Interference in Mammalian Cells. Proceedings of the National Academy of Sciences of the United States of America, 102, 13212-13217.
63. Dykxhoorn, D.M., Novina, C.D. and Sharp, P.A. (2003) Killing the Messenger: Short RNAs that Silence Gene Expression. Nature Reviews Molecular Cell Biology, 4, 457-467.
64. Dickins, R.A., Hemann, M.T., Zilfou, J.T., Simpson, D.R., Ibarra, I., Hannon, G.J. and Lowe, S.W. (2005) Probing Tumor Phenotypes Using Stable and Regulated Synthetic microRNA Precursors. Nature Genetics, 37, 1289-1295.
65. Silva, J.M., Li, M.Z., Chang, K., Ge, W., Golding, M.C., et al. (2005) Second-Generation shRNA Libraries Covering the Mouse and Human Genomes. Nature Genetics, 37, 1281-1288.
66. Sarnova, L., Malik, R., Sedlacek, R. and Svoboda, P. (2010) Shortcomings of Short Hairpin RNA-Based Transgenic RNA Interference in Mouse Oocytes. Journal of Negative Results in Biomedicine, 9, 8.
67. Chung, K.H., Hart, C.C., Al-Bassam, S., Avery, A., Taylor, J., Patel, P.D., Vojtek, A.B. and Turner, D.L. (2006) Polycistronic RNA Polymerase II Expression Vectors for RNA Interference Based on Bic/miR-155. Nucleic Acids Research, 34, e53.
68. Zeng, Y., Wagner, E.J. and Cullen, B.R. (2002) Both Natural and Designed Micro RNAs Can Inhibit the Expression of Cognate mRNAs when Expressed in Human Cells. Molecular Cell, 9, 1327-1333.
69. Du, G., Yonekubo, J., Zeng, Y., Osisami, M. and Frohman, M.A. (2006) Design of Expression Vectors for RNA Interference Based on miRNAs and RNA Splicing. FEBS Journal, 273, 5421-5427.
70. De Veer, M.J., Sledz, C.A. and Williams, B.R. (2005) Detection of Foreign RNA: Implications for RNAi. Immunology and Cell Biology, 83, 224-228.

71. Ossowski, S., Schwab, R. and Weigel, D. (2008) Gene Silencing in Plants Using Artificial microRNAs and Other Small RNAs. The Plant Journal, 53, 674-690.
72. Niu, Q.W., Lin, S.S., Reyes, J.L., Chen, K.C., Wu, H.W., Yeh, S.D. and Chua, N.H. (2006) Expression of Artificial microRNAs in Transgenic Arabidopsis Thaliana Confers Virus Resistance. Nature Biotechnology, 24, 1420-1428.
73. Schwab, R., Ossowski, S., Riester, M., Warthmann, N. and Weigel, D. (2006) Highly Specific Gene Silencing by Artificial microRNAs in Arabidopsis. The Plant Cell Online, 18, 1121-1133.
74. Zhang, X., Li, H., Zhang, J., Zhang, C., Gong, P., Ziaf, K., Xiao, F.M. and Ye, Z.B. (2011) Expression of Artificial microRNAs in Tomato Confers Efficient and Stable Virus Resistance in a Cell-Autonomous Manner. Transgenic Research, 20, 569-581.
75. Molnar, A., Bassett, A., Thuenemann, E., Schwach, F., Karkare, S., Ossowski, S., Weigel, D. and Baulcombe, D. (2009) Highly Specific Gene Silencing by Artificial microRNAs in the Unicellular Alga Chlamydomonas reinhardtii. The Plant Journal, 58, 165-174.
76. Zhao, T., Wang, W., Bai, X. and Qi, Y. (2009) Gene Silencing by Artificial microRNAs in Chlamydomonas. The Plant Journal, 58, 157-164.
77. Grimson, A., Farh, K.K.H., Johnston, W.K., Garrett-Engele, P., Lim, L.P. and Bartel, D.P. (2007) microRNA Targeting Specificity in Mammals: Determinants beyond Seed Pairing. Molecular Cell, 27, 91-105.
78. Joyce Tang, W., Chen, J.S. and Zeller, R.W. (2013) Transcriptional Regulation of the Peripheral Nervous System in Ciona intestinalis. Developmental Biology, 378, 183-193.
79. Lewis, B.P., Burge, C.B. and Bartel, D.P. (2005) Conserved Seed Pairing, Often Flanked by Adenosines, Indicates that Thousands of Human Genes Are microRNA Targets. Cell, 120, 15-20.
80. Clark, A.M., Goldstein, L.D., Tevlin, M., Tavare, S., Shaham, S. and Miska, E.A. (2010) The microRNA miR-124 Controls Gene Expression in the Sensory Nervous System of Caenorhabditis elegans. Nucleic Acids Research, 38, 3780-3793.
81. Kertesz, M., Iovino, N., Unnerstall, U., Gaul, U. and Segal, E. (2007) The Role of Site Accessibility in microRNA Target Recognition. Nature Genetics, 39, 1278-1284.
82. Brennecke, J., Stark, A., Russell, R.B. and Cohen, S.M. (2005) Principles of microRNA-Target Recognition. PLoS Biology, 3, Article ID: e85.
83. Shin, C., Nam, J.W., Farh, K.K.H., Chiang, H.R., Shkumatava, A. and Bartel, D.P. (2010) Expanding the microRNA Targeting Code: Functional Sites with Centered Pairing. Molecular Cell, 38, 789-802.
84. Helwak, A., Kudla, G., Dudnakova, T. and Tollervey, D. (2013) Mapping the Human miRNA Interactome by CLASH Reveals Frequent Noncanonical Binding. Cell, 153, 654-665.
85. Hafner, M., Landthaler, M., Burger, L., Khorshid, M., Hausser, J., et al. (2010) Transcriptomewide Identification of RNA-Binding Protein and microRNA Target Sites by PAR-CLIP. Cell, 141, 129-141.
86. Zeng, Y. and Cullen, B.R. (2005) Efficient Processing of Primary microRNA Hairpins by Drosha Requires Flanking Nonstructured RNA Sequences. Journal of Biological Chemistry, 280, 27595-27603.
87. Tang, R., Li, L., Zhu, D., Hou, D., Cao, T., Gu, H.W., Zhang, J., Chen, J.Y., Zhang, C.Y. and Zen, K. (2011) Mouse miRNA-709 Directly Regulates miRNA-15a/16-1 Biogenesis at the Posttranscriptional Level in the Nucleus: Evidence for a microRNA Hierarchy System. Cell Research, 22, 504-515.
88. Zisoulis, D.G., Kai, Z.S., Chang, R.K. and Pasquinelli, A.E. (2012) Autoregulation of microRNA Biogenesis by let-7 and Argonaute. Nature, 486, 541-544.
89. Wilbert, M.L., Huelga, S.C., Kapeli, K., Stark, T.J., Liang, T.Y., et al. (2012) LIN28 Binds Messenger RNAs at GGAGA Motifs and Regulates Splicing Factor Abundance. Molecular Cell, 48, 195-206.
90. Van Wynsberghe, P.M., Kai, Z.S., Massirer, K.B., Burton, V.H., Yeo, G.W. and Pasquinelli, A.E. (2011) LIN-28 Co-Transcriptionally Binds Primary let-7 to Regulate

miRNA Maturation in Caenorhabditis elegans. Nature Structural & Molecular Biology, 18, 302-308.

91. Neilsen, C.T., Goodall, G.J. and Bracken, C.P. (2012) IsomiRs-the Overlooked Repertoire in the Dynamic microRNAome. Trends in Genetics, 28, 544-549.

92. Zhou, H., Arcila, M.L., Li, Z., Lee, E.J., Henzler, C., Liu, J.Y., Rana, T.M. and Kosik, K.S. (2012) Deep Annotation of Mouse iso-miR and iso-moR Variation. Nucleic Acids Research, 40, 5864-5875.

CHAPTER 6

Healthy Ageing of Cloned Sheep

K.D. Sinclair[1] , S.A. Corr[1] , C.G. Gutierrez[1,2], P.A. Fisher[1] , J.-H. Lee[1], A.J. Rathbone[1] , I. Choi[1],w, K.H.S. Campbell[1], D.S. Gardner[1]

[1] Schools of Biosciences (KDS, CGG, PAF, J-HL, AJR, IC, KHSC), School of Veterinary Medicine and Science (SAC, DSG), University of Nottingham, Leicestershire LE12 5RD, UK.

[2]Universidad Nacional Autonoma de Mexico, Facultad de Medicina Veterinaria, Mexico City 04510, Mexico.

ABSTRACT

The health of cloned animals generated by somatic-cell nuclear transfer (SCNT) has been of concern since its inception; however, there are no detailed assessments of late-onset, non-communicable diseases. Here we report that SCNT has no obvious detrimental long-term health effects in a cohort of 13 cloned sheep. We perform musculoskeletal assessments, metabolic tests and blood pressure measurements in 13 aged (7–9 years old) cloned sheep, including four derived from the cell line that gave rise to Dolly. We also perform radiological examinations of all main joints, including the knees, the joint most affected by osteoarthritis in Dolly, and compare all health parameters to groups of 5-and 6-year-old sheep, and published reference ranges. Despite their advanced age, these clones are euglycaemic, insulin sensitive and normotensive. Importantly, we observe no clinical signs of degenerative joint disease apart from mild, or in one case moderate, osteoarthritis in some animals. Our study is the first to assess the long-term health outcomes of SCNT in large animals.

INTRODUCTION

The 20th anniversary of the birth of the first animal ('Dolly') to be derived from adult cells1, celebrated in July 2016, marks a milestone in the progress of somatic-cell nuclear transfer (SCNT), which has since been undertaken in more than 20 mammalian species including, quite recently, work involving human cells2,3,4. The pioneering study of Wilmut *et al.*1 was significant, however, because it demonstrated the potential to induce pluripotency in terminally differentiated cells. Indeed, although alternative approaches now exist for the generation of pluripotent stem cells5, it is generally recognized that SCNT remains the most effective way to reprogram somatic cells to a pluripotent state6,7. Nevertheless, despite technological advances in recent years8efficiency of SCNT remains low. This is best exemplified by the well-documented losses of partially reprogrammed embryos occurring throughout gestation and immediately following birth9,10,11,12. Although detailed clinical studies are lacking, the general consensus is that cloned offspring that survive beyond the neonatal period are healthy and reproductively fit13,14,15,16,17; although in some studies cloned mice tended to be obese, hyperinsulinaemic18 and were shorter lived19.

Longevity and healthy ageing among SCNT clones have been contentious issues from the outset20,21, with much made of the fact that Dolly had been undergoing treatment for osteoarthritis (OA) for some time before her death in 2003, aged 6½ years22,23,24. However, to the best of our knowledge, no formal, comprehensive assessment has been conducted of age-related, non-communicable diseases such as metabolic syndrome, hypertension and OA (three major comorbidities in aged human populations25) in cloned offspring of any species. This is significant because it is now well established that relatively subtle alterations to the periconceptional environment, either during natural mating or following assisted reproduction, can lead to heritable alterations to the epigenome linked to subsequent development and late-onset chronic diseases26,27,28,29. Given that SCNT requires the use of assisted reproductive procedures to generate donor oocytes and to culture reconstructed zygotes to the point of transfer, it is important to establish whether similar comorbidities exist in apparently healthy aged cloned offspring, using a long-lived and precocial mammalian species.

To this end we recently undertook detailed assessments of glucose tolerance, insulin sensitivity, ambulatory- and stimulated-blood pressure, together with musculoskeletal investigations, including clinical and radiological examination of all main joints, in 13 cloned sheep aged between 7 and 9 years (natural lifespan typically <10 years). In addition, magnetic resonance imaging (MRI) was undertaken of the stifle joints (the joint between the femur and tibia). These animals originated from studies undertaken between 2005 and 2007 that sought to improve the efficiency of SCNT by addressing issues related to mitochondrial heteroplasmy and chromatin remodelling at the point of nuclear transfer. Pregnancy outcomes for some of the clones derived from primary fetal fibroblasts were reported previously30,31. In addition to reporting the health status of aged clones derived from those studies we also report on the health of four 8-year-old Finn-Dorset clones (Figure 1) derived from the mammary gland cell line that gave rise to Dolly1. That is they are clones (genomic copies) of Dolly. Reference is made to the health status of contemporary groups of 5-and 6-year-old sheep, and published values for specific health-related parameters in sheep. Despite their advanced age, the clones were found to be euglycaemic, insulin sensitive and normotensive. Although most showed radiographic evidence of mild osteoathritis in one or two joints, no animal was lame and none have required treatment. We could find no evidence, therefore, of a detrimental long-term effect of cloning by SCNT on the health of aged offspring among our cohort.

Figure 1. Four 8-year old Finn-Dorset clones born in July 2007 and derived from the mammary-gland cell line that gave rise to Dolly.

RESULTS

Origin and genetic identity of cloned offspring

Microsatellite analyses confirmed that sheep within each clonal group were genetically identical to their respective parental-donor cell line for all 21 markers (Supplementary Table 1). The oldest clone (A089; male) was born in July 2006 and was derived from the primary fetal-fibroblast cell line SFF1 (Supplementary Table 1) as reported previously31. As part of that study the donor cell that gave rise to A089 had undergone mtDNA depletion before NT. Six of the female Lleyn clones were born in June 2008 and were derived from the primary fetal-fibroblast cell line LFF4 as reported previously30. Four of these clones (that is, 382, 385, 386 and 387) were derived from permeabilized fetal fibroblasts incubated with *Xenopus* oocyte extract with a view to enhancing epigenetic remodelling of chromatin before NT. The remaining two clones (that is, 388 and 389) were derived from permeabilized fibroblasts that were not exposed to extract. A seventh female Lleyn clone (2265) was also derived from LFF4 cells but was born in the previous year (August 2007) along with clone 2264 (derived from fetal-fibroblast cell line SFF5, breed unknown). These two clones formed part of a pilot study designed to assess the suitability of these two cell lines for SCNT using standard in-house protocols30,31.

The four female Finn-Dorset clones (2260, 2261, 2262 and 2263; Figure 1) were born in July 2007. Data from these animals have not been reported previously. They were derived from the mammary gland cell line that gave rise to Dolly1 (Supplementary Table 1). The study that produced these four ewes (Supplementary Table 6) was an extension of that of Choi *et al*32. which used fetal fibroblasts as nuclear donors. Both studies investigated the effects of treating ovine oocytes during the latter stages of *in vitro* maturation, before nuclear transfer, with caffeine. We now report that 10 lambs derived from these nuclear donor cells were born (derived from seven caffeine-treated and three control oocytes), seven lived beyond 1 week of age (derived from five caffeine-treated and two control oocytes), but only four survived to adulthood (all derived from caffeine treated oocytes).

Body composition and peripheral insulin sensitivity

In preparation for metabolic assessments animals were fed a nutritionally-balanced diet of hay and concentrates ((metabolizable energy=11.0 MJ kg^{-1} DM; crude protein=130 g kg^{-1} DM) *ad libitum* (equivalent to ~2 × maintenance33)) for a 28-week period to attain a body condition score (BCS) of 4.0 units ($1_{[thin]}$ to $5_{[obese]}$ scale34), equivalent to ~40% total body fat35 for female sheep. This was in line with our reference population of 6-year-old females conceived by embryo transfer (ET-controls). Mean (±s.e.m.) BCS of 4.0±0.1 and 4.1±0.2 units were attained for Lleyn and Finn-Dorset clones at the time of body composition assessment by dual-energy X-ray absorptiometry (DEXA). Body condition scores for clones 2264, 2265 and A089 at this time were 3.8, 3.8 and 3.0

units, respectively. Subsequent body fat content determined by DEXA was slightly greater (analysis of variance (ANOVA); $P<0.01$) in Finn-Dorset clones (~46%) than ET-Controls (~40%), but was lower (~21%; $P<0.001$) in Lleyn clones (Figure 2a). This was despite the fact that they were all fed the same diet *ad libitum* for an extended period (28 weeks), and BCS were similar. This probably reflects differences in body fat (that is, subcutaneous versus intra-abdominal fat) distribution between genotypes36. Corresponding levels of body fat for clones 2264, 2265 (both female) and A089 (male) were 42%, 38% and 23%, respectively (Supplementary Table 2).

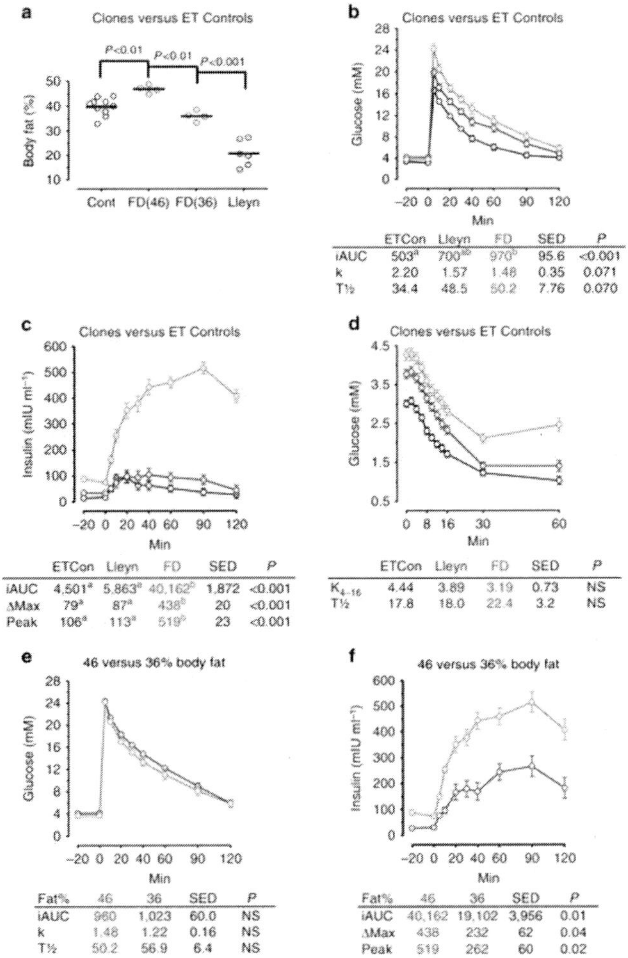

(b)

	ETCon	Lleyn	FD	SED	P
iAUC	503[a]	700[ab]	970[b]	95.6	<0.001
k	2.20	1.57	1.48	0.35	0.071
T½	34.4	48.5	50.2	7.76	0.070

(c)

	ETCon	Lleyn	FD	SED	P
iAUC	4,501[a]	5,863[a]	40,162[b]	1,872	<0.001
ΔMax	79[a]	87[a]	438[b]	20	<0.001
Peak	106[a]	113[a]	519[b]	23	<0.001

(d)

	ETCon	Lleyn	FD	SED	P
K_{4-16}	4.44	3.89	3.19	0.73	NS
T½	17.8	18.0	22.4	3.2	NS

(e)

Fat%	46	36	SED	P
iAUC	960	1,023	60.0	NS
k	1.48	1.22	0.16	NS
T½	50.2	56.9	6.4	NS

(f)

Fat%	46	36	SED	P
iAUC	40,162	19,102	3,956	0.01
ΔMax	438	232	62	0.04
Peak	519	262	60	0.02

Figure 2: Body fat, glucose tolerance and insulin sensitivity in aged cloned sheep and contemporary embryo transfer (ET) controls.

Percentage body fat (determined by DEXA) revealed differences between clonal groups (n=11 ET-Controls, 6 Lleyn and 4 FD clones) (**a**). Finn-Dorset (FD) clones were subsequently placed on a diet to reduce body fat percentage from 46% (FD46) to 36% (FD36). Initial glucose tolerance tests revealed modest differences in glucose tolerance (**b**) and large differences in insulinaemia (**c**) between clonal groups that probably reflected genotype differences in body fat percentage and body fat distribution. Insulin sensitivity did not differ between clonal groups and ET-controls (**d**). Although glucose tolerance was unaltered (**e**), insulinaemia was significantly reduced when Finn-Dorset clones lost body fat (**f**). Glucose was infused i.v. at 0.4 g·kg^{-1} body weight and insulin at 0.75 IU kg^{-1} body weight, both at time 0. iAUC, incremental area under curve; k, fractional clearance of glucose (% min^{-1}); $K_{4\text{-}16}$, % decline in glucose min^{-1} from 4 to 16 min; T½, half-life of glucose; Δ_{Max}, peak insulin—basal insulin. s.e.d., standard error of difference; NS, not significant. Different superscripts denote significance at P<0.05 following Bonferroni correction.

Non-fasted plasma glucose concentrations at the onset of the initial glucose tolerance test (GTT), 4 h after the morning meal, were all within the normal fasting range for well-fed sheep of moderate body condition (3.25 to 4.50 mmol/L (ref. 37)). The lower (ANOVA; P<0.001; SED=0.24) glucose concentrations (mmol/L) found in ET-Controls (3.20) than in either Finn-Dorset (3.70) or Lleyn (4.20) clones (Figure 2b) are therefore of no clinical significance. The Finn-Dorset and Lleyn clones, nevertheless, appeared less tolerant to glucose infusion (400 mg kg^{-1}) than the ET-Controls, as evident from the incremental area under the curve (Figure 2b). Resting plasma insulin (μIU ml^{-1}) was also greater (ANOVA; P<0.001) for the Finn-Dorset clones (81.5) than for either the Lleyn clones (34.1) or ET-controls (15.7) at the onset of the initial GTT, and a hyperinsulinaemic state was confirmed in these animals following glucose infusion (Figure 2c). Compared with fasted basal plasma insulin concentrations in lean and obese sheep from previous studies (Supplementary Table 3), these values for the Finn-Dorset clones appear high. However, neither percentage decline in plasma glucose per minute following insulin administration (0.75 IU kg^{-1}) ($K_{4\text{–}16}$), nor the half-time for glucose clearance t(1/2), differed between these animal groups (Figure 2d), indicating that peripheral tissue insulin sensitivity was similar. To get a better sense of the contribution of body fat to insulin sensitivity, the four Finn-Dorset ewes were subsequently placed on an energy-restricted diet (0.7 × maintenance (ref. 33)) for 20 weeks during which time their level of body fat declined from ~46 to ~36% (Figure 2a). Although glucose tolerance was not altered (Figure 2e), both resting (29 μIU ml^{-1}) and glucose-stimulated plasma insulin concentrations were markedly reduced (ANOVA; P<0.05) in the leaner Finn-Dorset clones (Figure 2f), indicating that a major factor contributing to hyperinsulinaemia in these animals was level of body fat.

Corresponding non-fasting plasma glucose concentrations for the other three clones were close to or slightly above the normal fasting range (Supplementary Table 2). At ~38% body fat, the 8-year-old clone 2265 was fatter than her 7-year-old fellow Lleyn clones, but otherwise her indices of glucose and insulin tolerance were similar. The 9-year-old male clone was also euglycaemic, with glucose indices similar to that of ET-control males during the GTT. Although glucose tolerance was normal for this animal, insulin secretion (Δ_{Max}) was low and peripheral tissue insulin sensitivity (that is, K_{4-16} and half-life) was reduced relative to ET-Control males of similar body composition. Parameters for this animal, however, do not indicate a diabetic state.

Ambulatory and stimulated blood pressure

After recovery from telemeter implantation, resting blood pressure and heart rate were monitored over the next 4 days. No sheep exhibited any clear circadian rhythm of pressure (for example, nocturnal dipping analysed by non-linear regression using Fourier curves). Hence, Figure 3 reports a composite pressure trace recorded over 4 days in which sheep were group-housed and exposed only to routine husbandry procedures. Mean arterial (MAP) and diastolic (DBP) blood pressure over the study period were similar among clones (MAP, 111±7 versus 102±8 mm Hg, P=0.18; DBP, 96±6 versus 90±7 mm Hg, ANOVA; P=0.24; Finn-Dorset versus Lleyns, respectively) and when compared with ET-controls (MAP, 114±6: DBP, 106±7 mm Hg). Systolic blood pressure was similar between clones and to ET-controls (Figure 3a,b), but resting heart rate in ET-controls was lower (average HR, 76±8 versus 91±11 (Finn-Dorset) and 92±6 (Lleyns) beats per min; ANOVA; P<0.001, Figure 2c,d). Spikes in pressure and heart rate were noted in all sheep at morning feed time (~08:00 hours, Figure 3a,c). Cardiovascular parameters recorded in the elderly clones were within the established normal range of pressures reported for aged sheep in the literature (Supplementary Table 4). The pressor response to incremental Angiotensin II (Ang II) was chosen as a cardiovascular challenge, as comparison with data from ET-controls and other studies conducted by us in sheep38 were available for reference. Also, Ang II is a potent vasoconstrictor, often targeted as a cardiovascular therapy through the use of angiotensin-converting enzyme inhibitors. Step-wise (every 10 min) doubling of intravenous (i.v.) Ang II delivery incrementally increased BP in all sheep (Figure 3e,f), the effect size being identical between clones (Finn-Dorset, 62±12 versus Lleyn, 62±12 mm Hg) and similar to ET-controls (58±10 mm Hg). Heart rate tended to decrease with increasing Ang II infusion in Lleyn clones, but was refractory to the depressor effect of high blood pressure in Finn-Dorset clones (Figure 3g), which was in keeping with the ET-controls (Figure 3h).

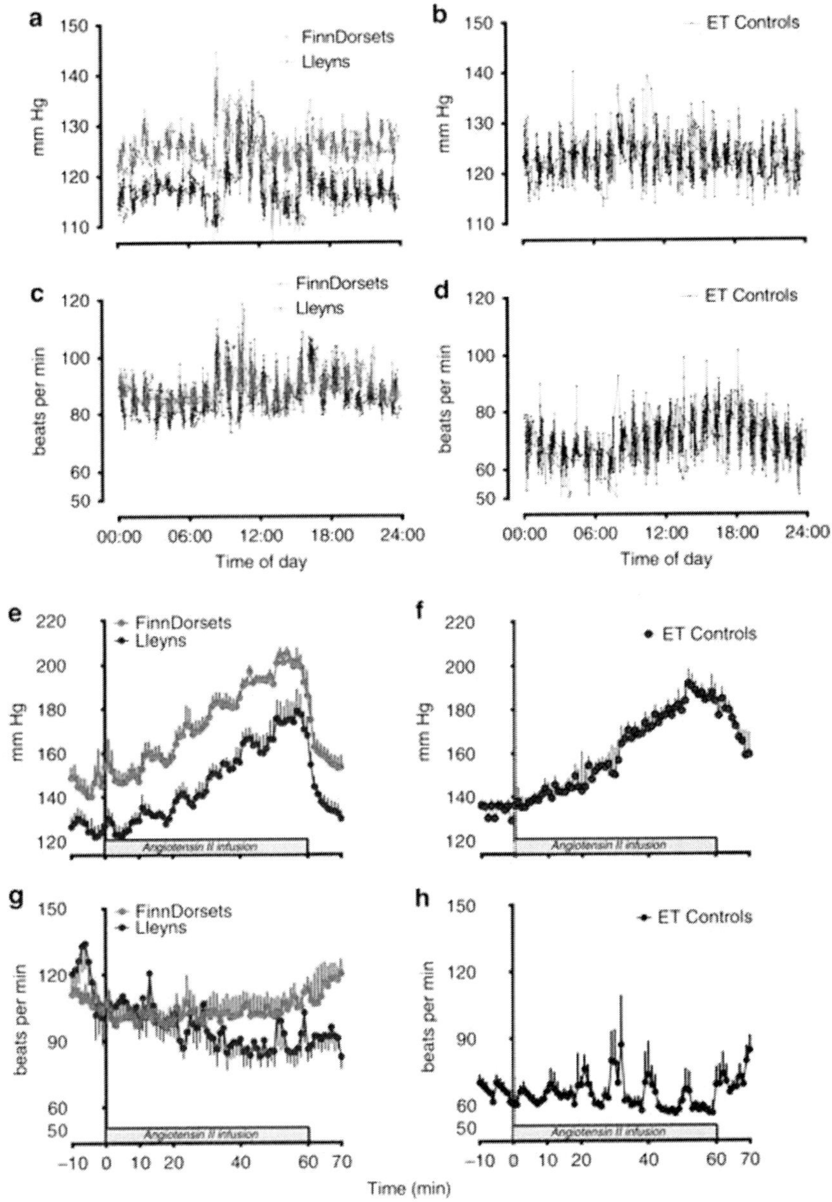

Figure 3. Ambulatory and Angiotensin-II stimulated systolic blood pressure and heart rate in Finn-Dorset (n=4) and Lleyn (n=6) clones, and embryo transfer controls (ET-controls; n=11).

Cardiovascular data were recorded using signals derived from radiotelemetry probes (Telemetry Research, distributed by Millar Ltd, Oxon, UK) implanted into the carotid artery. Data are presented as minute means±s.e.m. and represent either composite 24 h traces of ambulatory systolic blood pressure and heart rate recorded over 3–4 days with sheep group-housed in a floor pen (**a**,**b**,**c**,**d**) or the cardiovascular response to incremental i.v. angiotensin II infusion (0–60 ng kg^{-1} min^{-1}) (**e**,**f**,**g**,**h**). Wireless sampling was at 2 kHz (that is, beat-to-beat), data downloaded automatically to acquisition software on a laptop (LabChart 7; AD Instruments, Oxon, UK). Shaded box represents period of angiotensin II infusion.

MUSCULOSKELETAL ASSESSMENTS

Only two cloned sheep were identified as having an abnormal gait on clinical examination, scoring 1 and 2 on the six-point scale of Kaler *et al*39. (Supplementary Table 5). However no animal showed obvious signs of pain on manipulation of any joint. Radiographic assessment revealed that most sheep had either normal hip joints or mild grade 1 osteophytosis (Figure 4). The only sheep with significant degenerative joint disease was the Finn-Dorset clone 2262, with moderate to severe osteophytosis (scoring 2–3 on the modified scale of Kellgren and Lawrence40) of the majority of joints (Figs 5 and 6a). MRI of the stifles demonstrated the degenerative changes apparent in the radiographs, with periarticular osteophytosis primarily on the trochlear ridges and caudal tibia (Figure 6b,c). The severity of the changes was greatest in clone 2262. In addition, clones 2262 and 2263 had evidence of mild bilateral synovitis, while clones 2260 and 2261 had mild bilateral synovial effusions.

Both femoral heads appear normally seated within the acetabulae, with minimal remodelling and no obvious osteophytosis. A mild increase in bone density (white) is evident around both acetabular rims (black arrows), however there is no significant remodelling of the acetabulum or femoral head and neck regions. This sheep therefore has minimal radiographic evidence of osteoarthritis. The circular opacity is a 25 mm coin, placed as marker to indicate the right limb.

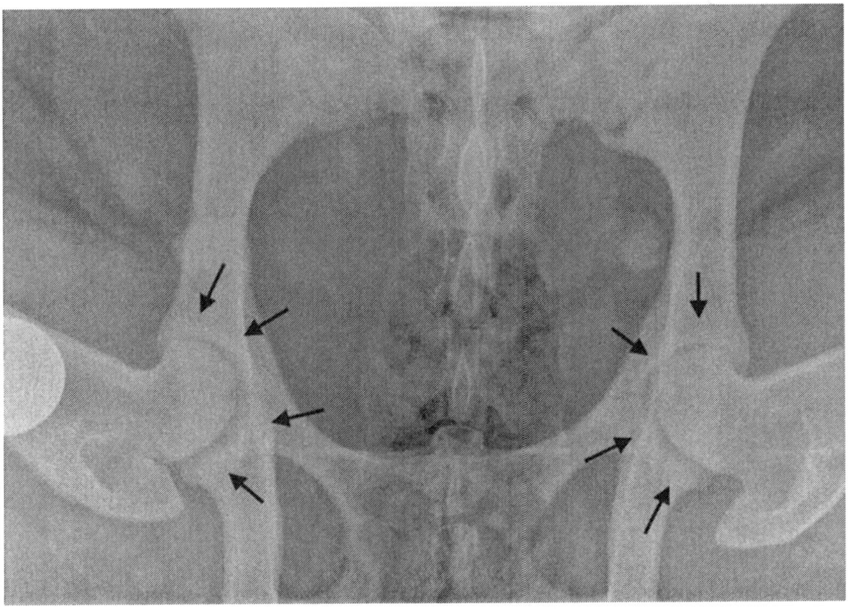

Figure 4: Image showing a ventrodorsal radiographic view of the hip joints of clone 2260.

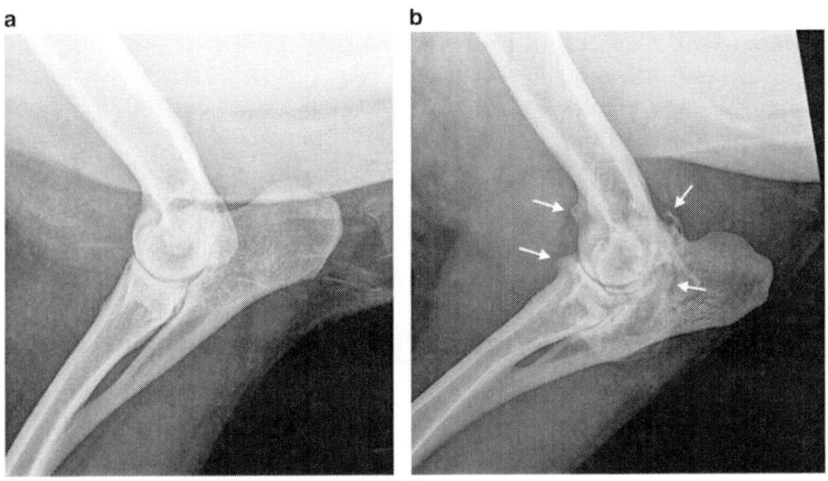

Figure 5: Mediolateral radiographic views of the elbow joint of two clones with and without osteoarthritis.

(**a**) The elbow joint of Clone 388 appears normal: there is no radiographic evidence of osteoarthritis. (**b**) The elbow joint of Clone 2262 shows evidence of severe osteoarthritis. Large osteophytes representing new bone formation are apparent at the joint margins (white arrows), and are associated with bone remodelling. This elbow was scored as Grade 3 on the modified scale of Kellgren and Lawrence40.

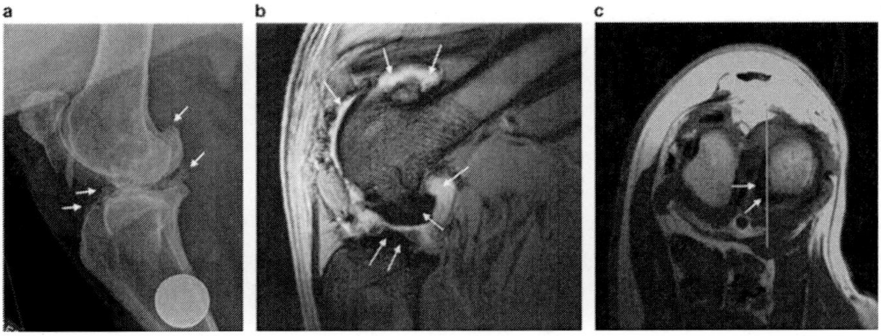

Figure 6: Images of the right stifle joint of Clone 2262.

(**a**) Mediolateral radiographic view of the right stifle of Clone 2262. White arrows indicate osteophytes/new bone formation at joint margins, with bone remodelling. This joint was scored as Grade 3 on the modified scale of Kellgren and Lawrence38. The circular opacity is a 25 mm coin, placed as marker to indicate the right limb. (**b**) Sagittal T2* MRI scan of the same stifle joint. (**c**) Transverse T1 MRI scan of the same stifle joint. The green lines on this image show the level corresponding to the sagittal image shown in **b**. In the MRI scans, bone appears hypointense (dark), while fat and fluid appear hyperintense (white). The irregular new bone formation on the femur and tibia are indicated by the yellow arrows. The white arrows indicate synovial fluid and fat within joint.

DISCUSSION

This study represents the first detailed and comprehensive assessment of age-related, non-communicable diseases in adult cloned offspring of large animals. Working with sheep, four of which were cloned from the cell line that gave rise to Dolly, we undertook assessments of metabolic, cardiovascular and musculoskeletal health. Importantly, despite their advanced age (7–9 years), none of the clones showed any clinical signs of disease, being euglycaemic, insulin sensitive and normotensive. No animal was lame despite most showing radiographic evidence of mild OA in one or two joints, as would be expected in sheep of that age. None have required treatment for OA. In the absence of perfectly matched (by age, genotype and environment), naturally conceived controls, these conclusions were reached with reference to (a) a contemporary group of six-year-old sheep that were managed alongside the clones, and underwent the same metabolic and cardiovascular assessments, (b) published reference ranges for blood glucose, (c) pelvic radiographs of eight 5-year-old healthy sheep, and (d) the available scientific literature. Whilst contemporaneous controls would have been ideal, our data are nevertheless compelling, indicating no detrimental long-term adverse effects of SCNT on the health of aged adult offspring.

Because cloning by SCNT can lead to obesity and hyperinsulinaemia in aged mice[18], and a single case of Type I (insulin dependent) diabetes was reported in a cloned calf[41], we decided to measure glucose tolerance and insulin sensitivity in our aged cloned offspring. There are (infrequent) reports of naturally occurring type I diabetes in sheep[42], where glucose concentrations for affected animals range between 16 to $19\ \text{mmol l}^{-1}$. However, we are unaware of cases of type II (non-insulin dependent) diabetes, although insulin resistance has been extensively studied in this species[43] where it is known, for example, that deficiencies in maternal periconceptional diet can lead to insulin resistance in adult offspring[44]. In humans, risk factors for insulin resistance leading to type II diabetes include advanced age and obesity; the latter risk factor is also related to enhanced insulin resistance in sheep[43,45]. Hence the current study sought to create an obese state in our experimental animals to fully test the extent of glucose tolerance and insulin sensitivity[46]. Hyperinsulinaemia in the four Finn-Dorset clones (Figure 2c) was certainly influenced by the level of body fat, as weight loss reduced plasma insulin concentrations (Figure 2f), consistent with improved peripheral insulin sensitivity which can be expected following weight loss in sheep[46]. However, peripheral insulin sensitivity is negatively related to visceral fat and intrahepatic lipid accumulation in humans[47,48], which varies between ethnic groups[49]. It follows that genotype differences in regional fat distribution in sheep[36] may at least partially account for the large difference in insulin response following glucose infusion observed between Finn-Dorset clones and ET-controls, which had a similar proportion of total body fat (Figure 2a). Although this awaits final confirmation following post-mortem examination, previous studies indicate that the Finn-Dorset genotype has larger intra-abdominal fat depots than other common genotypes such as our ET-controls[50,51].

Consistent with observations for cloned offspring from other species[9,14,52], perinatal losses of cloned sheep have been associated with structural defects of the kidneys and heart, and with pulmonary hypertension[10]. Enlarged and abnormally developed kidneys were also evident in some of the six live-born Finn-Dorset clones that failed to survive to 3 months of age in the current study (Supplementary Table 6). We hypothesized that if subtle renal defects in surviving offspring were to persist late into adult life then this could lead to increased blood pressure, known as reno-vascular hypertension53. Hypertension has been described to 'follow-the-kidney' in cross-transplantation studies, suggesting factors intrinsic to the kidney mediate increases in blood pressure54·55·56. The renin–angiotensin–aldosterone system is a major target for long-term anti-hypertensive treatment as elevated intra-renal renin–angiotensin–aldosterone system action and/or sensitivity may directly cause increased blood pressure57·58. Here, for the first time using radiotelemetry blood pressure probes, we recorded circadian cardiovascular parameters continuously for individual cloned and ET-control sheep kept as a group in a stress-free environment. Ambulatory blood pressure of these cloned animals was similar to ET-controls and within the normal range for other sheep reported in the literature (that is, ~120/80 mmHg; Supplementary Table 4). It was also similar to adult blood pressure in healthy humans. Classification of 'hypertension' in humans begins at a systolic blood pressure ≥140 mmHg (ref. 59) or ~20 mmHg above resting. The elderly cloned sheep in this study could not, therefore, be considered to be hypertensive. Direct i.v. infusion of angiotensin II, to further challenge cardiovascular responsivity, failed to induce a greater pressor response than in ET-controls. We therefore conclude that cloning by SCNT had no detrimental, latent effect on the cardiovascular system of sheep in this study that survived into late adulthood.

Detailed musculoskeletal investigations were undertaken of all sheep, as Dolly had required treatment for OA from around 5 years of age, raising concerns over premature ageing in cloned animals22·23·24. Radiographs taken of Dolly showed OA in the left stifle (knee), however OA was evident in both stifles at post-mortem examination 18 months later (TJ King, Roslin Institute, personal communication). The consensus at that time was that, because the OA was localized rather than generalized, it was more likely of traumatic aetiology. However, the aetiopathogenesis of OA is accepted to be multifactorial, involving both genetic and acquired factors (such as joint trauma/overloading, and obesity). The prevalence increases with age (the strongest risk factor), but the course of the disease varies greatly between individuals60·61.

Clinical examination of cloned sheep in the present study revealed only mild lameness in the left foreleg of one animal, which may have been associated with her slightly abnormal forelimb conformation. None of the Finn-Dorset clones had an abnormal gait or showed lameness. Several clones had soft tissue swelling or fibrous thickenings of various parts of their limbs, as would be expected in a random sample of sheep. However, these did not appear to be associated with any discomfort. Subsequent radiography revealed mild OA in various joints in several sheep. It is not surprising that no sheep showed marked lameness, as radiographic changes indicative of OA do not necessarily correlate with the extent of clinical disease62. Mild OA (scored by two independent reviewers) was seen most often in the hip (13 of 13 sheep) and stifle (12 of 13 sheep) joints. Scott63 reported that the stifle and elbow joints are most commonly affected by joint trauma. However in the present study, only 5 of 13 sheep showed elbow OA, which was bilateral in most cases. Only Clone 2262 had significant OA in multiple, but not all, joints. The majority of sheep appeared to have a mild increase in bone density of the acetabular rim, however none had significant osteophytosis or remodelling of the acetabulum or femoral head and neck as would be expected with OA. Similar findings were apparent in pelvic radiographs of eight healthy 5-year-old sheep (Supplementary Figure 2), enroled in an unrelated study, made available courtesy of Professors Allen Goodship and Gordon Blunn (University College London). Thus it may be a normal radiographic finding in healthy ageing sheep, and not, in isolation, indicative of OA. This will be further assessed when post-mortem examination of the present cohort is undertaken at a future time.

Radiographic changes appear relatively late in the disease process and so MRI of the stifles (knees) of the four Finn-Dorset clones was undertaken, as MRI is more sensitive in detecting early structural changes60. To minimize the duration of sedation required, MRI scans were taken only of the stifle joints, because these were among the most commonly affected joints in our population, and were affected in Dolly23. T2* and proton density, fat-saturated MRI images provided the most useful diagnostic information, showing periarticular osteophytosis primarily on the trochlear ridges and caudal tibia. However, the extent of the disease remained mild, and not unexpected in aged sheep.

Reports of OA in non-experimental sheep are rare in the literature. A case series was reported by Scott64, describing elbow OA in 10 sheep (aged 2 to 5 years) that had shown forelimb lameness of at least 3 months duration. A post-mortem study of the stifle joints of 65 clinically normal sheep aged 6 months to 11 years by Vandeweerd *et al.*65identified cartilage defects in 66% of the sheep, the severity of the lesions increasing with age. These authors proposed that OA exists in normal ageing sheep, and is not always associated with clinical signs. Pathological changes were also found in the femoral head cartilage of clinically normal 2.5- to 3-year-old sheep by Zilkens *et al*66. Thus there is no evidence of an increased incidence or severity of OA in the clones in this study. However, a single cloned sheep in the present study had obvious OA in multiple joints, although it varied in severity between joints, and not all joints were affected. It is not surprising that the joints were variably affected, as studies in humans have shown, for example, that a factor such as obesity increases the risk of knee OA by a factor of three, but has less effect on hip OA60. As discussed previously, a more homogenous distribution might have been expected, both within and between sheep, if cloning by SCNT was a direct causative factor in the development of OA.

From the current series of assessments we conclude that there are no long-term detrimental health effects of cloning by SCNT for a long-lived species such as the sheep. This conclusion is consistent with less detailed longevity studies in cattle17, and suggests that the ageing process in surviving clones of large animal species is not accelerated. On initial inspection, our data in sheep may appear at odds with the health status of Dolly and predictions of premature ageing, which were based on terminal restriction fragment analyses of her genomic DNA20. While telomere length was reduced in SCNT clones relative to age-matched controls in that and subsequent studies in sheep67, these effects did not manifest following SCNT in cattle21. Further inconsistent reports of shorter telomeres in cloned offspring from other species68have led to the consensus that telomere length is generally restored during nuclear reprogramming69. The extent of telomere restoration in turn is dictated by intricate epigenetic alterations to telomeric and sub-telomeric chromatin, variation in which could explain discrepancies between species and donor cell types within species. It follows that the relationship between telomere length, health and longevity in multicellular organisms is complex and, for our current cohort of animals, awaits organ-specific cell enrichment and analysis following post-mortem at a future date.

In conclusion, although the efficiency of SCNT has improved in recent years, its overall efficiency remains low, with high embryonic and gestational losses compared to natural mating and assisted reproduction. A relatively high proportion of clones also fail to successfully make the transition to extra-uterine life, some harbouring congenital defects, such as observed in the kidney. For those clones that survive beyond the perinatal period, however, the emerging consensus, supported by the current data, is that they are healthy and seem to age normally.

METHODS

Animal work

Animal work was conducted under the authority of the United Kingdom Animal (Scientific Procedures) Act 1986, and approval was obtained from the University of Nottingham Animal Welfare and Ethical Review Board before commencement of these studies.

Genetic identity of cloned sheep

This was confirmed using 21 microsatellite markers including those recommended by the International Society of Animal Genetics (Supplementary Table 1). Sheep blood samples and cell lines were submitted for PCR and fragment analysis to an independent genetic Company (VHL Genetics, The Netherlands). PCR amplification and fragment analysis was conducted in an ABI PRISM 3130XL automatic sequencer (Applied Biosystems), and the amplified fragments were classified using the GENESCAN 3.7 software (Applied Biosystems). We undertook a blinded pairing of cloned sheep and parental cells, and all clones matched their cells of origin for all 21 markers (CERVUS v.3.0 software70).

Reference population (ET-controls)

This consisted of a group of 6-year-old virgin male (n=6) and female (n=11) Suffolk x Scottish Blackface sheep generated as part of the study of Sinclair *et al.*44. These animals were the offspring of the control group of normally fed ET donors in that study, and were accommodated in pens adjacent to the cloned offspring where they were fed the same diet. Health assessments were originally reported for these animals at two years of age44. Further assessments of body composition, glucose and insulin tolerance, and blood pressure were undertaken again for these animals at 6 years of age using protocols described herein.

Whole animal body composition

Glucose tolerance and insulin sensitivity are each challenged in the obese state43, which is why we endeavoured to create this condition in both our reference population of ET-controls and clones. Animals were fed a proprietary concentrate (Manor Farm Feeds, Owston, Leicestershire, UK) consisting 60 g kg^{-1} oil, 145 g kg^{-1} crude protein, 80 g kg^{-1} fibre and 65 g kg^{-1} ash, in a mixed diet with meadow hay (dry matter (DM), 830 g kg^{-1}; metabolizable energy, 8.5 MJ kg^{-1} DM; crude protein, 95 g kg^{-1} DM). This diet was fed on an *ad libitum* basis. To standardize body composition, body condition scores (BCS) ($1_{[thin]}$ to $5_{[obese]}$ scale34), were obtained at fortnightly intervals for these animals with the aim of attaining a BCS of 4.0 units at the time of metabolic assessments, equivalent to approximately 40% total body fat in mature female ewes35. Formal assessments of body composition were assessed by DEXA using a validated in-house protocol71. Animals were sedated (intramuscular (i.m.) 0.1 mg kg^{-1} Xylazine; i.v. 4 mg kg^{-1} Ketamine) and scanned transversally along the longitudinal axis using a Norland XR-800 DXA scanner. The scan lasted ~15 min after which animals were returned to the barn.

Glucose and insulin tolerance tests

Glucose tolerance and insulin sensitivity were determined using standard in-house protocols44,71. Briefly, for the GTT, basal plasma insulin and glucose concentrations were established 4 h after the animals' morning meal. Infusion of sterile glucose (i.v. 0.4 g·kg^{-1} body weight given as a single bolus) was followed by blood sample collection from indwelling jugular catheters into K^{+}EDTA vacutainers at 5, 10, 20, 30 40, 60, 90 and 120 min. The insulin tolerance tests were carried out before the morning meal. Infusion of insulin (0.75 IU kg^{-1}) was followed by blood collection into K+EDTA vacutainers at two-minute intervals up to 16 min, and then again at 30 and 60 min. Plasma glucose concentrations were determined on a Cobas Miras Plus autoanalyzer (ABX Diagnostics); the kit for glucose was supplied by Randox Laboratories (catalogue no. GL2623). Plasma insulin concentrations were determined using an ovine specific insulin ELISA kit (Mercodia AB, Uppsala, Sweden).

Radiotelemetric assessments of heart rate and blood pressure

Sheep were anaesthetised for general surgery using i.v. ketamine (2.0 mg kg^{-1}) and midazolam (0.3 mg kg^{-1}), dosed to effect for intubation. Anaesthesia was maintained with isoflurane (1–2% in O_2). Solid-state pressure radiotelemeters designed for large animal use (TRM84PB; Telemetry Research, NZ and distributed in the UK by Millar Inc, Oxon, UK) were inserted (10 cm) into the carotid artery with the body of the transmitter secured subcutaneously in the neck (Supplementary Figure 1). The incision site was closed and the animal recovered to a pen. Post-operative antibiotics were administered (i.m. long-acting amoxycillin) and analgesia maintained with a fentanyl (0.2 mg kg^{-1}) patch fixed to the forelimb. Sheep (4 on each occasion) were recovered to a floor pen and wireless recording of individual cardiovascular parameters (for example, blood pressures, heart rate) began immediately for 3–4 days for each sheep. Continuous recording at a sampling rate of 2 kHz was pre-programmed for 10 min of each hour, all day and night for all four sheep in each batch using an automated scheduler pod linked to data acquisition software (LabChart 7; AD Instruments, Oxon, UK). On the final day, all sheep were challenged with a step-wise incremental Angiotensin II (0–60 ng kg^{-1} min^{-1}; Sigma-

Aldrich, UK) infusion, delivered into the jugular vein after earlier venipuncture and restraint in a metabolic crate. Recording at this time was of individual sheep on a second-by-second basis. Sampled data were stored, analysed and minute means derived (after exclusion of artefacts) using LabChart 7 and MS Excel. Telemeters were surgically removed from all sheep under general anaesthesia the following day, following similar protocols as described above. The solid-state telemeters were used (after recalibration and sterilization) on three separate occasions, to obtain cardiovascular data for ten sheep (n=4, Finn-Dorset clones; n=6, Lleyn clones). Cardiovascular protocols were successfully completed in all sheep.

Clinical examination and diagnostic imaging of joints

All sheep underwent a full musculoskeletal examination by a veterinary orthopaedic specialist (co-author SC) involving visual assessment of gait, followed by palpation and manipulation of the limbs and spine. Palpation assessed for evidence of muscle atrophy, swelling, heat or pain in the limb. Manipulation assessed individual joints for range of motion, pain or crepitus. The feet were also examined for pathology. Assessment of gait was made using the seven-point scale (0 to 6) developed by Kaler *et al*39.

All sheep were subsequently sedated (i.m. 0.1 mg kg^{-1} Xylazine; i.v. 4 mg kg^{-1} Ketamine) for 10 to 15 minutes to obtain mediolateral radiographs of both hocks, stifles, elbows and carpi, and a ventrodorsal view of the pelvis, including both hips and the lumbosacral junction. Shoulders were radiographed but the images found to be of inconsistent quality for valid interpretation. Radiographs were taken using a portable Tru-DR Digital Radiography System, with laptop (R8-51D45) and plate (50-C907-08B) (BCF Technology Ltd, Strathclyde, UK), and analysed using OsiriX Lite (http://www.osirix-viewer.com/download_form/ download_form.php). Images were assessed for evidence of OA, and graded using a modified Kellgren and Lawrence40 scale. Assessment of joint space width was not undertaken as the sheep were sedated and recumbent (that is, non-weight-bearing) when the radiographs were taken. Radiographs were scored independently by two veterinary orthopaedic specialists (co-author SAC, and Malcolm Ness, independent) who were unaware of sheep identity. OA was graded based on the presence of osteophytes and bone remodelling, where 0=no evidence, 1=questionable, mild osteophytosis, 2=moderate osteophytosis and minor bone remodelling, and 3=severe osteophytosis with definite bone remodelling. The four Finn-Dorset clones underwent MRI of their stifle joints, using a mobile MRI scanner (Burgess Diagnostics Ltd. Earnshaw Business Centre, Hugh Lane Leyland PR26 6PD), with a high-field 1.5Tesla magnet (Philips Medical Systems, Achieva). Sheep were induced using i.v. ketamine (2.0 mg kg^{-1}) and midazolam (0.3 mg kg^{-1}), dosed to effect for intubation. Anaesthesia was maintained with isoflurane (1–2% in O_2). Scans were obtained using a SENSE Flex-S receiver coil and Body Transmit Coil, with a 130mm Field of View. Transverse T1, Saggital T2*, PD and PDFS protocols used to obtain images extending from 2 cm proximal to the patella, to 5 cm distal to the tibial plateau. Images were transferred in DICOM format on CD and subsequently analysed using OsiriX Lite.

Statistical analyses

Because the six Lleyn clones were derived from a study that assessed the effects of *Xenopus* oocyte extract before nuclear transfer on pregnancy outcomes30, we initially investigated if this had any influence on our measurements of body composition, glucose metabolism and blood pressure by ANOVA. As the extract had no effect we treated these animals as a single cohort in all subsequent analyses. Glucose and insulin tolerance tests were subsequently analysed by repeated-measures ANOVA (with Bonferroni correction for comparisons between means) where clone identity (Finn-Dorset, Lleyn) and time formed the fixed effects. Single values for incremental areas under the response curves were generated for each individual using the trapezoid rule in Prism 6 (GraphPad Software, La Jolla, CA, USA) and were analysed by one-way ANOVA with Bonferroni correction for comparisons between means. For cardiovascular analyses, data were either analysed as summary measures (e.g. minute means of sec-sec radiotelemetry pressures or heart rate) and analysed similarly to metabolic data or, to test for any circadian effects, by incorporating all recorded cardiovascular datapoints (~2,500–4,500 for each animal) into a non-linear regression model fitting a Fourier-curve ($Y=\alpha+\beta sin(2\pi(X+\varepsilon)/w)$) to derive four parameters α, set-point; β, amplitude; w, wavelength and ε, offset. No sheep exhibited any circadian cardiovascular rhythm (pressure or heart rate) and therefore only baseline (average for each variable) or minute means of responses (for example, to angiotensin II infusion) are presented. All data were analysed using Genstat v18 (VSNi, Rothampsted, UK) and are presented as means±s.e.m. or s.e.d. (standard error of the differences between means, for a more conservative estimate of the contrast variance).

DATA AVAILABILITY

The data that support the findings of this study are available from the corresponding author (KDS) on reasonable request.

ADDITIONAL INFORMATION

How to cite this article: Sinclair, K. D. *et al.* Healthy ageing of cloned sheep. *Nat. Commun.* 7:12359

ACKNOWLEDGEMENTS

This article is dedicated to the memory of co-author Keith H.S. Campbell (1954 to 2012). The University of Nottingham provided financial support for this study. Professors Gordon Blunn (University College London) and Allen Goodship (University College London and Royal Veterinary College) for making available pelvic radiographs of 5-year-old healthy sheep. Malcolm Ness, DipECVS, FRCVS (Longframlington, Northumberland, UK) for independent assessment of radiographs. Professor CG Gutierrez received financial support from DGAPA, UNAM, towards a sabbatical visit to the University of Nottingham.

REFERENCES

1. Wilmut, I., Schnieke, A. E., McWhir, J., Kind, A. J. & Campbell, K. H.Viable offspring derived from fetal and adult mammalian cells. Nature 385, 810–813 (1997).
2. Tachibana, M. et al. Human embryonic stem cells derived by somatic cell nuclear transfer. Cell 153, 1228–1238 (2013).
3. Yamada, M. et al. Human oocytes reprogram adult somatic nuclei of a type 1 diabetic to diploid pluripotent stem cells. Nature 510, 533–536 (2014).
4. Chung, Y. G. et al. Human somatic cell nuclear transfer using adult cells. Cell Stem Cell 14, 777–780 (2014).
5. Takahashi, K. & Yamanaka, S. Induction of pluripotent stem cells from mouse embryonic and adult fibroblast cultures by defined factors. Cell 126, 663–676 (2006).
6. Narbonne, P., Miyamoto, K. & Gurdon, J. B. Reprogramming and development in nuclear transfer embryos and in interspecific systems. Curr. Opin. Genet. Dev. 22, 450–458 (2012).
7. Han, F. et al. SCNT versus iPSCs: proteins and small molecules in reprogramming. Int. J. Dev. Biol. 59, 179–186 (2015).
8. Ogura, A., Inoue, K. & Wakayama, T. Recent advancements in cloning by somatic cell nuclear transfer. Philos. Trans. R. Soc. Lond. B Biol. Sci. 368, 20110329 (2013).
9. Hill, J. R. et al. Clinical and pathologic features of cloned transgenic calves and fetuses (13 case studies). Theriogenology 51, 1451–1465 (1999).
10. Rhind, S. M. et al. Cloned lambs-lessons from pathology. Nat. Biotechnol. 21, 744–745 (2003).
11. Heyman, Y. Nuclear transfer: a new tool for reproductive biotechnology in cattle. Reprod. Nutr. Dev. 45, 353–361 (2005).
12. Panarace, M. et al. How healthy are clones and their progeny: 5 years of field experience. Theriogenology 67, 142–151 (2007).

13. Wakayama, T. et al. Cloning of mice to six generations. Nature407, 318–319 (2000).
14. Lanza, R. P. et al. Cloned cattle can be healthy and normal. Science294, 1893–1894 (2001).
15. Cibelli, J. B., Campbell, K. H., Seidel, G. E., West, M. D. & Lanza, R. P. The health profile of cloned animals. Nat. Biotechnol. 20, 13–14 (2002).
16. Wells, D. N., Forsyth, J. T., McMillan, V. & Oback, B. The health of somatic cell cloned cattle and their offspring. Cloning Stem Cells6, 101–110 (2004).
17. Watanabe, S. Effect of calf death loss on cloned cattle herd derived from somatic cell nuclear transfer: clones with congenital defects would be removed by the death loss. Anim. Sci. J. 84, 631–638 (2013).
18. Tamashiro, K. L. et al. Cloned mice have an obese phenotype not transmitted to their offspring. Nat. Med. 8, 262–267 (2002).
19. Ogonuki, N. et al. Early death of mice cloned from somatic cells. Nat. Genet. 30, 253–254 (2002).
20. Shiels, P. G. et al. Analysis of telomere length in Dolly, a sheep derived by nuclear transfer. Cloning 1, 119–125 (1999).
21. Lanza, R. P. et al. Extension of cell life-span and telomere length in animals cloned from senescent somatic cells. Science 288, 665–669 (2000).
22. Dyer, O. Dolly's arthritis dents faith in cloning. Brit. Med. J. 324, 67 (2002).
23. Williams, N. Dolly clouds cloning hopes. Curr. Biol. 12, R79–R80 (2002).
24. Williams, N. Death of Dolly marks cloning milestone. Curr. Biol. 13, R209–R210 (2003).
25. Forouzanfar, M. H. et al. Global, regional, and national comparative risk assessment of 79 behavioural, environmental and occupational, and metabolic risks or clusters of risks in 188 countries, 1990-2013: a systematic analysis for the Global Burden of Disease Study 2013. Lancet 386, 2287–2323 (2015).
26. Young, L. E. et al. Epigenetic change in IGF2R is associated with fetal overgrowth after sheep embryo culture. Nat. Genet. 27, 153–154 (2001).
27. Gluckman, P. D., Hanson, M. A., Buklijas, T., Low, F. M. & Beedle, A. S. Epigenetic mechanisms that underpin metabolic and cardiovascular diseases. Nat. Rev. Endocrinol. 5, 401–408 (2009).
28. Steegers-Theunissen, R. P., Twigt, J., Pestinger, V. & Sinclair, K. D.The periconceptional period, reproduction and long-term health of offspring: the importance of one-carbon metabolism. Hum. Reprod. Update 19, 640–655 (2013).
29. Lazaraviciute, G., Kauser, M., Bhattacharya, S., Haggarty, P. & Bhattacharya, S. A systematic review and meta-analysis of DNA methylation levels and imprinting disorders in children conceived by IVF/ICSI compared with children conceived spontaneously. Hum. Reprod. Update 20, 840–852 (2014).
30. Rathbone, A. J., Fisher, P. A., Lee, J. H., Craigon, J. & Campbell, K. H. Reprogramming of ovine somatic cells with Xenopus laevis oocyte extract before SCNT improves live birth rate. Cell Reprogram. 12, 609–616 (2010).
31. Lee, J. H. et al. Generation of mtDNA homoplasmic cloned lambs. Cell Reprogram. 12, 347–355 (2010).
32. Choi, I., Lee, J.-H., Fisher, P. & Campbell, K. H. Caffeine treatment of ovine cytoplasts regulates gene expression and foetal development of embryos produced by somatic cell nuclear transfer. Mol. Reprod. Dev. 77, 876–887 (2010).
33. AFRC. Energy and Protein Requirements of Ruminants. An Advisory Manual Prepared by the AFRC Technical Committee on Responses to Nutrients CAB INTERNATIONAL (1993).
34. Russel, A. J. F., Doney, J. M. & Gunn, R. G. Subjective assessment of body fat in live sheep. J. Agric. Sci. Camb. 72, 451–454 (1969).
35. Guerra, J. C., Thwaites, C. J. & Edney, T. N. Assessment of the proportion of chemical fat in the bodies of live sheep. J. Agric. Sci. Camb. 78, 147–149 (1972).
36. Lister, D. Effects of nutrition and genetics on the composition of the body. Proc. Nutr. Soc. 35, 351–356 (1976).

37. Jackson, P. G. G. & Cockcroft, P. D. Clinical Examination of Farm Animals. Appendix 3, Laboratory Reference Values: BiochemistryClin Exam Farm Animals, Blackwell Sci Ltd (2002).
38. Gardner, D. S. et al. Peri-implantation undernutrition programs blunted angiotensin II evoked baroreflex responses in young adult sheep. Hypertension 43, 1290–1296 (2004).
39. Kaler, J., Wassink, G. J. & Green, L. E. The inter- and intra-observer reliability of a locomotion scoring scale for sheep. Vet. J.180, 189–194 (2009).
40. Kellgren, J. H. & Lawrence, J. S. Radiological assessment of osteoarthrosis. Ann. Rheumatic Dis. 16, 494e502 (1957).
41. Hill, J. R. et al. Development rates of male bovine nuclear transfer embryos derived from adult and fetal cells. Biol. Reprod. 62, 1135–1140 (2000).
42. Mattheeuws, D., Derijcke, J., Rottiers, R. & Hoorens, J. Diabetes mellitus in two twin male lambs. Vet. Q. 4, 135–138 (1982).
43. Bergman, E. N., Reulein, S. S. & Corlett, R. E. Effects of obesity on insulin sensitivity and responsiveness in sheep. Am. J. Physiol. 257, E772–E781 (1989).
44. Sinclair, K. D. et al. DNA methylation, insulin resistance, and blood pressure in offspring determined by maternal periconceptional B vitamin and methionine status. Proc. Natl Acad. Sci. USA 104, 19351–19356 (2007).
45. McCann, J. P., Ullmann, M. B., Temple, M. R., Reimers, T. J. & Bergman, E. N. Insulin and glucose responses to glucose injection in fed and fasted obese and lean sheep. J. Nutr. 116, 1287–1297 (1986).
46. Adam, C. L., Findlay, P. A., Aitken, R. P., Milne, J. S. & Wallace, J. M.In vivo changes in central and peripheral insulin sensitivity in a large animal model of obesity. Endocrinology 153, 3147–3157 (2012).
47. Kirchhoff, K. et al. Impact of different fat depots on insulin sensitivity: predominant role of liver fat. J. Diabetes. Sci. Technol.1, 753–759 (2007).
48. Lalia, A. Z. et al. Predictors of Whole-Body Insulin Sensitivity Across Ages and Adiposity in Adult Humans. J. Clin. Endocrinol. Metab. 101, 626–634 (2016).
49. Eastwood, S. V. et al. Ethnic differences in associations between fat deposition and incident diabetes and underlying mechanisms: the SABRE study. Obesity 23, 699–706 (2015).
50. Robinson, J. J., McDonald, I., McHattie, I. & Pennie, K. Studies on reproduction in prolific ewes 4. Sequential changes in the maternal body during pregnancy. J. Agric. Sci. Camb. 91, 291–304 (1978).
51. Cowan, R. T., Robinson, J. J., McDonald, I. & Smart, R. Effects of body fatness at lambing and diet in lactation on body tissue loss, feed intake and milk yield of ewes in early lactation. J. Agric. Sci. Camb. 95, 497–514 (1980).
52. Lai, L. et al. Production of alpha-1,3-galactosyltransferase knockout pigs by nuclear transfer cloning. Science 295, 1089–1092 (2002).
53. Crowley, S. D. & Coffman, T. M. The inextricable role of the kidney in hypertension. J. Clin. Invest. 124, 2341–2347 (2014).
54. Bianchi, G., Fox, U., Di Francesco, G. F., Giovanetti, A. M. & Pagetti, D. Blood pressure changes produced by kidney cross-transplantation between spontaneously hypertensive rats and normotensive rats. Clin. Sci. Mol. Med. 47, 435–448 (1974).
55. Dahl, L. K., Heine, M. & Thompson, K. Genetic influence of the kidneys on blood pressure. Evidence from chronic renal homografts in rats with opposite predispositions to hypertension. Circ. Res. 34, 94–101 (1974).
56. Curtis, J. J. et al. Remission of essential hypertension after renal transplantation. N. Engl. J. Med. 309, 1009–1015 (1983).
57. Gonzalez-Villalobos, R. A. et al. Intrarenal angiotensin-converting enzyme induces hypertension in response to angiotensin I infusion. J. Am. Soc. Nephrol. 22, 449–459 (2011).
58. Gonzalez-Villalobos, R. A. et al. The absence of intrarenal ACE protects against hypertension. J. Clin. Invest. 123, 2011–2023 (2013).

59. Kearney, P. M. et al. Global burden of hypertension: analysis of worldwide data. Lancet 365, 217–223 (2005).
60. Glyn-Jones, S. et al. Oesteoarthritis. Lancet 386, 376–387 (2015).
61. Thysen, S., Luyten, F. P. & Lories, R. J. U. Targets, models and challenges in osteoarthritis research. Dis. Model. Mech. 8, 17–30 (2015).
62. Gordon, W. J. et al. The relationship between limb function and radiographic osteoarthrosis in dogs with stifle osteoarthrosis. Vet. Surgery 32, 451–454 (2003).
63. Scott, P. R. Sheep Medicine. 2nd Edn Chapter 9, Musculoskeletal System 266CRC Press (2006).
64. Scott, P. R. Osteoarthritis of the elbow joint in adult sheep. Vet. Rec. 149, 652–654 (2001).
65. Vandeweerd, J.-M. et al. Prevalence of naturally occurring cartilage defects in the ovine knee. Osteoarthritis Cartilage 21, 1125–1131 (2013).
66. Zilkens, C. et al. Magnetic resonance imaging and histology of ovine hip joint cartilage in two age populations: a sheep model with assumed healthy cartilage. Skeletal Radiol. 42, 699–70 (2013).
67. Alexander, B. et al. Telomere length status of somatic cell sheep clones and their offspring. Mol. Reprod. Dev. 74, 1525–1537 (2007).
68. Niemann, H. Epigenetic reprogramming in mammalian species after SCNT-based cloning. Theriogenology 86, 80–90 (2016).
69. Marion, R. M. & Blasco, M. A. Telomere rejuvenation during nuclear reprogramming. Curr. Opin. Genet. Dev. 20, 190–196 (2010).
70. Kalinowski, S. ,T., Taper, M. L. & Marshall, T. C. Revising how the computer program CERVUS accommodates genotyping error increases success in paternity assignment. Mol. Ecol. 16, 1099–1106 (2007).
71. Rhodes, P. et al. Adult-onset obesity reveals prenatal programming of glucose-insulin sensitivity in male sheep nutrient restricted during late gestation. PLoS ONE 14, e7393 (2014).

CHAPTER 7

Rapid Generation of Hypomorphic Mutations

Laura L. Arthur[1], Joyce J. Chung[2], Preetam Janakirama[3,4], Kathryn M. Keefer[1], Igor Kolotilin[5], Slavica Pavlovic-Djuranovic[1], Douglas L. Chalker[2], Vojislava Grbic[3], Rachel Green[6], Rima Menassa[4], Heather L. True[1,7], James B. Skeath[8] & Sergej Djuranovic[1]

[1]Department of Cell Biology and Physiology, Washington University School of Medicine, St Louis, Missouri 63110, USA.

[2]Department of Biology, Washington University, St Louis, Missouri 63105, USA.

[3]Department of Biology, The University of Western Ontario, 1151 Richmond Street, London, Ontario, Canada N6A5B7.

[4] Science and Technology Branch, Agriculture and Agri-Food Canada, 1391 Sandford Street, London, Ontario, Canada N5V4T3.

[5] Scattered Gold Biotechnology Inc. 14 Denali Terrace, London, Ontario, Canada N5X 3W2.

[6]Department of Molecular Biology and Genetics, Howard Hughes Medical Institute, Johns Hopkins University School of Medicine, 725 North Wolfe Street, Baltimore, Maryland 21205, USA.

[7] The Hope Center for Neurological Diseases, Washington University School of Medicine, St Louis, Missouri 63110, USA.

[8]Department of Genetics, Washington University School of Medicine, St Louis, Missouri 63110, USA.

ABSTRACT

Hypomorphic mutations are a valuable tool for both genetic analysis of gene function and for synthetic biology applications. However, current methods to generate hypomorphic mutations are limited to a specific organism, change gene expression unpredictably, or depend on changes in spatial-temporal expression of the targeted gene. Here we present a simple and predictable method to generate hypomorphic mutations in model organisms by targeting translation elongation. Adding consecutive adenosine nucleotides, so-called polyA tracks, to the gene coding sequence of interest will decrease translation elongation efficiency, and in all tested cell cultures and model organisms, this decreases mRNA stability and protein expression. We show that protein expression is adjustable independent of promoter strength and can be further modulated by changing sequence features of the polyA tracks. These characteristics make this method highly predictable and tractable for generation of programmable allelic series with a range of expression levels.

INTRODUCTION

Manipulation of gene activity is a standard genetic approach to gain insight into gene function in single and multicellular organisms1. In many cases, complete loss of gene function (null allele or knockout of the locus) will provide the most valuable information about gene function. However, for essential genes, complete loss of function leads to lethality, which usually precludes obtaining functional information for later cellular or developmental stages. Similarly, for genes that function in multiple cellular/developmental processes and have pleiotropic null mutant phenotypes, it can be difficult to distinguish primary from secondary effects. In many of these cases, however, partial loss of function or hypomorphic mutations can overcome lethality and pleiotropy, allowing later stage cells and organisms to be examined for phenotypic consequences. Furthermore, hypomorphic mutations, because they retain residual gene activity and partial phenotypes, are used in suppressor or enhancer genetic screens to identify other genes that act in the same biological process. One example of a hypomorphic condition is a 50% reduction in gene activity from heterozygosity for a null allele, which for some genes can display mutant phenotypes (called a haploinsufficency). However, for other genes a 50% reduction in gene activity is sufficient for normal function, and thus hypomorphic mutations with a further reduction are required to observe a phenotype. For these reasons, it is important to have the ability to generate hypomorphic mutations with a range of loss of gene activity.

Hypomorphic mutations are traditionally obtained in forward genetic chemical mutagenesis screens. The hypomorphic allele needs to be isolated, identified, and then genetically and biochemically characterized in order to be further used in analysis to deduce gene function. As a consequence, this process is time consuming. In addition, because of evolution, discovered alleles can be species specific. These difficulties and the importance of hypomorphic alleles have prompted the development of several methods to generate hypomorphic mutations directly in various model organisms[2,3,4,5,6,7,8,9,10,11]. However, these methods again are usually specific to one organism, can have unpredictable alterations in gene activity (separation of function, gain of function), or can change other aspects of regulation that affect interpretation of phenotype, such as spatial-temporal gene expression. With the rise of gene editing systems, such as CRISPR/Cas9 (ref. 12) or TALEN technology[13], and more extensive use of synthetic and systems biology approaches[14,15], there is an increasing interest in generating hypomorphic mutations of a target gene through simpler, more systematic and rapid approaches.

Here we present a method for the generation of hypomorphic mutations that produces a range of reduced levels of nearly wild type protein through the use of disrupted translational elongation. This method is based on polyA tracks, a novel cis regulatory element that decreases gene expression by disrupting messenger RNA (mRNA) translation[16,17]. Insertion of consecutive adenosine nucleotides into the open reading frame of an mRNA will decrease protein expression by decreasing the efficiency of the translation elongation phase leading to diminished production of protein and mRNA destabilization, and thus to diminished mRNA levels.

RESULTS

Use of polyA tracks to generate hypomorphic mutants

We have recently identified polyA tracks as a regulator of gene expression[16,17]. This mechanism is used endogenously in most eukaryotic genomes and regulates ~2% of human genes[16,18]. The polyA track causes ribosomal stalling and frameshifting during translation elongation, leading to mRNA instability and degradation of nascent protein products[16,17]. The translation elongation cycle is an ideal target for a universal method of gene regulation because it is the most highly conserved step in protein biosynthesis between bacteria and eukaryotes[19]. Therefore, we reasoned that polyA tracks, because of their versatility in lengths and sequence composition, can be used as a system to create programmable hypomorphic mutants and regulate gene expression in a wide variety of model organisms (Figure 1).

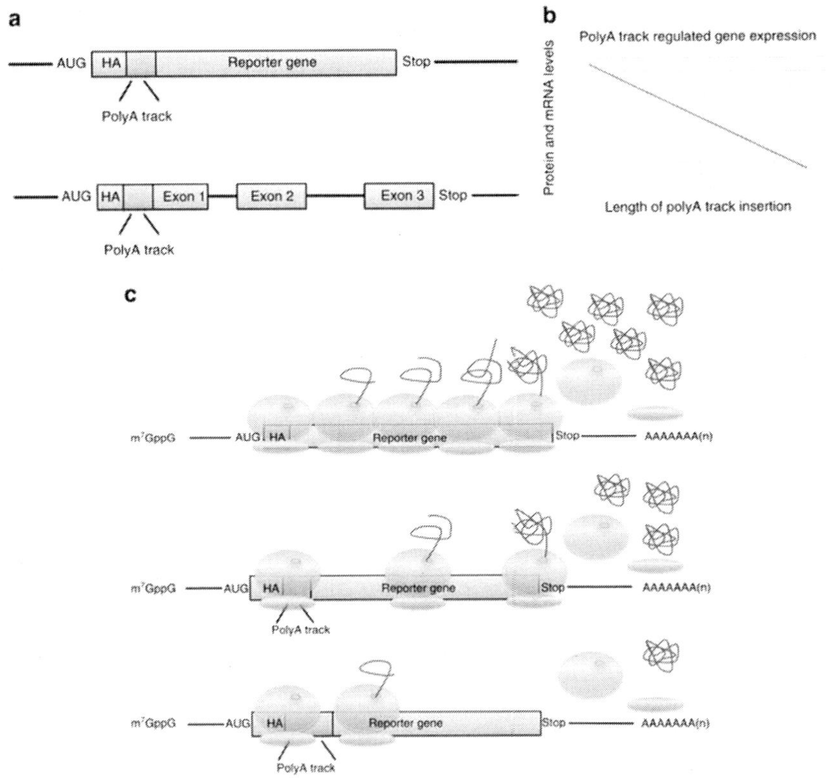

Figure 1: Design and mechanism of polyA track tag regulated gene expression.

(**a**) Scheme of inserted polyA tracks in the reporter genes used in this study. Hemagglutinin (HA) tag (grey) and polyA tracks (red) were introduced in the coding region of the reporter genes next to the start AUG codon. Exon boundaries as well as termination codon (Stop) are indicated. (**b**) Proposed correlation between gene products levels, mRNA and protein, and the length of inserted polyA track tags. The reduction in levels of both reporter protein and mRNA is dependent on increasing length of consecutive adenosine nucleotides in the coding sequence. (**c**) Scheme of translation of eukaryotic reporter mRNA with or without inserted polyA tracks. The length of inserted polyA track tag determines the protein output of the regulated reporter gene as indicated by the number of globular protein structures. Features of the eukaryotic mRNAs (m7GpppG—cap, AUG—start codon, Stop—termination codon and polyA tail), as well as HA-tag, position of the polyA track tag, ribosome and nascent polypeptide chain are illustrated in the scheme.

We have generated a fluorescent reporter gene that has an insertion of defined polyA tracks in order to control the amount of expression (Figure 1a). The reporter consists of either a constitutive or inducible promoter driving expression of the mCherry fluorescence and other reporter proteins. A double HA-tag was added at the beginning of the coding sequence for detection through western blot analysis. The polyA track is inserted directly after the HA-tag. The length of the polyA track varies from 9 to 36 consecutive adenosine nucleotides, adding 3–12 lysine residues to the protein sequence (Figure 1a). To assess the effects of polybasic peptide arising from sequential lysine residues20,21, we also generated control reporters with consecutive lysine AAG codons. We hypothesized that as the length of the polyA track is increased, expression of the reporter gene products will decrease (Figure 1b,c). These reporters can be transiently transfected, recombined or inserted into the genome of cell cultures or whole organisms. Likewise, endogenous genes can be edited to include a polyA track in their open reading frames (ORF's) using genome editing methodology.

PolyA tracks can regulate gene expression in *Escherichia coli*

We first tested whether polyA tracks can be used in single-cell model organisms to attenuate gene expression from a defined reporter gene. Previous attempts to control gene expression in *Escherichia coli* cells have used degeneration of ribosome binding sequence2 (rbs) or spacer sequence between rbs sequence and starting AUG codon22, but did not focus on the coding sequence. To show that polyA tracks can be used to control gene expression in *E. coli* cells, we created a set of reporters with increasing length of polyA tracks under the arabinose-inducible promoter pBAD (Supplementary Figure 1). We transformed *E. coli* cells with plasmids expressing HA-mCherry, HA-$(AAG)_n$-mCherry or HA-$(AAA)_n$-mCherry. All *E. coli* cell cultures were induced at the same optical density and monitored for both cell growth and fluorescence of the mCherry constructs during induction. While *E. coli* containing wild type and Lys_{AAG} controls (6 × , 9 × and 12 × lysine AAG codons) show no significant differences in the amount of mCherry fluorescence, cell cultures containing constructs with polyA tracks show progressively less fluorescence with increasing length of the polyA track (Figure 2a). Addition of 9 and 12As in a row (3 or 4 Lys_{AAA} codons) consistently reduced fluorescence of mCherry reporter by 15–35%. Further additions in the length of the polyA track resulted in continuing decreases in mCherry reporter fluorescence to the point where

36 consecutive adenosine nucleotides resulted in barely visible expression of the reporter (<5% of wild type). Western blot analyses of equal amounts of *E. coli* cell lysates expressing different polyA track and control reporters were consistent with the mCherry fluorescence data and indicated again that protein abundance of reporter proteins strongly depends on the length of the polyA track (Figure 2b). Reporters with 9 and 12As in a row (3 and 4Lys$_{AAA}$ codons, respectively) show reduction in protein abundance in the range of 20–40% of the wild type mCherry, and constructs with >27As in a row (9 and more Lys$_{AAA}$codons) were nearly undetectable by western blot analyses.

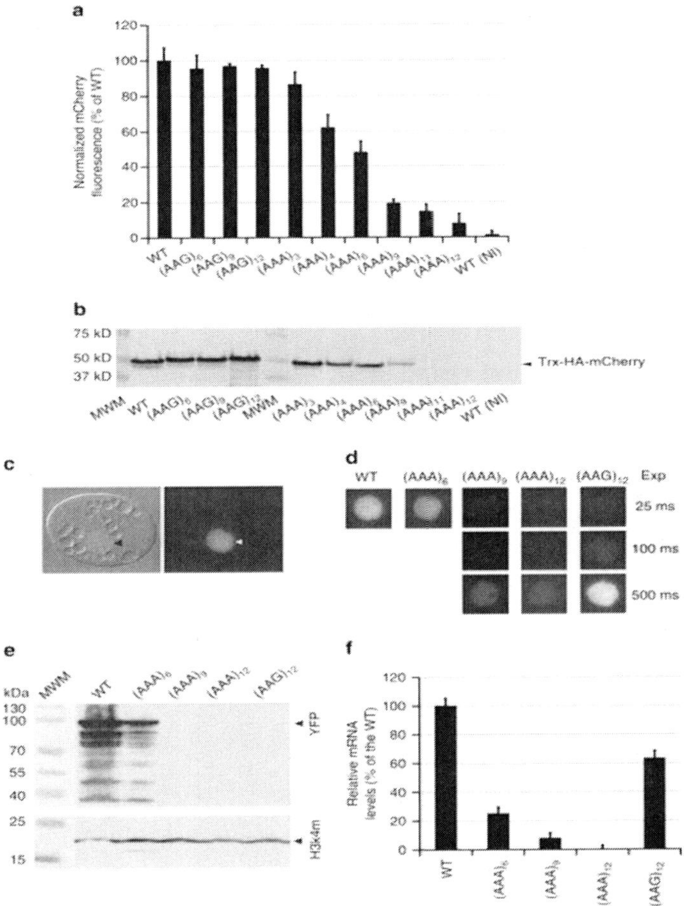

Figure 2: Regulation of reporter gene by polyA tracks in the single-cell prokaryotic and eukaryotic organisms.

(**a**) Percentage of mCherry fluorescence of tested Lys_{AAG} ($(AAG)_{6-12}$) and Lys_{AAA}($(AAA)_{3-12}$) insertion constructs compared with wild type fluorescence (WT, no insertion construct). mCherry fluorescence was assayed at excitation wavelength of 475±9 nm and emission was detected at 620±9 nm. Error bars indicate mean mCherry fluorescence values±s.d. for three individual *E. coli* colonies for each construct. Background levels of mCherry expression can be estimated from the fluorescence of the non-induced wild type construct (WT(NI)). (**b**) Western blot analysis of mCherry constructs expressed in *E. coli* cells. Equal amounts of *E. coli* cell lysates with Thioredoxin(Trx) fusion proteins were used for analysis. Fusion proteins were detected using HA-tag specific antibody. Positions of the fusion protein (Trx-HA-mCherry) and sizes of molecular weight markers (MWM) are indicated. (**c**) Representative differential interference contrast microscopy (left panel) and the corresponding fluorescence image (right panel −25 ms exposure) of a *T. thermophila* cell expressing the wild type (WT) MLP1-HA-YFP fusion. Arrowheads denote the position of the macronucleus. (**d**) *MPL1*-HA-YFP accumulation within macronuclei of live *T. thermophila* cells expressing an allelic series of fusion proteins—WT, $(AAA)_{6-12}$, and $(AAG)_{12}$—was visualized by epifluorescence microscopy. Different exposures times are indicated on the right to demonstrate the relative accumulation of each variant. (**e**) Western blot analysis was performed with whole cell lysates made from *T. thermophila* cells expressing the MLP-HA-YFP fusion proteins. Protein from equivalent cell numbers was loaded in each lane and detected using YFP specific antibody (top panel) and normalized to the nuclear histone species, histone H3 trimethyl-lysine 4 (H3K4m) (bottom panel). Positions of the full-length fusion protein (YFP), normalization control (H3k4m), and sizes of molecular weight markers (MWM) are indicated. Degradation of excess fusion protein is readily apparent as faster migrating species below the full-length MLP1-HA-YFP. (**f**) Steady state levels of fusion gene constructs measured by qRT-PCR. Relative levels of the mRNA for $(AAG)_{12}$ and $(AAA)_{6-12}$ are presented as percentage of the wild type (WT) construct mRNA levels. Error bars represent mean±s.d. values (*n*=3).

To further fine tune expression of the reporter gene with polyA tracks, we tested whether insertion of non-lysine codons in the middle of long polyA tracks (33As, 11 AAA codons) would result in change of the reporter accumulation in *E. coli* cells. Indeed, addition of such codons did not change levels of the fluorescent reporter drastically but fine-tuned them to the fluorescent values between constructs with 30As or 33As in a row (Supplementary Figure 2). Additional analysis of XAA and AAY codons, where X and Y denote C/G or T/C/G nucleotides respectively, for programming the length of polyA tracks revealed that gene attenuation by polyA tracks does not depend necessarily on lysine codons but other codons can contribute to the overall length and effects of polyA tracks (Supplementary Figure 2). In a similar manner endogenous genes with different arrangement of A–rich codons (mainly lysine encoding AAA and AAG codons) have shown differential gene regulation in our previous study17.

PolyA tracks can be used in protozoan *T. thermophila*

We previously showed that polyA tracks can influence expression of the reporter genes in *Saccharomyces cerevisiae* cells17. To test whether polyA tracks can regulate gene expression in another model, single-cell eukaryotic organism, we monitored the effect of various length tracks on YFP expression in the protozoan *Tetrahymena thermophila*. The genome of *T. thermophila* has extremely high AT content (>75%) and has been extensively used as a microbial animal model23. Our *T. thermophila* reporter contained the coding sequence of a *M*acronucleus-*L*ocalized *P*rotein of unknown function (*MLP1*, TTHERM_00384860) fused to eYFP protein (Supplementary Figure 3). The fusion with MLP1 directed YFP to *Tetrahymena* macronuclei to allow easier quantification of YFP levels (Figure 2c). These two proteins were fused, separated by linkers containing an HA-tag (MLP1-HA-YFP (WT)) and polyA tracks of 18, 27 or 36As, $(AAA)_6$, $(AAA)_9$ or $(AAA)_{12}$, respectively, or 12 Lys_{AAG} $(AAG)_{12}$ codons inserted as a control. All constructs were expressed upon cadmium-induction of the upstream *MTT1* promoter. Just as in our *E. coli* experiments with mCherry reporter, the YFP gene containing increasing lengths of polyA tracks exhibited a progressive decrease in total protein accumulation, measured by fluorescence, relative to the HA-linker fusion or Lys_{AAG} insertion controls (Figure 2d). The construct with 18As in a row (6Lys_{AAA}) showed ~50% reduction in protein fluorescence, while constructs with 27 and 36As exhibited nearly undetectable levels of fluorescence and required 20 times longer exposure for detection of YFP by microscopy. The construct with 12Lys_{AAG} codons showed fluorescence that was 4–5 fold lower than the WT construct. This effect was comparable to the polybasic peptide stalling that was observed earlier in *S. cerevisiae* cells17,20,21. We further supported our fluorescence results by western blot analyses (Figure 2e). The MLP1-HA-YFP-fusion (WT) was readily visible, whereas the polyA track-YFP fusions exhibited attenuated expression. Insertion of 18A's (6Lys_{AAA}) showed expression levels that were ~35% of the HA-YFP control, while 27 and 36A's (9 and 12Lys_{AAA}) constructs were at the limit of detection (Figure 2e). Insertion of 12Lys_{AAG} codons in the fusion protein resulted in a 6- to 8-fold reduction in expression of fusion protein. Since the different fusion proteins were all transcribed from the *MTT1* promoter under identical induction conditions we reasoned that the amount of mRNA produced for each construct should be equivalent. We used qRT-PCR to quantify the steady-state level of YFP mRNA for each fusion (Figure 2f). The steady-state mRNA levels of polyA track-YFP constructs reflected the decreasing YFP protein accumulation relative to the WT consistent with earlier reports in *S. cerevisae*17,20. Insertion of 18As (6Lys_{AAA} codons) reduced mRNA levels to approximately 30–35% of WT levels, while insertion of 27 and 36As (9 and 12Lys_{AAA} codons, respectively) reduced mRNA levels to <5% of the HA-YFP construct (Figure 2d). Interestingly, while attenuation at the protein level was stronger for the insertion of 12Lys_{AAG} codons than for 6Lys_{AAA} codons, the trend was not consistent at the mRNA level. The observed discrepancies between the protein levels and mRNA degradation in the case of polybasic peptide (12 AAG codons) are likely attributed to the two distinct pathways employed in regulation of polybasic peptide stalling and polyA track induced stalling and frameshifting in eukaryotic cells16,17,24. Nevertheless, the effects of polyA tracks on regulation of the reporter gene showed a consistent relationship between mRNA levels and protein accumulation levels. Most importantly, we were able to control expression of reporter genes using polyA tracks in *T. thermophila*, a single-cell AT-rich protozoan, as we previously reported in *S. cerevisiae* and the above-mentioned *E. coli* cells17.

PolyA tracks can regulate gene expression in plant tissues

To test whether polyA tracks can attenuate gene expression in plants, we transiently co-expressed HA-$(AAA)_n$-mCherry with an YFP construct as an internal control in the model plant *Nicotiana benthamiana* (Supplementary Figure 4). The expression of mCherry and YFP was assessed by fluorescence imaging (Figure 3a). Like cell cultures, *N. benthamiana* epidermal cells showed attenuated mCherry fluorescence proportional to the length of the polyA tracks (6,9 and 12 Lys_{AAA} codons) compared with the HA-mCherry and 12 Lys_{AAG} control constructs (Figure 3a). The fluorescence data for each construct revealed the same trend of gene expression regulation as in *T. thermophila* cells (Figure 2d). As fluorescence in this assay was not quantifiable, protein abundance was determined by semi-quantitative western blot analysis of *N. benthamiana* leaves infiltrated with the HA-$(AAA)_n$-mCherry. The levels of HA-mCherry proteins were normalized to levels of the cis-linked selectable marker phosphinotricin acetyl transferase (BAR) in the same sample (Figure 3b,c). The addition of a polyA track with 18As (6 Lys_{AAA}) decreased protein accumulation to ~70% of HA-mCherry levels. Further reduction of mCherry protein accumulation, to 30% and below the detection limit was observed in 9 Lys_{AAA} and 12 Lys_{AAA} constructs, respectively (Figure 3c). Insertion of 12 Lys_{AAG} codons displayed ~50% reduction in the reporter expression compared with WT construct. Parallel analyses of steady-state mRNA levels of transcripts with increasing lengths of polyA tracks showed progressively reduced levels of polyA track mRNAs when compared with transcript levels of the HA-mCherry and AAG-containing control constructs (Figure 3d). mRNA levels were reduced to ~50–55% of WT expression for 6Lys$_{AAA}$ transcripts, while 9 and 12Lys$_{AAA}$ constructs had reduced mRNA levels to ~30 and 20% of control, respectively (Figure 3d). Insertion of 12 Lys$_{AAG}$ codons exhibited marginal effects on mRNA levels and again showed the already observed discrepancy between mRNA levels and protein accumulation as a result of polybasic peptide stalling. However, these results indicate once again that polyA tracks affect both mRNA and protein levels more consistently and can be used to regulate gene expression in plants.

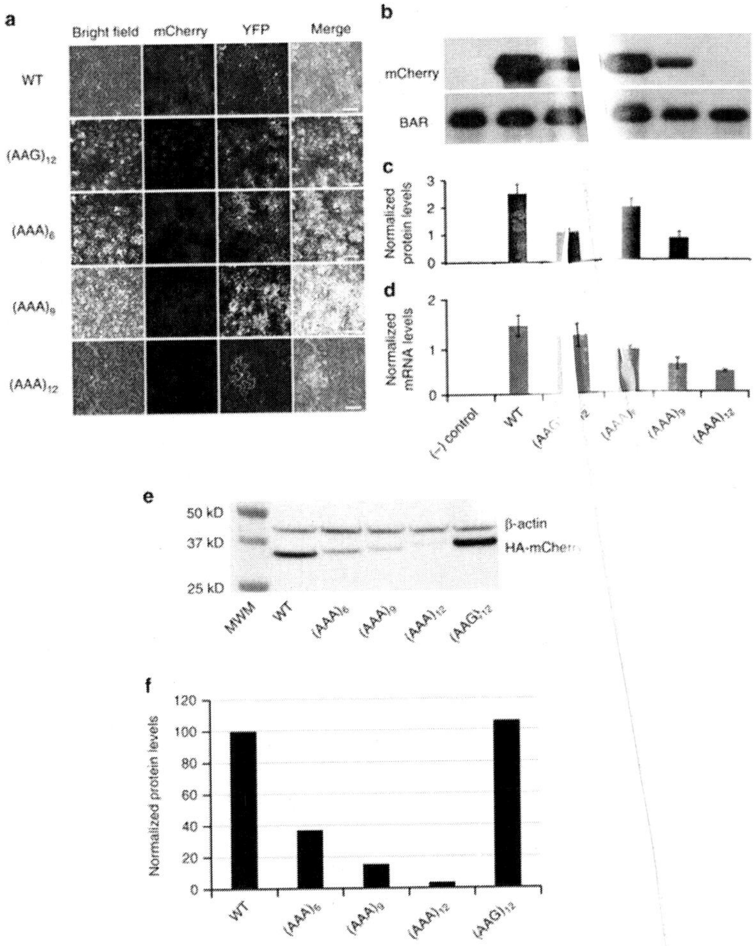

Figure 3: Regulation of reporter gene by polyA tracks in the eukaryotic tissue cultures.

(**a**) Fluorescence images of *N. benthamiana* epidermal cells transiently expressing wild type (WT), $(AAG)_{12}$ and $(AAA)_{6\text{-}12}$ mCherry constructs. YFP expression was used as a transfection control. The scale bars in images are 100 μm. (**b**)—Western blot analysis, (**c**)—protein level estimate and (**d**)—mRNA levels for transfected (−) insert control and WT, $(AAG)_{12}$ and $(AAA)_{6\text{-}12}$ mCherry constructs expressed transiently in *N. benthamiana* epidermal cells. (**b**) Primary HA-tag antibody was used for detection of HA-mCherry constructs (molecular weight 34 kDa). Phosphinotricin acetyl transferase (BAR) specific antibody was used as a loading and normalization control (molecular weight 22 kDa) (**c**). Levels of mCherry protein from different constructs were derived from detected band intensities normalized for BAR accumulation detected in the same sample. Error bars represent mean values±standard error from biological replicates (n=8). (**d**) mRNA levels for different mCherry constructs were calculated as cycle threshold (Ct) values and normalized to *BAR* gene mRNA values. Error bars represent mean values±s.e. from biological replicates (n=3). (**e**) Western blot analysis of transient mCherry constructs expression in HeLa cells. WT, 12 Lys_{AAG} ($(AAG)_{12}$) and 6–12 Lys_{AAA} ($(AAA)_{6\text{-}12}$) mCherry proteins were detected using HA-tag specific primary antibody. β-actin was used as a loading control and was detected using specific antibody. Positions of the fusion protein (HA-mCherry), normalization control (β-actin) and sizes of molecular weight markers (MWM) are indicated. (**f**) Quantification of the mCherry protein levels from detected western blot intensities. Levels of mCherry were normalized to β-actin band intensities and represented as a percentage of the wild type construct values.

PolyA tracks can regulate genes in human tissue cultures

To further assess the universality of polyA tracks on protein expression, we tested our reporter series in human tissue cultures using HeLa cells. Plasmids with HA-mCherry, HA-$(AAG)_{12}$-mCherry and HA-$(AAA)_n$-mCherry reporters, driven by the constitutively active CMV promoter, were electroporated into HeLa cells for transient expression. Protein abundances were assessed by western blot analyses 24 h after electroporation (Figure 3e). As in our previous study on expression of endogenous and synthetic polyA tracks in various human tissue cultures16, constructs with increasing length of polyA tracks (6, 9 and 12 Lys_{AAA}) were expressed at lower levels than control constructs and the reductions in protein expression were proportional to the length of polyA track. The construct with 18As (6Lys_{AAA}) displayed an ~3-fold reduction in expression compared with the WT construct. Insertion of 27 and 36As (9 and 12Lys_{AAA}, respectively), exhibited a 6- and 25-fold reduction of HA-mCherry expression compared with WT

(Figure 3f). The control construct with 12Lys$_{AAG}$ codons did not show any reduction in protein levels compared with the WT construct (Figure 3e,f). These results again argue for differences between translational stalling induced by polybasic peptides[16,17,20,21], which seems to be cell- or organism-specific and unpredictable, and polyA track-induced ribosomal stalling and frameshifting[16,17], which is clearly dependent on the length of polyA tracks. Importantly, this latter pathway appears to be conserved between multiple organisms. Altogether with our previous study[16], our results indicate that polyA tracks can readily be used to regulate expression of reporters or genes transiently transfected in diverse eukaryotic tissues and cultured cell systems, such as *N. benthamiana* and human cell cultures, as well as other mammalian or insect tissue culture systems[16].

PolyA tracks can regulate gene expression in model organisms

We next sought to test whether polyA tracks can be used to regulate reporter gene expression in complex, multicellular organisms. We chose the fruit fly, *Drosophilla melanogaster*, because of the well-developed tools in the manipulation of endogenous genetic loci, as well as for the ready assessment of the mCherry reporter activity. Using the PhiC31-integrase approach[25], we generated single transgene insertions of the HA-mCherry and HA-(AAG)$_{12}$-mCherry controls, and HA-(AAA)$_n$-mCherry (6, 9 and 12Lys$_{AAA}$) constructs in the identical genomic location in the third chromosome (see methods; Supplementary Figure 5). All constructs contained an Upstream Activation Sequence (UAS) followed by the *HSP70* promoter which actively transcribes mCherry reporter mRNAs in response to expression of GAL4 protein[26]. To drive expression of mCherry in all tissues, each transgenic line was crossed to a driver line with Tub-GAL4 to express GAL4 protein in all tissues. In addition, the driver line carried a UAS-linked GFP transgene, which allowed us to use GFP expression for normalization of the mCherry reporter genes (Supplementary Figure 5).

Expression of mCherry was assessed by fluorescence imaging of formaldehyde fixed salivary glands (SG), central nervous system (CNS) and proventriculus (PV) dissected from otherwise wild type third instar larvae (Figure 4a). Wild type HA-mCherry expressed well in all imaged tissues. Addition of a polyA track with 18A's (6Lys$_{AAA}$) reduced mCherry expression to ~30% of the wild type construct in all three tissues (Figure 4b–d; Supplementary Figure 6). Constructs with 27As and 36As (9 and 12Lys$_{AAA}$ codons, respectively) reduced expression of mCherry in all assayed tissues to ~20 and 10% of wild type levels, respectively (Figure 4b–d; Supplementary Figure 6). Western blot analyses on cell lysates produced from five fruit fly larvae for each independent construct were consistent with our quantification of fluorescence imaging data (Supplementary Figure 7). As in the previous experiments with *T.*

thermophila and tissue culture systems (Figs 2 and 3), mRNA stability of polyA track constructs in fruit fly larvae exhibited an inverse correlation with the length of polyA track (Supplementary Figure 8) and concordance with protein abundances measured by western blot analyses. Insertion of 12Lys$_{AAG}$ codons had a moderate effect on levels of mCherry mRNA and protein and was similar to the expression levels for the 18As (6Lys$_{AAA}$) insertion construct (Figure 4b–d; Supplementary Figs 7 and 8). Our data indicate that individual tissues of a complex multicellular organism, such as fruit fly, are equally sensitive to gene expression attenuation mediated by polyA tracks. Therefore, one can use polyA track constructs to create hypomorphic alleles and allelic series, in complex multicellular organisms with similar relative gene expression attenuation efficiencies as observed in the isolated tissues.

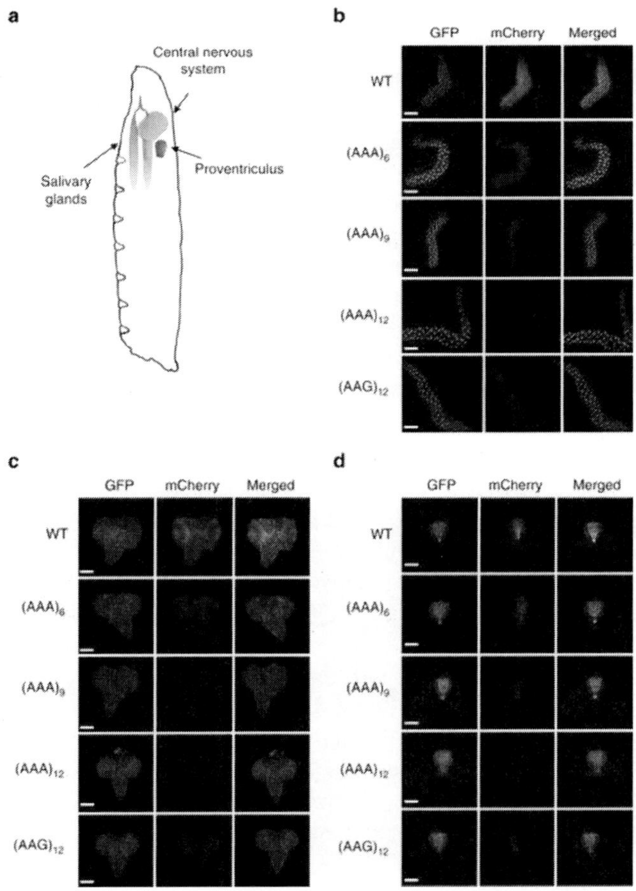

Figure 4: PolyA tracks regulate mCherry reporter gene expression in different organs of *D. melanogaster*.

(**a**) Diagram of third instar fruit fly larva showing approximate location of salivary glands (SG, blue), central nervous system (CNS, green) and proventriculus (PV, red). Fluorescence imaging of formaldehyde fixed SG (**b**), CNS (**c**) and PV (**d**), dissected from larvae expressing wild type (WT), $(AAG)_{12}$ and $(AAA)_{6-12}$ mCherry constructs. mCherry and GFP indicate images acquired by selective fluorescence filter setting. Overlay of mCherry and GFP fluorescence is shown in the merged panel. The scale bars in images are 200 μm.

One of the potential limitations of polyA tracks could be their homopolymeric nature and observed hypermutability of short tandem repeats (STRs) and homopolymeric sequences that was previously observed in different systems22,27,28. To control for the cell population heterogeneity that may arise from hypermutability of polyA tracks, we have analysed the mutation rate of a single locus insertion of our longest polyA track insertion (36As) in the *D. melanogaster* genome. The sequence of the mCherry reporter gene from genomic DNA isolated from a whole adult fly after more than a year of homozygote crosses, approximately 30 generations of fruit flies, was compared with the original DNA construct that was used for generation of the transgenic insect. Both Sanger and Illumina sequencing of the amplified polyA track (36As) regions from the transgenic fruit fly genome and original DNA vector revealed differences between two data sets (Supplementary Figure 9). In ~8% of the cases by Illumina sequencing we have observed loss of the polyA track because of the possible recombination event indicated by the similarity of the amplified sequence to the genomic sequence located on the chromosome X and 3L of the fruit fly genome. Illumina sequencing was not reliable enough to show differences in the polyA tracks length because of the difficulty associated with sequencing of the homopolymers. The Sanger sequencing of the amplified polyA track regions from plasmid and genomic DNA used for the Illumina sequencing revealed a low frequency of mutations in the polyA tracks. We observed insertions as well as polyA track shortening, which is in good concordance with previously published data from different cellular systems22,27,28,29. This is in agreement with our previous analysis of endogenous polyA track genes, which showed rather strong conservation of polyA track sequences in multiple analysed genomes. Taken together, these data indicate that polyA tracks can be used for stable control of gene expression over a multiple generations of model organisms and for an extended period of time without a strong mutational drift in cell mutability.

PolyA track control is independent of promotor strength

Our data from the fruit fly experiment indicated that the ratio between reporters with polyA track insertion and control is maintained in all tissues

(Figure 4b–d; Supplementary Figs 6–8). This suggests that inserted polyA tracks maintain their capacity for gene regulation independent of the strength of DNA transcription, which is known to have a large dynamic range across genes and cell types30.

To systematically evaluate how differences in the strength of transcription would affect gene regulation and hypomorphic expression of reporters with polyA track insertion, we used human Flp-In T-REx 293 cell lines (Thermo Fisher Scientific). Using a protocol for generation of stable and inducible expression cell lines, we generated cells with a single insertion of our HA-mCherry control and HA-$(AAA)_{12}$-mCherry polyA construct in a defined chromosomal locus (Supplementary Figure 10). The strength of transcription in these cell lines was varied by use of increasing concentrations of doxycycline (0.001–0.1 mg μl^{-1} of Dox) added to the growth media, and levels of transcription were assayed in relation to constitutively expressed hygromycin B phosphotransferase (Supplementary Figure 10). The dose-dependent response of the doxycycline-inducible CMV promoter for both the polyA track and control mCherry transcript ranged over two orders of magnitude. At the same time, relative expression of the polyA track construct was constant; 12–17% expression relative to the control construct based on the western blot analysis (Figure 5a,b). Moreover, relative mRNA levels of control and polyA track constructs did not change under different transcriptional regimes (Figure 5c). The steady-state amount of the polyA track construct mRNA was consistently in a range between 1 and 3% of the normalized control construct. The same results are obtained using stable cell lines that express HA-tagged human haemoglobin (delta chain, WT-HBD) and an 18As HBD construct (HBD-$6Lys_{AAA}$) with polyA track inserted in the second exon of the HBD coding sequence (Supplementary Figure 11). Expression of the HBD-$6Lys_{AAA}$ protein was 2- to 3-fold reduced compared with the WT-HBD construct, based on western blot analysis (Supplementary Figure 12), and mRNA levels were ~5-fold lower than HBD-WT mRNA levels, measured by qRT-PCR (Supplementary Figure 13). The relative ratios of WT-HBD and HBD-$6Lys_{AAA}$ protein and mRNA levels were constant for all doxycycline induction levels. Altogether with previous data, showing regulated expression of mCherry reporter in different tissues of the transgenic fruit fly, these data demonstrate that polyA tracks can control gene expression independently of the promoter strength associated with the assayed gene.

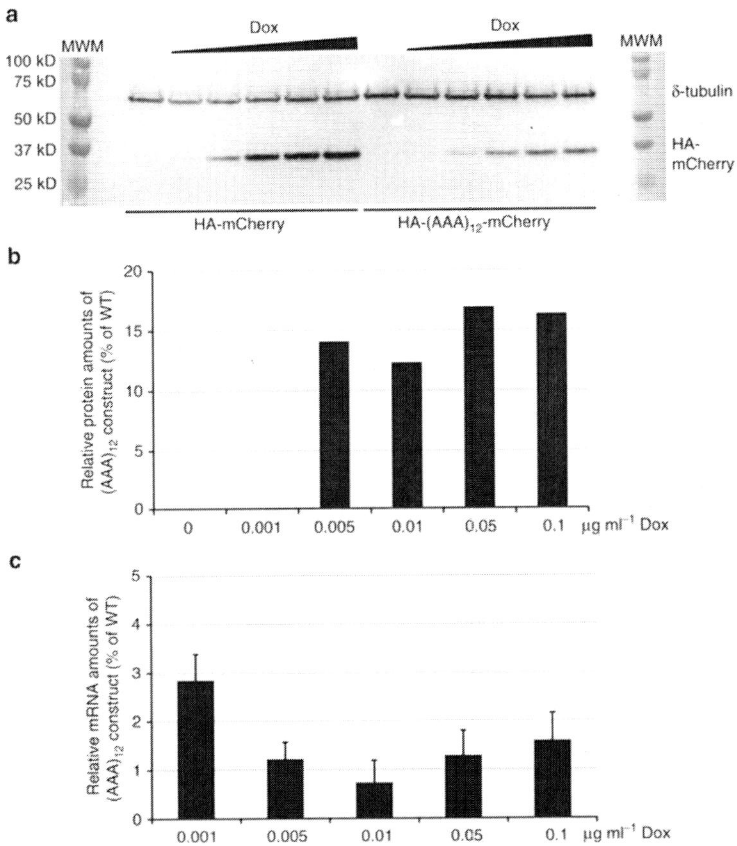

Figure 5: PolyA tracks regulate mCherry reporter expression independently of the promoter strength.

(**a**) Western blot analysis of the cell lysates from Flp-In T-REx 293 stable cell lines expressing doxycycline (Dox) inducible wild type (HA-mCherry) and 12 Lys_{AAA} insertion construct (HA-$(AAA)_{12}$-mCherry) from a single locus. Dox concentration in the media was varied from 0 to 0.1 µg ml^{-1}. Constitutively expressed δ-tubulin was used as a loading control and was detected using specific antibody. Positions of the fusion protein (HA-mCherry), normalization control (δ-tubulin) and sizes of molecular weight markers (MWM) are indicated. (**b**) Quantification of the mCherry protein levels from detected western blot intensities. Levels of mCherry were normalized to δ-tubulin band intensities and represented as a percentage of the wild type construct values at each Dox concentration. Numbers indicate concentration of Dox in the media. (**c**) Steady state mRNA levels of the 12 Lys_{AAA} insertion construct ($(AAA)_{12}$) measured by qRT-PCR. Relative levels of the mRNA for $(AAA)_{12}$ are presented as percentage of the wild type (WT) construct mRNA level at each Dox concentration. Error bars represent mean±s.d. values (n=3). Numbers indicate final concentration of Dox in the media.

PolyA tracks create hypomorphic mutants in functional genes

The polyA tracks are mainly composed of lysine residues, AAA or AAG codons, which can be problematic when expressed as tags because of their charge and propensity for acquiring specific modifications (ubiquitination, acetylation, SUMOylition and hydroxylation). These features of poly-lys chains can potentially influence protein function as well as cell homoeostasis[24,31,32,33].

We tested our ability to regulate gene expression of functional proteins in both the bacterial and eukaryotic cell systems. In *E. coli*, the chloramphenicol acetyltransferase (*CAT*) gene confers resistance to the broad spectrum antibiotic chloramphenicol (CAM) in a dose-dependent manner[34]. Moreover, the CAT protein is functional only as a homotrimer which has a rather complex protein structure (PDB: 3CLA)[35]. As such, regulation of this gene and subsequent protein folding could be challenging because of the additional lysine residues introduced either by polyA track or control AAG constructs. To show that we can regulate expression of CAT protein by insertion of polyA tracks, we assessed *E. coli* survival under increasing concentrations of CAM in comparison with the wild type *CAT* gene. To control for the influence of additional lysine residues in the N-terminus of the CAT protein, we also inserted 10Lys_{AAG} codons in the N-terminus of *CAT*. Expression of WT-CAT, $(AAG)_{10}$-CAT and $(AAA)_n$-CAT constructs was driven by the inducible arabinose promoter (*pBAD*, Supplementary Figure 14). All *E. coli* cultures were pulse induced, with addition of 0.1% arabinose, and growth was monitored on LB plates using different concentrations of CAM in the media. WT-CAT and AAG_{10}-CAT control constructs were able to survive CAM selection to the same extent (75 mg ml^{-1} CAM, Figure 6a). Therefore, the function of CAT protein is not affected by the addition of 10 consecutive Lys residues. By contrast, polyA track constructs led to increased CAM sensitivity of *E. coli* cells which correlated nicely with the length of the polyA tracks inserted in the *CAT* gene (Figure 6a). While the majority of constructs could grow on minimal addition of CAM in the media (15 mg ml^{-1}), constructs with 24, 27 or 30As (8, 9 and 10Lys_{AAA}) were unable to grow on LB-plates with a CAM concentration of 30 mg ml^{-1}. Furthermore, survivability of *E. coli* cells with CAT constructs having 15, 18 and 21As (5, 6 and 7Lys_{AAA}) on one hand, and 9 and 12As (3 and 4Lys_{AAA}) on the other hand, was impaired when cells were grown on LB-plates with final CAM concentrations of 50 mg ml^{-1} or 75 mg ml^{-1}, respectively (Figure 6a). The survivability of *E. coli* cultures with different CAT constructs was in concordance with expression levels of CAT protein assayed by western blot analyses (Supplementary Figure 15). The insertion of 10Lys$_{AAG}$ codons in the *CAT* gene did not affect *E. coli* cell growth on CAM selective media or levels of CAT protein expression, arguing that insertion of multiple lysine residues in the N-terminus is not detrimental for the function, structure and stability of CAT protein. These data demonstrate that polyA tracks can regulate levels of certain enzyme expression (CAT) in *E. coli* cells proportionally with their length.

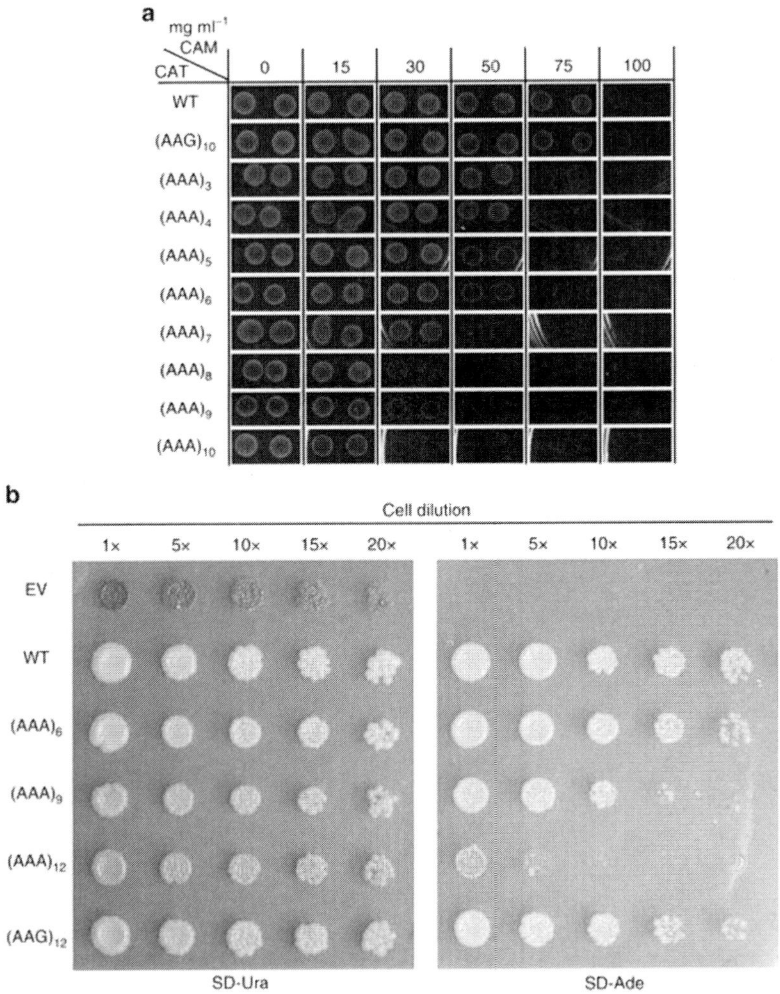

Figure 6: Regulation of drug resistance and metabolic survival by insertion of polyA track tags in genes from *E. coli* and *S. cerviseae*.

(**a**) Survival of *E. coli* cells expressing wildtype (WT), 10Lys$_{AAG}$ ((AAG)$_{10}$) and 3–10 Lys$_{AAA}$(AAA)$_{3-10}$ chloramphenicol acetyltransferase (CAT) constructs on chloramphenicol (CAM) selective media. Pulse induced *E. coli* cells, expressing different CAT constructs, were plated on selective antibiotic plates with varying amounts of CAM in the media (0–100 mg ml^{-1}). Two independent clones were assessed for each construct. *E. coli* colonies were imaged 16 h after plating. (**b**) Assays for *ADE1* gene regulation by polyA tracks ((AAA)$_{6-12}$). Ability of *S. cerevisiae ade1Δ* cells to produce sufficient levels of functional Ade1 protein were assayed by reintroduction of single copy vector with wild type (WT), 12 Lys$_{AAG}$ ((AAG)$_{12}$) and 6–12 Lys$_{AAA}$ ((AAA)$_{6-12}$) Ade1 construct. Empty vector (EV) served as a negative control. Yeast colonies show differential red coloration, on the selective SD-Ura media, which is proportional to the activity of Ade1 protein. Adenine dropout media (SD-Ade) selects for yeast cells expressing sufficient amounts of functional Ade1 protein. Dilutions of the yeast cultures showing relative survival and growth are indicated.

To test the ability of polyA tracks to regulate expression and function of protein in a eukaryotic cell we monitored how polyA tracts affect expression of N-succinyl-5-aminoimidazole-4-carboxamide ribotide synthetase (Ade1, SAICAR) in *S. cerevisiae* (Figure 6b). The Ade1 protein is a monomeric protein with multiple domains in the structure36 (PDB: 1A48) and requires strict folding of an active site for catalysis of the seventh step in the purine biosynthesis pathway. Disruption of the *ADE1* gene results in the storage of a red pigment because of the buildup of a metabolic byproduct of the adenine biosynthesis pathway. Colonies of yeast that are *ade1Δ* are a dark red colour; reintroduction of functional Ade1 protein restores the wild type white coloration in a dose-dependent manner. Such differences in colony colour and ability to grown on adenine dropout media (SD-Ade) have been utilized to differentiate strains of yeast prions37, assess mitotic stability38 and monitor gene expression39.

To survey how polyA tracts affect expression of Ade1, we transformed *ade1Δ* strains of *S. cerevisiae* with single copy plasmids (p416) encoding polyA-*ADE1*-FLAG, with the polyA tracks containing 18, 27 or 36As (6, 9 or 12 Lys$_{AAA}$, Supplementary Figure 16). Control plasmids contained no insertions (WT) or 12 Lys$_{AAG}$ codons, which showed moderate attenuation of reporter expression in our previous study17. Transformants were spotted onto plates to monitor colour phenotypes and growth on media lacking adenine (SD-Ade, Figure 6b). The empty vector control exhibited a dark red coloration and inability to grow on SD-Ade, consistent with disruption of the *ADE1* locus, while the wild type Ade1-FLAG restored both the white phenotype on media containing adenine (SD-Ura) and growth on SD-Ade. Yeast containing constructs with polyA track lengths of

18, 27 and 36As showed progressively pinker coloration and poorer growth on SD-Ade; however, the control 12 Lys_{AAG} construct conferred a nearly-WT white colour and strong growth on SD-Ade (Figure 6b). Dot blot analysis of Ade1 protein expression, normalized to total protein, was in accordance with our phenotypic results and revealed visibly reduced amounts of expression for constructs with insertion of 9 and 12 Lys_{AAA} codons (Supplementary Figure 16). Expression of Ade1 protein with 12 lysine residues at the N-terminus, as in the case of 12 Lys_{AAG}, did not impair folding or function of the assayed protein (Ade1) and exhibited similar phenotypes relative to the insertion of 6 Lys_{AAA} codons. Therefore, addition of polyA tracks to functional genes in both *E. coli* and *S. cerevisiae* preserved protein structure and function but regulated protein abundance. While behaviour of some proteins will of course be sensitive to the inclusion of multiple lysines within their structure, polyA tracks can potentially be quite broadly used in the creation of hypomorphic gene mutants with fixed levels of protein expression.

DISCUSSION

We have presented a rapid method of generating hypomorphic mutations in a reporter or gene of interest. Insertion of a polyA track into a coding sequence exhibits predictable and robust attenuation of gene expression in all tested cell culture and model organism systems. The length and the sequence of the polyA track can be manipulated to achieve full range of expression levels, allowing for the generation of an allelic series from the level of a complete knockout to wild type expression for the study of gene function. This method can also be used in synthetic biology applications that require precise gene control and modelling of metabolic and signalling networks[14,15].

The use of polyA tracks overcomes many of the challenges present in current methods of generating hypomorphic mutations and controllable gene expression. For instance, an approach tested for attenuating gene expression in *E. coli* involves mutagenesis of the Shine-Dalgarno sequence (ribosome binding sequence, rbs) in the gene of interest[2,22]. The expression levels from all possible six-mer Shine-Dalgarno sequences were experimentally determined and the information is available in the EMOPEC database[2]. However, this valuable resource would have to be generated anew to use this approach in other bacteria and it cannot be applied to eukaryotic systems. In addition, many orthogonal translation systems that are used to perturb gene networks rely on modified rbs (refs 40, 41, 42, 43). Use of an orthogonal translation system would prohibit use of the rbs for expression regulation. A similar method has overcome this problem by using tandem repeats to change the distance between the rbs and starting ATG codon and, thereby, controlling gene expression levels22. However, the restricted use in one system, *E. coli* cells, still limits application of this method. The polyA track system of gene regulation that we describe here for the creation of hypomorphic mutations overcomes these issues because of its dependency on regulation of translation elongation cycle which is well conserved between bacteria and eukaryotes19.

Hypomorphic mutations have been generated in eukaryotic cell systems by insertion of an antibiotic resistance gene into an intron10 or the 3′-untranslated region of a gene3. Insertion of the neomycin resistance gene (neo) into an intron introduces a cryptic splice site that results in aberrant splicing of transcripts, effectively reducing gene expression10. The reliance on stochastic cryptic splicing events leads to unpredictable changes in transcript expression and is rather gene dependent. Insertion of neo in various genes have resulted in expression of a functionally null allele44, hypomorphic expression10·45 or no change in expression46. Our system of polyA tracks gives predictable gene expression attenuation in a variety of different eukaryotic systems, and furthermore shows similar relative gene expression attenuation efficiencies in different tissues of the same organism.

We have primarily introduced polyA tracks at the N-terminal regions of reporter genes because of the uniformity of the construct design and to reduce potential frameshifting effects16·17. We do not anticipate this to be a major limitation of this method. Our Tetrahymena reporters place the polyA tracts at the N-terminus of YFP, but at the C-terminus of the linked Tetrahymena gene (Supplementary Figure 3). Furthermore, insertion of polyA track in the second exon of the human beta globin gene (HBD) gene, an unstructured loop of the protein, argue that polyA tracks can be effectively introduced at various positions in the gene (Supplementary Figs 11–13). In addition, we have shown previously that naturally occurring polyA track sequences exist in the human genome and that potential frameshifted products are efficiently degraded by non-sense mediated decay mechanisms16.

The addition of a polyA track to the target gene will result in additional lysine residues in the protein product. Like any protein tag, it is important to consider the effects of the additional residues when studying the functionality of the protein. We have shown that the function and stability of two structurally diverse proteins, CAT and Ade1, are not affected by up to 12 additional lysine residues. To control for possible effects of the poly-lysine tracks, investigators can create an allele with the same number of lysine residues encoded by AAG codons. The AAG codons will have minimal effect on expression levels while encoding a synonymous protein. Furthermore, the flexibility in polyA track placement within the coding sequence allows investigators to choose the most suitable insertion site for the protein of interest.

The conservation of the polyA track sequences in the multiple genes across vertebrates as well as our analysis of mutation rates of polyA tracks (36As) inserted in the defined locus of *D. melanogaster* genome argue that polyA tracks can be used to create stable hypomorphic gene alleles. Our results are in the range of already described hypermutability (~8%) of the short tandem repeats (STRs) and BAT-40 microsatellite (40As) located in the second intron of the 3-beta-hydroxysteroid dehydrogenase gene28. The distinction is that our data show general mutation rates for the whole fruit fly after >30 generations, while in the case of the mentioned study28 the mutation rate is dependent on the cell type. An additional study found that the mutation rate in polyA region, 10As in this case, is in the range of 10^{-4}per cell per generation29. As such, the authors argue that ~1% of cells will be affected by a polyA region mutation in 100 generations. Similar rates were observed in the other studies with ~10^{-6}–10^{-2} mutation rate for the different lengths of homopolymeric regions or STRs (refs 22, 47, 48). PolyA tracks used in our study tend to operate on the shorter side of the length distribution of STRs and as such should have similar if not even lower rates of mutations.

PolyA tracks that are used endogenously in eukaryotic genomes are typically interrupted by other nucleotides at various positions within the A-rich sequence, which further reduces potential hypermutability effects. We have observed that the position of the interrupting nucleotide or codon, in combination with the length of the A-rich sequence, modulates gene expression (Supplementary Figure 2)16. These observations suggest that polyA-mediated regulation can be further developed for even more precise control of gene expression. Last, ~2% of human genes are endogenously regulated by polyA tracks, including many well-studied, disease-associated genes, such as BRCA1, TCOF1 and MTDH among others16,18. As we showed in our previous study, synonymous mutations of the internal polyA track of such genes can allow investigators to dramatically change expression levels of these genes without manipulation of protein sequence or gene regulatory elements such as promoters and enhancers16. The use of our method is not restricted only to these genes, and we feel that the synthetic biology field will benefit from this application. Control of biosynthetic pathways for production of useful secondary metabolites49, antibiotics50 or recombinant antibodies51, as well as introduction of controllable retrosynthetic and fully engineered pathways52 or ultimate control of metabolic pathways in the modelling of diseases are just a few among the multiple possible applications of this method in the future.

METHODS

Escherichia coli experiments

mCherry reporter constructs used for expression in *E.coli* cells were subcloned using LR clonase recombination (Thermo Fisher Scientific) from pENTR/D-Topo constructs used in this study or in previous studies16. The resulting pBAD-DEST49 vector constructs express Thioredoxin (Thrdx) fusion protein as Thrdx-HA tag-polyA-mCherry. For assaying expression of mCherry reporter all constructs were expressed in 2 ml *E.coli* Top10 strain grown in LB-Carbencilin (LB-Carb; final concentration 100 μg ml^{-1}). The cells were grown to optical density at 600 nm (OD_{600}) of 0.4 at 37 °C and induced with addition of arabinose (0.5% w v^{-1}). Fluorescence of mCherry reporter for each construct was measured in triplicates 2–4 h after induction using Biotek Synergy H4 plate reader (Excitation 475±9, Emission 620±9). The amount of fluorescence was normalized to number of cells measured by OD_{600}. To additionally check for expression of fusion proteins, 200 μl of the cells was harvested 2 h post-induction, resuspended in 100 μl of 2 × SDS sample buffer and analysed by SDS–polyacrylamide gel electrophoresis (SDS–PAGE) followed by western blot analysis using HA-tag specific probe (Santa Cruz Biotechnology Inc.). Images of western blot analyses were generated by Bio-Rad Molecular Imager ChemiDoc XRS System with Image Lab software for chemi-luminescence detection (Supplementary Figure 17).

The chloramphenicol acetyltransferase (CAT) constructs used for functional protein studies were created by amplification of the CAT gene from pENTR/D-Topo vector (Thermo Fisher Scientific) using primers listed in Table 1. Constructs were subcloned into pBAD-DEST49 vector for use in functional assays. *E. coli* Top10 cells freshly transformed with pBAD-DEST49 plasmids expressing CAT reporters with different polyA tracks, as well as CAT control reporters were grown in liquid LB-Carb media (100 μg ml^{-1}). For the chloramphenicol (CAM) survivability assay *E. coli* cells were grown to OD_{600}=0.4 and non-induced fractions were spotted on LB-Carb plates (Carb 100 μg ml^{-1}) without chloramphenicol. The residual amount of the cells was induced for 1 h with arabinose (final concentration 0.1% (w v^{-1})). Fraction of cells, 100 μl, was harvested after 30 min of induction, resuspended in 50 μl of 2 × SDS sample buffer and analysed by SDS–PAGE followed by western blot analysis using HA-tag specific probe (Santa Cruz Biotechnology Inc., SC-7392HRP). Reminder of cells were washed twice in M9 minimal media, resuspended in the starting volume of LB-Carb media and 5 μl of cells was spotted as induced fraction on LB-Carb or to LB-Carb/CAM plates with raising amount of chloramphenicol in the media (CAM 15–100 μg ml^{-1}). Plates were incubated overnight at 37 °C and imaged 24 h post-induction using Bio-Rad Molecular Imager ChemiDoc XRS System (Supplementary Figure 21).

Table 1: Oligos used for generation of *E. coli*expressing CAT constructs.

Construct/ oligo name	Primer sequence
CAT WT For	CACCATGCACCATCACCATCACCATGAAAAAAAAATCACTGGATATACCACCGTTGATATATCCC
CAT 10xAAG	CACCATGCACCATCACCATCACCATGAGAAGAAGAAGAAGAAGAAGAAGAAGAAGAAGATCACTGGATATACCACCGTTGATATATCCC
CAT 3xAAA	CACCATGCACCATCACCATCACCATGAAAAAAAAAAAATCACTGGATATACCACCGTTGATATATCCC
CAT 4xAAA	CACCATGCACCATCACCATCACCATGAAAAAAAAAAAAAAATCACTGGATATACCACCGTTGATATATCCC
CAT 5xAAA	CACCATGCACCATCACCATCACCATGAAAAAAAAAAAAAAAAAATCACTGGATATACCACCGTTGATATATCCC
CAT 6xAAA	CACCATGCACCATCACCATCACCATGAAAAAAAAAAAAAAAAAAAAATCACTGGATATACCACCGTTGATATATCCC
CAT 7xAAA	CACCATGCACCATCACCATCACCATGAAAAAAAAAAAAAAAAAAAAAAAATCACTGGATATACCACCGTTGATATATCCC
CAT 8xAAA	CACCATGCACCATCACCATCACCATGAAAAAAAAAAAAAAAAAAAAAAAAAAATCACTGGATATACCACCGTTGATATATCCC
CAT 9xAAA	CACCATGCACCATCACCATCACCATGAAAAAAAAAAAAAAAAAAAAAAAAAAAAAATCACTGGATATACCACCGTTGATATATCCC
CAT 10xAAA	CACCATGCACCATCACCATCACCATGAAAAAAAAAAAAAAAAAAAAAAAAAAAAAAAAAATCACTGGATATACCACCGTTGATATATCCC
CAT Rev	CATTACAGATCTTCTTCAGAAATAAGTTTTTGTTCCGCCCCGCCCTGCCACTCATCGCAG

Saccharomyces cerevisiae experiments

To conduct functional studies with polyA track-induced hypomorphic attenuation in *S. cerevisiae* cells the *ADE1* locus was deleted from 74-D964 yeast strain via homologous recombination. Resultant *ade1Δ* strains (Table 2) were transformed with an empty vector or a plasmid-based *ADE1* containing variable length of polyA tracks, as well as WT and 12Lys$_{AAG}$ insertion constructs. Constructs were generated by performing PCR on *ADE1* isolated from the yeast genomic tiling library (Open Biosystems) with primers listed in Table 3. PCR products were digested and ligated into a p416 vector backbone containing the *ADE1* endogenous promoter. Clones were verified via sequencing and correct constructs were transformed into *ade1Δ* deletions strains via the PEG-LiOAc method (Table 2). To generate dilution spottings, three colonies were picked from each transformation plate and grown overnight in selective media. In the morning, cultures were normalized to OD_{600}=1.0 and 10 µl of cells were spotted onto rich media, SD-Ura and SD-Ade for phenotypic analysis.

Table 2: *S. cerevisiae* strains used in this study.

Name	Strain background	Genotype
HT 971	74-D694	Mat A, ade1::KANMX4, trp1-289(UAG), his3Δ-200, ura3-52, leu2-3, 112
HT 972	74-D694	Mat A, ade1::KANMX4, trp1-289(UAG), his3Δ-200,ura3-52, leu2-3, 112

Table 3: Primers used for generation of ADE1 constructs.

Name	Sequence
FwdAde1SpelWT	5'GGactagtATGTCAATTACGAAGACTGAACTGG
FwdAde1Spel6AAA	5'GGactagtATGAAAAAAAAAAAAAAAAAATCAATTACGAAGACTGAACTGG
FwdAde1Spel9AAA	5'GGactagtATGAAAAAAAAAAAAAAAAAAAAAAAAAAATCAATTACGAAGACTGAACTGG
FwdAde1Spel12AAA	5'GGactagtATGAAAAAAAAAAAAAAAAAAAAAAAAAAAAAAAAAAAATCAATTACGAAGACTGAACTGG
FwdAde1Spel12AAG	5'GGactagtATGAAGAAGAAGAAGAAGAAGAAGAAGAAGAAGAAGAAGTCAATTACGAAGACTGAACTGG
RevAde1ClaIFLAG	5'GGatcgatTTACTTGTCGTCATCGTCCTTGTAGTCGTGAGACCATTTAGACCC
FwdAdelPromSacI	5'GGgagctcACACGATAGCAAAGCAG
RevAdelPromXbaI	5' GGtctagaTATCGTTAATATTTCG

Relative protein abundance was determined via dot blot. Briefly, yeast transformants were picked from selection plates to inoculate 10 ml of SD-Ura and grown overnight to ~OD=0.6. In the morning, cells were harvested and lysed in buffer (25 mM Tris–HCl pH 7.5, 50 mM KCl, 1 mM EDTA, Roche protease inhibitor cocktail) via mechanical disruption with acid-washed glass beads (Sigma). Total protein was normalized to 1 mg ml^{-1} via Bradford assay, and 20 μg of total protein was spotted onto a nitrocellulose membrane. Western blotting was performed by overnight incubation with anti-Flag (Sigma M2, F3165, 1:1,000 in 5% milk) and goat anti-rabbit (Sigma A0545, 1:10,000 in 5% milk) antibodies, followed by detection with chemiluminescence (Amersham ECL).

Tetrahymena thermophila experiments

T. thermophila strain B2086 (II) was used for all experiments reported. Similar results were obtained with strain CU428 ((VII) mpr1-1/mpr1-1). To assess the effect of polyA-tracks on protein accumulation, we modified a fluorescent protein tagging vector, pBSICY-gtw (ref. 53) so as to fuse YFP to the carboxyl-terminus of a Macronucleus-Localized Protein 1 of unknown function (MLP1, TTHERM_00384860), separated by a Gateway recombination cassette (Invitrogen/Life Technologies, Inc), and expressed from the cadmium inducible *MTT1* promoter54. The *MLP1* gene coding region was amplified with oligonucleotides 5′ ALM Bsi' 5′–CAC CCG TAC GAA TAA AAT GAG CAT TAA TAA AGA AGA AGT–3′ and 3′ ALM RV 5′–GAT ATC TTC AAT TTT AAT TTT TCT TCG AAG TTG C 3′ and cloned into pENTR-D in a topoisomerase mediated reaction before digesting with BsiWI and EcoRV and inserting into *BsiWI/PmeI* digested pBSICY-gtw. Subsequently, LR Clonase II was used to insert a linker containing the sequence coding for an HA epitope tag alone (WT) or the tag plus different length of polyA tracks or AAG insertions in place of the Gateway cassette.

The expression cassette is located within the 5′ flanking region of a cycloheximide resistant allele of the rpL29 gene to direct its integration into this genomic locus. These constructs were linearized with *PvuI* and *SacI* in the region flanking the *Tetrahymena* rpl29 sequences and introduced into starved *Tetrahymena* cells by biolistic transformation55,56. Transformants were selected in 1x SPP medium containing 12.5 μg ml^{-1} cycloheximide. To control for copy number, PCR assays with primers MTT2386 5′–TCTTAGCTACGTGATTCACG–3′ and Chx-117, 5′- ATGTGTTAT TAATCGATTGAT–3′ and Chx85r, 5′–TCTCTTTCATGC ATGCTAGC–3′ verified that all rpL29 loci contained the integrated expression construct.

Transgene expression was induced by addition of 0.6 μg ml^{-1} $CdCl_2$ and cells were grown 12–16 h before monitoring protein accumulation. YFP accumulation was visualized by epifluorescence microscopy as previously described57. Whole cells extracts were generated by boiling concentrated cell pellets (2×10^6) in 120 µl of 1 × Laemmli sample buffer, followed by fractionation on 10% SDS polyacrylamide gels and transfer to nitrocellulose. YFP accumulation was a monitored with mouse anti-GFP antisera (G28R anti-GFP (OAEA00007) antibody, Aviva Systems Biology) and normalized to acetylated Rabbit anti-Histone H3 trimethyl-lysine (Upstate Biotechnologies/Millipore, NY, 07-473). qPCR analysis was done using 5′–AGGCCTACAAGACCAAGGGT–3′ and 5′–AGAGCGGTT TTGACG TTGGA–3′ primers for *T. thermophila* ribosomal protein L21 (rpl21) which was used for normalization. Primers 5′–CCCGTATGACGTACCGGATTATG–3′ and 5′–ACTTCAGGGTCAGCTTGCC–3′ were used for detection and estimation of fusion protein transcript levels using SybrGreen master mix and CFX96 Touch Real time PCR Detection System (BioRad). Normalized ΔCt values were used to calculate fold ratio between WT, $12Lys_{AAG}$ and polyA track constructs.

Nicotiana benthamiana experiments

Constructs for expression of HA-tagged mCherry reporters that were already cloned in pEntryD-TOPO vector were sub-cloned to pEarleyGate 100 (ABRC stock number CD3-724) through LR reaction using LR clonase (Invitrogen). The mCherry reporter constructs, pEARLY100 and pBIN61 plasmids were individually electroporated into *Agrobacterium tumefaciens* strains GV3101 (ref. 58). The strain carrying pBIN61 construct expressing p19 protein from tomato bushy stunt virus was co-infiltrated with the reporter constructs to suppress post-transcriptional gene silencing59. The *Agrobacterium* suspensions carrying the reporter constructs were infiltrated into the leaves of 5- to 6-week-old *N. benthamiana* plants as described in Joensuu *et al.*, 2010 (ref. 60). Briefly, saturated over-night cultures were spun-down and resuspended in the infiltration solution (3.2 g l^{-1} Gamborg's B5 plus vitamins, 20 g l^{-1} sucrose, 10 mM MES pH 5.6, 200 µM 4′-Hydroxy-3′,5′-dimethoxyacetophenone) to a final OD_{600}=1.0; *Agrobacterium* suspensions carrying the reporter constructs were individually mixed with suspensions carrying the pBIN61 construct in 1:1 ratio before infiltrations. These suspensions were infiltrated into separate segments of two young leaves on each of eight different *N. benthamiana* plants, which served as biological replicates. For control, 1:1 suspension of *A. tumefaciens* carrying pEARLY 100 with no insert along with pBIN61 was used. The infiltrated plants were maintained in a controlled growth chamber conditions at 22 °C, with a 16 h photoperiod.

Samples of the abaxial epidermis of *N. benthamiana* leaves infiltrated with different mCherry reporter constructs were collected 6 days post-infiltration. Infiltration was performed as described in the previous section, with the addition of an YFP-expressing construct pEARLY104 (ABRC stock number CD3-686), which served as infiltration control. The samples were visualized for fluorescence by confocal laser-scanning microscopy using a Leica TCS SP2 confocal microscope. Samples for RNA and total soluble protein (TSP) extraction were separately collected from the infiltrated plants 6 days post-infiltration using a cork borer (7.1 mm in diameter); each sample contained equal amounts of leaf tissue (TWO leaf discs) collected from each of the segments on the two leaves infiltrated with the same construct.

Analysis of mCherry protein accumulation was carried out by western blot as described in Gutiérrez *et al.* and Conley *et al.*[61,62]. Briefly, phosphate-buffered saline (PBS: 8 g l^{-1} NaCl, 1.16 g l^{-1} Na_2HPO_4, 0.2 g l^{-1} KH_2PO_4, 0.2 g l^{-1} KCl, pH 7.4), supplemented with 1 mM EDTA, 1 mM phenylmethanesulfonylfluoride (PMSF), 1 μg ml^{-1} leupeptin 0.1% Tween-20 and 100 mM sodium L-ascorbate was used for total soluble protein (TSP) extractions. Bradford assay (Biorad) was used to quantify TSP in the extracts using a standard curve (r^2=0.99) of known concentrations of Bovine Serum Albumin (BSA). Sample extracts (25 μg TSP for mCherry and 5 μg TSP for phosphinotricin acetyl transferase (BAR) protein detection) were separated by SDS–PAGE, blotted onto nitrocellulose membrane and probed with a primary anti-HA tag antibody (Genscript, A01244-100) for mCherry or anti-Phosphinotricin acetyl transferase antibody (Abcam, Ab50504) for BAR, both at 1:2,000 dilution, followed by HRP-conjugated secondary antibody (Biorad, 170-6516 and 170-6515) at 1:5,000 dilution. The blots were washed (3 times × 10 min) in 1 × Tris-buffered Saline (TBS, 50 mM Tris, 150 mM NaCl, pH 7.5) containing 0.1% Tween (Sigma) and images were obtained after 1 min incubation with the enhanced chemiluminescence detection system (GE Healthcare) (Supplementary Figure 18). Numerical values for protein accumulation were derived from the detected band intensities on the analysed images using TotalLab TL 100 software (Nonlinear Dynamics, Durham, USA). The mCherry accumulation values were normalized for BAR accumulation detected in the same sample. Normalized values of the mCherry protein accumulation for each reporter construct were presented as the mean of eight biological replicates±s.e.; Tukey's honest significance test (JMP software, SAS Institute Inc.) was used to identify significantly different means (α=0.05).

For quantitative RT–PCR (qPCR), total RNA was extracted using an RNeasy plant mini kit coupled with DNase treatment (Qiagen). The purified RNA (500 ng) was reverse-transcribed using the Maxima first-strand complementary DNA (cDNA) synthesis kit (Thermo Fisher Scientific). The resulting cDNA (2 ng μl^{-1}) was quantified by qPCR using the Maxima SYBR Green/ROX qPCR master mix (Thermo Fisher Scientific) and CFX384 Touch

Real-Time PCR Detection System (Biorad). Cycle threshold (Ct) values were normalized to phosphinothricin N-acetyltransferase (*BAR*) gene expressed from the same plasmid used for transient expression. Primer sequences used: For mCherry—mCherryFWD: 5′–GGCTACCCATACGATGTTCC–3′; mCherryREV: 5′–CCTCCATGTGCACCTTGAAG–3′; for BASTA–BAR-F3: 5′–TCAAGAGCGTGGTCGCTG–3′ and BAR-R3: 5′–CAAATCTCGGTGACGGGCAG–3′.

Drosophila melanogaster experiments

Reporter gene expression was achieved with the GAL4/UAS system. The UAS-mCherry transgene plasmids were constructed from the phiC31 integrase plasmid, pJFRC28-10XUAS-IVS-GFP-p10 (Addgene plasmid #36431)63. GFP was removed by digestion with KpnI and XbaI and replaced with HA-mCherry and HA-polyA-mCherry. Transgenic fly lines were obtained by injecting P{CaryP}attP2 embryos with each pJFRC28 mCherry construct to achieve site-specific, single insertion on the third chromosome at the attP2 landing site (Rainbow Transgenic Flies, Inc.). Injected G_0 adult flies were backcrossed to w^{1118} flies. Red-eyed progeny indicated successful germline integration of the UAS-mCherry expression cassette. Male red-eye progeny were crossed to female w;TM3 Sb/TM6 Tb flies followed by sib-crosses of the F_1progeny to generate homozygous UAS-mCherry transgenic lines. Insertion was confirmed by Sanger sequencing of PCR amplified mCherry from genomic DNA of individual flies. Each mCherry transgenic fly line was crossed to a TubGal4 UAS-GFP driver line (derived from BSC42734) to achieve mCherry expression in all tissues. GFP expression was used for normalization. All flies were maintained at 25 °C.

Third instar larvae from each cross were fixed in formaldehyde and dissected to recover the salivary glands (SG), intact central nervous system (CNS) and proventriculus (PV). The tissues were mounted on glass cover slips and confocal images were taken on a Zeiss Imager 2 upright microscope using identical parameters for all images of each tissue type. Fluorescence intensity of the mCherry and GFP were quantified with Zen 9 software.

Total RNA was extracted from each cross by pooling five third instar larvae in 1.5 ml RNase-free Eppendorf tubes which were then frozen in dry ice. Frozen samples were homogenized using 1.5 ml pestles (Fisherbrand, RNase- and DNase-free). After homogenization, 1 ml RiboZol reagent (Amresco) was added and extraction was completed according to manufacturer's instructions. Total RNA samples were treated with TURBO DNA-free kit (Ambion) to remove potential genomic DNA. cDNA synthesis was performed with iScript Reverse Transcription Supermix (Bio-Rad) with 1 µg of total RNA in a 20 µl reaction. RT-qPCR was performed in the Bio-Rad CFX96 Real-Time System with iQ SYBR Green Supermix (Bio-Rad). The mCherry transcript was detected with the following primers: 5′–TGACGTACCGGATTATGCAA–3′ and 5′–ATATGAACTGAGGGGACAGG–3′. Cycle threshold (Ct) values were normalized to EF1 with the following primers: 5′–GCGTGGGTTTGTGATCAGTT–3′ and 5′–GATCTTCTCC TTGCCCATCC–3′)64.

For western blot analysis, five third instar larvae from each cross were collected and frozen in dry ice. Frozen samples were homogenized using 1.5 ml pestles (Fisherbrand, RNase- and DNase-free). After homogenization, SDS sample buffer was added and the samples were boiled for 10 min. Anti-HA tag (Santa Cruz Biotechnology Inc., SC7392HRP) was used to detect mCherry expression. Samples were normalized with anti-GFP (Clontech, 632381).

For Illumina sequencing analysis, genomic DNA was isolated from a single adult HA-(AAA)12-mCherry fly approximately 30 generations after transgene insertion. The QIAGEN DNeasy Blood and Tissue Kit were used, following the supplementary protocol for total DNA extraction from insect cells. The polyA track of the mCherry transgene was amplified from both the genomic DNA and the plasmid DNA that was used to generate the transgenic fly line. The primers (5′–GGCTGCTC GAGCCGTATGA CGTACCGGATTATGC–3′ and 5′–CCTTGTGATCAGGCCATGTTATCCTCCTCGC–3′) added XhoI and SpeI restriction enzyme sites flanking the polyA track region. PCR products were purified with NucleoSpin Gel and PCR Clean-up (Macherey-Nagel) and digested with XhoI and SpeI to generate overhangs for ligation of Illuminia adaptor sequences. To enrich for amplicons with both adaptor sequences, the product was amplified with the following primers: 5′- -3′ and 5′- -3′. Samples were run as a spike-in on an Illumina HISeq machine. Reads that matched the first 32 or 33 expected nucleotides, depending on the indexing barcode used, were counted. These resulted in ~670,000 genomic reads and 1,000,000 plasmid reads after removing sequences that had fewer than 50 reads.

For Sanger sequencing analysis, the polyA track from the genomic DNA extracted for Illumina sequencing and the original plasmid were amplified with the following primers: 5′-GGCTGCTCGAGCCGTATGACGTACCGG-3′ and 5′-ACATGAACTGAGGGGACAGG-3′. The PCR products were purified with NucleoSpin Gel and PCR Clean-up kit (Macherey-Nagel) and ligated into the pCR-Blunt vector using the Zero Blunt PCR Cloning Kit (Thermofisher). The vector was transformed into Max Efficiency DH5α competent cells (Thermofisher) and 60 colonies from both the genomic DNA and plasmid DNA were sent for sequencing (Genewiz). Sequences with good quality scores were used for downstream analysis.

Homo sapiens cell culture experiments

mCherry reporter constructs used for transient expression in human cells were subcloned using LR clonase recombination (Thermo Fisher Scientific) from pEntryD-Topo constructs used in other experiments or in previous studies16. DNA fragments for constructs used for creation of inducible and stable cell lines were PCR amplified, purified and ligated into pcDNA 5/FRT/TO vector (Thermo Fisher Scientific).

HeLa cells were cultured in Dulbecco's modified Eagle's medium (DMEM) (Gibco) and supplemented with 10% fetal bovine serum, 5% minimum essential medium nonessential amino acids (100 × , Gibco), 5% penicillin and streptomycin (Gibco) and L-glutamine (Gibco). Flp-In T-REx 293 cells were grown in the same media with addition of 5 μg ml^{-1} of blastocidin and 100 μg ml^{-1} of Zeocin for non-recombined cells, or 5 μg ml^{-1} of blastocidin and 100 μg ml^{-1} of hygromycin for growth of stable cell lines expressing mCherry or HBD constructs.

Plasmids were introduced to the tissue culture cells by the Neon Transfection System (Thermo Fisher Scientific) using 100-μl tips according to cell-specific protocols (www.lifetechnologies.com/us/en/home/life-science/cell-culture/transfection/transfection---selection-misc/neon-transfection-system/neon-protocols-cell-line-data.html). Hela cells, used for transient expression, were electroporated with 1.5 μg of DNA plasmids and were harvested 24 h after the electroporation. Flp-In T-REx 293 cells were electroporated with plasmids, selected for positive clones as described by protocol (https://tools.thermofisher.com/content /sfs/manuals/flpinsystem_man.pdf). Expression of polyA track and control constructs was induced by addition of various amounts of doxycycline from a common stock (1 mg ml^{-1}) and harvested 24 or 48 h after induction, if not indicated differently.

Total RNA was extracted from cells using the RiboZol RNA extraction reagent (Amresco) according to the manufacturer's instructions or using GenEluteTM Direct. RiboZol reagent (500 μl) was used in each well of 6- or 12-well plates for RNA extraction. Precipitated nucleic acids were treated with Turbo deoxyribonuclease (Ambion), and total RNA was dissolved in ribonuclease-free water and stored at −20 °C. RNA concentration was measured by NanoDrop (OD260/280). iScript Reverse Transcription Supermix (Bio-Rad) was used with 1 μg of total RNA following the manufacturer's protocol. RT-qPCR was performed in the Bio-Rad CFX96 Real-Time System with iQ SYBR Green Supermix (Bio-Rad). For both transient expression samples and stable cell line samples, the mCherry transcript was detected with the following primers: 5′-TGACGTACCGGATTATGCAA-3′ and 5′-ATATGAACTGAGGGGACAGG-3′. Cycle threshold (Ct) values were normalized to the neomycin resistance gene expressed from the same plasmid for transient expression (5′-CTGAATGAACTGCAGGACGA-3′ and 5′-ATACTTTCTCGGCAGGAGCA-3′) or hygromycin (5′-GATGTAGGAGGGCGTGGATA-3′ and 5′-ATAGGTCA GGCTCTCGCTGA-3′ or actin gene for stable cell lines (5′-AGAAAATCTGGCACCACACC-3′ and 5′-AGAGGCGTACAGGGATAGCA-3′.

Total cell lysates were prepared with passive lysis buffer (Promega). Blots were blocked with 5% milk in 1 × tris-buffered saline with 0.1% Tween 20 (TBST) for 1 h. Horseradish peroxidase–conjugated or primary antibodies (anti-β-actin (Santa Cruz Biotechnology Inc., SC1616HRP), anti δ-tubulin (Sigma-Aldrich, T3950)) were diluted according to the manufacturer's recommendations and incubated overnight with the membranes. The membranes were washed four times for 5 min in TBST and prepared for imaging, or secondary antibody (Cell Signaling Technology, 7076S) was added for additional 1 h incubation. Images were generated by Bio-Rad Molecular Imager ChemiDoc XRS System with Image Lab software by chemiluminescence detection or by the LI-COR Odyssey Infrared Imaging System (Supplementary Figs 19 and 20). Blots imaged by the LI-COR system were first incubated for 1 h with Pierce DyLight secondary antibodies.

DATA AVAILABILITY

The data that support the findings of this study are available from the authors on reasonable request; see author contributions for specific data sets.

ACKNOWLEDGEMENTS

We thank D. Piston, M. Jovanovic, T. Schedl and P. Szczęsny and members of Djuranovic lab for helpful comments. This work was supported by the NIH grant T32 GM007067 to LLA and KMK, National Science Foundation grant # 1412336 (MCB, PI: Chalker) to JCC and DLC, the NSERC-DG to VG, NIH Grant R01 NS036570 to JBS, NIH Grant R01 GM072778 to HT, NIH Grant R01 GM112824 and the American Cancer Society (grant IRG-58-010-58-2) to SD. Subject to the U.S. Provisional Patent Application, Serial No. 62/361,307.

REFERENCES

1. Muller, H. J. Further studies on the nature and causes of gene mutations. Proc. 6th Int. Congr. Genet. 1, 213–255 (1932).
2. Bonde, M. T. et al. Predictable tuning of protein expression in bacteria. Nat. Methods 13, 233–236 (2016).
3. Breslow, D. K. et al. A comprehensive strategy enabling high-resolution functional analysis of the yeast genome. Nat. Methods 5, 711–718 (2008).
4. Chappell, J., Watters, K. E., Takahashi, M. K. & Lucks, J. B. A renaissance in RNA synthetic biology: New mechanisms, applications and tools for the future. Curr. Opin. Chem. Biol. 28, 47–56 (2015).
5. Dawlaty, M. M. & van Deursen, J. M. Gene targeting methods for studying nuclear transport factors in mice. Methods 39, 370–378 (2006).
6. Ferri, A. L. et al. Sox2 deficiency causes neurodegeneration and impaired neurogenesis in the adult mouse brain. Development 131, 3805–3819 (2004).
7. Garí, E., Piedrafita, L., Aldea, M. & Herrero, E. A set of vectors with a tetracycline-regulatable promoter system for modulated gene expression in Saccharomyces cervisiae. Yeast 13, 837–848 (1997).
8. Goto, T., Hara, H., Nakauchi, H., Hochi, S. & Hirabayashi, M. Hypomorphic phenotype of Foxn1 gene-modified rats by CRISPR/Cas9 system. Transgenic Res. 25, 533–544 (2016).
9. LaFave, M. C. & Sekelsky, J. Transcription initiation from within P elements generates hypomorphic mutations in Drosophila melanogaster. Genetics 188, 749–752 (2011).
10. Meyers, E. N., Lewandoski, M. & Martin, G. R. An Fgf8 mutant allelic series generated by Cre- and Flp-mediated recombination. Nat. Genet. 18, 136–141 (1998).
11. Redden, H., Morse, N. & Alper, H. S. The synthetic biology toolbox for tuning gene expression in yeast. FEMS Yeast Res. 15, 1–10 (2015).
12. Doudna, J. A. & Charpentier, E. The new frontier of genome engineering with CRISPR-Cas9. Science 346, 1258096–1258096 (2014).
13. Joung, J. K. & Sander, J. D. TALENs: a widely applicable technology for targeted genome editing. Nat. Rev. Mol. Cell. Biol. 14, 49–55 (2013).
14. Chuang, H.-Y., Hofree, M. & Ideker, T. A decade of systems biology. Annu. Rev. Cell Dev. Biol. 26, 721–744 (2010).
15. Xie, M., Haellman, V. & Fussenegger, M. Synthetic biology—application-oriented cell engineering. Curr. Opin. Biotechnol. 40, 139–148 (2016).
16. Arthur, L. L. et al. Translational control by lysine-encoding A-rich sequences. Sci. Adv. 1, e1500154 (2015).
17. Koutmou, K. S. et al. Ribosomes slide on lysine-encoding homopolymeric A stretches. Elife 4, 1–18 (2015).
18. Habich, M., Djuranovic, S. & Szczesny, P. PATACSDB—the database of polyA translational attenuators in coding sequences. PeerJ Comput. Sci. 2, e45 (2016).
19. Melnikov, S. et al. One core, two shells: bacterial and eukaryotic ribosomes. Nat. Struct. Mol. Biol. 19, 560–567 (2012).
20. Kuroha, K. et al. Receptor for activated C kinase 1 stimulates nascent polypeptide-dependent translation arrest. EMBO Rep. 11, 956–961 (2010).
21. Brandman, O. et al. A ribosome-bound quality control complex triggers degradation of nascent peptides and signals translation stress. Cell 151, 1042–1054 (2012).
22. Egbert, R. G. & Klavins, E. Fine-tuning gene networks using simple sequence repeats. Proc. Natl Acad. Sci. 109, 16817–16822 (2012).
23. Collins, K. & Gorovsky, M. A. Tetrahymena thermophila. Curr. Biol. 15, R317–R318 (2005).
24. Brandman, O. et al. A ribosome-bound quality control complex triggers degradation of nascent peptides and signals translation stress. Cell 151, 1042–1054 (2012).

25. Groth, A. C., Fish, M., Nusse, R. & Calos, M. P. Construction of transgenic Drosophila by using the site-specific integrase from phage phiC31. Genetics 166, 1775–1782 (2004).
26. Duffy, J. B. GAL4 system in Drosophila: a fly geneticist's Swiss army knife. Genesis 34, 1–15 (2002).
27. Brinkmann, B., Klintschar, M., Neuhuber, F., Hühne, J. & Rolf, B. Mutation rate in human microsatellites: influence of the structure and length of the tandem repeat. Am. J. Hum. Genet. 62, 1408–1415 (1998).
28. Ellegren, H. Microsatellites: simple sequences with complex evolution. Nat. Rev. 5, 435–445 (2004).
29. Chung, H. et al. Mutation rates of TGFBR2 and ACVR2 coding microsatellites in human cells with defective DNA mismatch repair. PLoS ONE 3, 2–10 (2008).
30. Li, J. & Zhang, Y. Relationship between promoter sequence and its strength in gene expression. Eur. Phys. J. E 37, 1–6 (2014).
31. Dimitrova, L. N., Kuroha, K., Tatematsu, T. & Inada, T. Nascent peptide-dependent translation arrest leads to Not4p-mediated protein degradation by the proteasome. J. Biol. Chem. 284, 10343–10352 (2009).
32. Choe, Y.-J. et al. Failure of RQC machinery causes protein aggregation and proteotoxic stress. Nature 531, 191–195 (2016).
33. Yonashiro, R. et al. The Rqc2/Tae2 subunit of the Ribosome-Associated Quality Control (RQC) complex marks ribosome-stalled nascent polypeptide chains for aggregation. Elife 5, e11794 (2016).
34. Shaw, W. V. & Leslie, A. G. W. Chloramphenicol acetyltransferase. Annu. Rev. Biophy. Biophys. Chem. 20, 363–386 (1991).
35. Leslie, A. G. W. Refined crystal structure of type III chloramphenicol acetyltransferase at 1.75 A resolution. J. Mol. Biol. 213, 167–186 (1990).
36. Levdikov, V. M. et al. The structure of SAICAR synthase: an enzyme in the de novo pathway of purine nucleotide biosynthesis. Structure 6, 363–376 (1998).
37. Liebman, S. W. & Chernoff, Y. O. Prions in yeast. Genetics 191, 1041–1072 (2012).
38. Hieter, P., Mann, C., Snyder, M. & Davis, R. W. Mitotic stability of yeast chromosomes: a colony color assay that measures nondisjunction and chromosome loss. Cell 40, 381–392 (1985).
39. Mano, Y., Kobayashi, T. J., Nakayama, J., Uchida, H. & Oki, M. Single cell visualization of yeast gene expression shows correlation of epigenetic switching between multiple heterochromatic regions through multiple generations. PLOS Biol. 11, e1001601 (2013).
40. Hui, A. & de Boer, H. A. Specialized ribosome system: preferential translation of a single mRNA species by a subpopulation of mutated ribosomes in Escherichia coli. Proc. Natl Acad. Sci. USA 84, 4762–4766 (1987).
41. Hui, A. et al. Directing ribosomes to a single mRNA Species: a method to study ribosomal RNA mutations and their effects on translation of a single messenger in Escherichia coli. Methods Enzymol. 153, 432–452 (1987).
42. Lee, K., Holland-Staley, C. A. & Cunningham, P. R. Genetic analysis of the Shine-Dalgarno interaction: Selection of alternative functional mRNA-rRNA combination. RNA 2, 1270–1285 (1996).
43. Rackham, O. & Chin, J. W. A network of orthogonal ribosome·mRNA pairs. Nat. Chem. Biol. 1, 159–166 (2005).
44. Nagy, A. et al. Dissecting the role of N-myc in development using a single targeting vector to generate a series of alleles. Curr. Biol. 8, 661–664 (1998).
45. Hirotsune, S. et al. Graded reduction of Pafah1b1 (Lis1) activity results in neuronal migration defects and early embryonic lethality. Nat. Genet. 19, 333–339 (1998).
46. Wolpowitz, D. et al. Cysteine-rich domain isoforms of the neuregulin-1 gene are required for maintenance of peripheral synapses. Neuron 25, 79–91 (2000).
47. Blake, C., Tsao, J. L., Wu, A. & Shibata, D. Stepwise deletions of polyA sequences in mismatch repair-deficient colorectal cancers. Am. J. Pathol. 158, 1867–1870 (2001).

48. Zhivotovsky, L. A. et al. The effective mutation rate at Y chromosome short tandem repeats, with application to human population-divergence time. Am. J. Hum. Genet. 74, 50–61 (2004).
49. O'Connor, S. E. Engineering of Secondary Metabolism. Annu. Rev. Genet. 49, 71–94 (2015).
50. Braff, D., Shis, D. & Collins, J. J. Synthetic biology platform technologies for antimicrobial applications. Adv. Drug Deliv. Rev. 105, 35–43 (2015).
51. Wu, C.-Y., Rupp, L. J., Roybal, K. T. & Lim, W. A. Synthetic biology approaches to engineer T cells. Curr. Opin. Immunol. 33, 123–130 (2015).
52. Carbonell, P., Planson, A.-G., Fichera, D. & Faulon, J.-L. A retrosynthetic biology approach to metabolic pathway design for therapeutic production. BMC Syst. Biol. 5, 122 (2011).
53. Motl, J. A. & Chalker, D. L. Zygotic expression of the double-stranded RNA binding motif protein Drb2p is required for DNA elimination in the ciliate Tetrahymena thermophila. Eukaryot. Cell 10, 1648–1659 (2011).
54. Shang, Y. et al. A robust inducible-repressible promoter greatly facilitates gene knockouts, conditional expression, and overexpression of homologous and heterologous genes in Tetrahymena thermophila. Proc. Natl Acad. Sci. USA 99, 3734–3739 (2002).
55. Cassidy-Hanley, D. et al. Germline and somatic transformation of mating Tetrahymena thermophila by p bombardment. Genetics 146, 135–147 (1997).
56. Bruns, P. J. & Cassidy-Hanley, D. Biolistic transformation of macro- and micrnuclei. Methods Cell Biol. 62, 303–305 (2000).
57. Matsuda, A., Shieh, A. W. Y., Chalker, D. L. & Forney, J. D. The conjugation-specific Die5 protein is required for development of the somatic nucleus in both Paramecium and Tetrahymena. Eukaryot. Cell 9, 1087–1099 (2010).
58. Koncz, C. & Schell, J. The promoter of TL-DNA gene 5 controls the tissue-specific expression of chimaeric genes carried by a novel type of Agrobacterium binary vector. MGG Mol. Gen. Genet. 204, 383–396 (1986).
59. Voinnet, O., Rivas, S., Mestre, P. & Baulcombe, D. An enhanced transient expression system in plants based on suppression of gene silencing by the p19 protein oftomato bushy stunt virus. Plant J. 33, 949–956 (2003).
60. Joensuu, J. J. et al. Hydrophobin fusions for high-level transient protein expression and purification in Nicotiana benthamiana. Plant Physiol. 152, 622–633 (2010).
61. Gutiérrez, S. P., Saberianfar, R., Kohalmi, S. E. & Menassa, R. Protein body formation in stable transgenic tobacco expressing elastin-like polypeptide and hydrophobin fusion proteins. BMC Biotechnol. 13, 40 (2013).
62. Conley, A. J., Joensuu, J. J., Jevnikar, A. M., Menassa, R. & Brandle, J. E. Optimization of elastin-like polypeptide fusions for expression and purification of recombinant proteins in plants. Biotechnol. Bioeng. 103, 562–573 (2009).
63. Pfeiffer, B. D., Truman, J. W. & Rubin, G. M. Using translational enhancers to increase transgene expression in Drosophila. Proc. Natl Acad. Sci. USA 109, 6626–6631 (2012).
64. Ponton, F., Chapuis, M. P., Pernice, M., Sword, G. A. & Simpson, S. J. Evaluation of potential reference genes for reverse transcription-qPCR studies of physiological responses in Drosophila melanogaster. J. Insect Physiol. 57, 840–850 (2011).

CHAPTER 8

Artificial Rearing Influences The Morphology, Permeability and Redox State of The Gastrointestinal Tract of Low and Normal Birth Weight Piglets

Hans Vergauwen[1†], Jeroen Degroote[2†], Sara Prims[1], Wei Wang[2,3], Erik Fransen[4], Stefaan De Smet[3], Christophe Casteleyn[1], Steven Van Cruchten[1], Joris Michiels[2] and Chris Van Ginneken[1]

[1] *Laboratory of Applied Veterinary Morphology, Department of Veterinary Sciences, Faculty of Biomedical, Pharmaceutical and Veterinary Sciences, University of Antwerp, Campus Drie Eiken, Universiteitsplein 1, D.U.015, 2610 Wilrijk, Belgium.*

[2] *Department of Applied Biosciences, Faculty of Bioscience Engineering, Ghent University, Ghent, Belgium.*

[3] *Laboratory for Animal Nutrition and Animal Product Quality (LANUPRO), Department of Animal Production, Faculty of Bioscience Engineering, Ghent University, Melle, Belgium.*

[4] *StatUa Center for Statistics, University of Antwerp, Antwerp, Belgium.*

ABSTRACT

Background

In this study the physiological implications of artificial rearing were investigated. Low (LBW) and normal birth weight (NBW) piglets were compared as they might react differently to stressors caused by artificial rearing. In total, 42 pairs of LBW and NBW piglets from 16 litters suckled the sow until d19 of age or were artificially reared starting at d3 until d19 of age. Blood and tissue samples that were collected after euthanasia at 0, 3, 5, 8 and 19 d of age. Histology, ELISA, and Ussing chamber analysis were used to study proximal and distal small intestine histo-morphology, proliferation, apoptosis, tight junction protein expression, and permeability. Furthermore, small intestine, liver and systemic redox parameters (GSH, GSSG, GSH-Px and MDA) were investigated using HPLC.

Results

LBW and NBW artificially reared piglets weighed respectively 40 and 33% more than LBW and NBW sow-reared piglets at d19 ($P < 0.01$). Transferring piglets to a nursery at d3 resulted in villus atrophy, increased intestinal FD-4 and HRP permeability and elevated GSSG/GSH ratio in the distal small intestine at d5 ($P < 0.05$). GSH concentrations in the proximal small intestine remained stable, while they decreased in the liver ($P < 0.05$). From d5 until d19, villus width and crypt depth increased, whereas PCNA, caspase-3, occludin and claudin-3 protein expressions were reduced. GSH, GSSG and permeability recovered in artificially reared piglets ($P < 0.05$).

Conclusion

The results suggest that artificial rearing altered the morphology, permeability and redox state without compromising piglet performance. The observed effects were not depending on birth weight.

Keywords: Milk replacer Oxidative stress Small intestine Suckling period Tight junction proteins

BACKGROUND

The neonatal period in the pig's life is accompanied with high morbidity and mortality [1, 2]. In addition, increasing litter sizes in modern swine production have led to higher rates of piglets born with a low birth weight (LBW) [3]. Both newborn LBW human [4, 5, 6] and LBW piglets [7, 8, 9, 10] seem to have a lower capacity to mount an antioxidant response. Newborns transitioning from maternal mediated respiration to autonomous pulmonary respiration outside the uterus are suddenly exposed to O_2-derived free radicals [11]. This increased the production of reactive species in various organs [7]. A redox imbalance affects cellular signaling, protein synthesis, and enhances proteolysis that can ultimately lead to a dysfunctional intestinal barrier and suboptimal regenerative potential as shown in vitro [12, 13, 14, 15]. Consequently, the observed redox imbalance and the downstream effects could explain the abnormal absorption and metabolism of nutrients, reduced growth and impaired development of the small intestine, liver, and muscle observed in LBW piglets [10, 16, 17, 18, 19, 20, 21]. This redox imbalance appears to persist beyond weaning [9]. Wang et al. observed that mRNA expression of occludin, heme oxygenase 1, catalase, thioredoxin reductase genes and occludin protein expression continued to be lower in LBW pigs during the suckling period [21]. This apparently conflicts with the observation that LBW piglets that survive the critical first days after birth show an intestinal morphology, digestive capacity, cytokine production, intestinal motility and permeability that is comparable with those seen in normal birth weight (NBW) littermates [22, 23, 24].

The increasing incidence of supernumerary and LBW piglets has raised an urgent need for innovative rearing strategies [1]. Next to cross-fostering [25], supplementing piglets [26] and split nursing [27], piglets can be transferred to a nursery and artificially reared [28, 29, 30]. Similar to conventional weaning, the artificially reared piglet encounters psychological and physical stressors including maternal and littermate separation, abrupt changes in diet composition and environment, lower intake of bioactive substances, as well as unfamiliar drinking nipples, increased exposure to pathogens and antigens, comingling and establishment of social hierarchy with unfamiliar pigs from different litters. The physiological responses to the strategy of full artificial rearing are largely unknown. Conventional weaning is associated with the induction of intestinal oxidative stress [31, 32, 33] and LBW piglets have more difficulties to maintain a balanced redox state when exposed to weaning stressors [9]. It is unknown at present if the response to artificial rearing, since it includes similar stressors as conventional weaning, is different in LBW piglets—which have a lower antioxidant capacity [21]—compared to NBW piglets.

Therefore, we aimed to investigate the impact of artificial rearing on piglet performance, proximal and distal small intestinal (SI) morphology, mitosis, apoptosis, and tight junction protein expression, permeability, and SI, liver and systemic redox state development compared to conventional rearing. Given the similarities between conventional weaning stressors and artificial rearing stressors we hypothesized that artificial rearing results in a redox imbalance and negatively affects intestinal morphology and functionality. Secondly, given the differences observed between NBW and LBW piglets during the suckling period, we hypothesized that in view of their affected redox state, the morphology and functionality of the small intestine is suboptimal in LBW piglets and that artificial rearing has a greater negative impact in this birth weight category.

METHODS

Pig model and tissue collection

Eighty-four piglets (Topigs hybrid × Piétrain) were selected during two consecutive farrowing rounds. Only sows with 14 or more live-born piglets were selected and no cross-fostered piglets were included in the experiment. Piglets were tagged and weighed within 12 h after parturition. A LBW piglet was defined as a pig having a birth weight between 0.75 and 0.90 kg and belonging to the lower quartile of litter birth weights. A NBW piglet had a birth weight within 1 SD unit of the mean birth weight of the whole litter. The average birth weight of all the piglets used in this study was 0.81 ± 0.01 kg and 1.26 ± 0.02 kg for LBW and NBW piglets, respectively. Forty-two pairs of LBW and NBW gender-matched littermates were either left to suckle the sow or transferred to a nursery at the age of 3 d and had ad libitum access to a milk replacer that was refreshed twice a day (Table 1). Milk replacer was made by mixing 200 g of milk replacer powder into 1 L of water, preheated at 42 °C. This resulted in a dry matter (DM) content of 193 g/kg milk. This is comparable with the DM content of sow milk, which is around 170 to 190 g/kg milk between 3 d and 42 d of lactation [34]. In Table 1, the calculated nutrient composition of some important nutrients is shown, expressed on DM basis. When account for the dilution factor, it is possible to compare this milk formula with sow milk. For instance, it is clear that the crude protein level in the formula (49.8 g/kg milk) is low but comparable with what can be expected in sow milk (51–64 g/kg milk) ([34, 35, 36]). The crude fat content was calculated at 22 g/kg milk, which is considerably lower than the crude fat content of sow milk (5.3–6.5 g/kg milk) [34, 35, 36]. In contrast, the total amount of lactose in the liquid milk replacer was 80 g/kg milk. Values for sow milk range from 48 to 59 g/kg milk, and thus are lower than what was offered to the formula fed piglets. From this, it might be clear that the milk formula was not completely comparable to the nutrient composition of porcine milk. However, this tailor made milk formula was meant as a compromise of many existing, commercially available milk replacer formulas, and as a copy of sow milk as such. A maximum of 4 NBW-LBW gender matched littermate pairs were co-housed. The starting ambient temperature was 32 °C and linearly decreased to 28 °C towards d19 post-natal. Heating and ventilation was automatically controlled in function of the temperature settings. Six pairs of sow-reared LBW and NBW piglets were sampled on d0, d3, d8 and d19 of age, and 6 pairs of artificially reared LBW and NBW piglets were sampled on d5, d8 and d19 of age. Piglets sampled on d0 were removed from the sow between 12 and 24 h after parturition. At d3, intramuscular iron injections were given (Iron(III) Dextran, 200 mg/piglet, Uniferon). The selected animals did not receive any antibiotic treatment prior or during the experiment.

Table 1. Composition of the milk replacer used for piglets from 3 d of age until weaning at d 19

Ingredient composition, %	
Coco fat filled whey 50/50	42.00
Skimmed milk powder	17.61
Whey permeate	8.29
Soy protein concentrate Soycomil K	10.00
Cheddar whey powder	8.29
Whey protein concentrate80, DVN	7.00
Spray dried blood plasma P80	4.00
Dicalciumphosphate 18% P	0.32
DL-Methionine	0.31
Citric acid	0.30
L-Tryptophan	0.08
Vitamin and mineral premix[a]	1.8
Calculated nutrient levels	
NEv(1997) MJ/kg	15.48
CP, g/kg	249
CF, g/kg	110
dLYS, g/kg	18.2
dMET + CYS, g/kg	11.5
dTHR, g/kg	11.2
dTRY, g/kg	4.1

[a]The mineral and vitamin premix supplied as the following (per kg diet): Vitamin A, 30, 000 IU; Vitamin D_3, 5000 IU; Vitamin E; 75 mg; Fe^{2+}, 120 mg; Zn^{2+}, 35 mg; Cu^{2+}, 135 mg;Mn^{2+}, 45 mg; Se^{6+}, 350 µg; I^-, 1 mg, BHT, 75 mg/kg

Pigs were killed by exsanguination by severing the carotid arteries and jugular veins following induction of terminal anesthesia by intramuscular injection of ketamine (15 mg/kg BW) combined with xylazine (2 mg/kg BW). All piglets were weighed prior to euthanasia.

Blood was collected in EDTA and heparinized tubes containing supplemental bathophenanthroline disulfonate sodium salt. Erythrocytes were isolated by centrifuging (3000 × g, 15 min) 0.5 mL of unclotted, heparinized blood. After removing the plasma, erythrocytes were lysed by adding 100 μL of a 70% metaphosphoric acid solution, 600 μL milli Q and intense vortexing. These extracts were then centrifuged (3000 × g, 15 min), and 0.5 mL of the remaining acid extract was transferred to a vial containing 50 μL of a γ-glu-glu internal standard solution. After opening the abdomen, the liver was isolated and samples of the left lateral lobe were dissected for acid and phosphate buffered aqueous extraction, as described for the small intestinal mucosa. Subsequently, the small intestine (SI), defined as the part of the gastrointestinal tract between the pylorus and the ileocecal valve, was dissected and its length was measured. A 10 cm segment of proximal and distal SI (5 and 75% of total SI length, respectively) was taken for Ussing chamber measurements. In addition, 20 cm segments at 5 and 75% of the total SI length were emptied and carefully flushed with saline. The tissue of these 20 cm segments was placed on an ice-cold surface and the mucosa was retrieved by gently scraping the mucosal surface with a glass slide. Aliquots of the mucosa were either used instantaneously for acid and phosphate buffered aqueous extracts or transferred to plastic 2 mL screw-capped tubes, snap-frozen in liquid nitrogen and stored at -80 °C pending redox state analysis. Furthermore, 5 cm segments at 5 or 75% of the total SI length were taken, flushed with saline, snap-frozen in liquid nitrogen and stored at -80 °C pending protein expression analysis. Finally, a 5 cm segment at 5 or 75% of the total SI length was flushed with saline, divided in smaller pieces of max 1.5 cm in length and fixated for 2 h in 4% freshly prepared paraformaldehyde (in 0.01 mol/L phosphate-buffered saline) (volume tissue/volume fixative: 1/5) and routinely processed for paraffin-embedding [37].

Small intestinal histo-morphological measurements

In brief, 4 μm sections of paraffin-embedded samples were mounted on slides and stained with hematoxylin-eosin. Villus height, mid-villus width, and crypt depth were measured at 10× magnification using an Olympus BX61 microscope and image analysis software (analySIS Pro, Olympus Belgium, Aartselaar, Belgium) in 1–3 well-oriented villi and associated crypts in at least 12–15 sections per tissue sample, to yield 30 measurements per small intestinal region.

Small intestinal protein expression profile analysis

The concentration of specific tight junction proteins and markers for apoptosis and mitosis of the proximal and distal intestinal tissue samples was investigated using commercially available enzyme-linked immunosorbent assays (ELISA) of occludin (SEC228Hu), claudin-3 (SEF293Hu), proliferating cell nuclear antigen (PCNA) (SEA591Hu) and caspase-3 (SEA626Hu) (Cloud-Clone Corporation®, Houston, TX, USA). All tissue samples were crushed, dissolved in phosphate-buffered saline solution (PBS, pH 7.4, 0.01 mol/L), sonicated 6 times for 5 s at 4 °C (Sonics Vibracell™, VCX130, Newtown CT, USA), and kept on ice for 30 min. Subsequently the samples were centrifuged for 2 min at 13,400 rpm at 4 °C (Heraeus X3R with TX-750, Thermo Scientific, Rockford, USA), after which the supernatant was isolated, total protein concentration was determined using a Pierce TM BCA Protein Assay Kit (Thermo Scientific, Rockford, USA) and finally the samples were diluted to a total protein concentration of 10 ng/μL. Then, samples were processed on a sandwich ELISA plate and the experiment was performed according to the manufacturer's instructions. Absorbance was measured using an Infinite M200 Pro spectrophotometer with X-Fluor software at 450 nm at 25 °C (Tecan Group Ltd., Männedorf, Switzerland). Values of protein expression were determined per gram of total protein in a sample, measured using a Pierce™ BCA Protein Assay Kit (ThermoFisher Scientific, Belgium), and expressed as fmol/mg.

Ex vivo measurement of small intestinal permeability

Intestinal mucosal permeability was assessed ex vivo by measuring the translocation of macromolecular markers using the Ussing chamber technique. The segments were first rinsed with saline. The mucosal layer was stripped from the seromuscular layer and pinned onto 1.07 cm^2 sliders that were mounted into modified Ussing chambers (Dipl.-Ing. Mußler Scientific Instruments, Aachen, Germany). All tissues were mounted within 10 min following euthanasia. Tissues were immersed in 6.5 mL Ringer solution (115 mmol/L NaCl, 5 mmol/L KCl, 25 mmol/L $NaHCO_3$, 2.4 mmol/L Na_2HPO_4, 0.4 mmol/L NaH_2PO_4, 1.25 mmol/L $CaCl_2$, 1 mmol/L $MgSO_4$) with 6 mmol/L of mannitol or glucose in the luminal and serosal side, respectively. The system was water-jacketed to 37 °C and oxygenated with 95% O_2 and 5% CO_2. After an equilibration period of 20 min, 4 kDa fluorescein isothiocyanate-dextran (FD-4, Sigma-Aldrich, Bornem, Belgium) and 40 kDa horseradish peroxidase (HRP, type IV, Sigma-Aldrich, Bornem, Belgium) were added to the mucosal side to a final concentration of 0.8 mg/mL of FD-4 and 0.4 mg/mL of HRP. Samples of the buffer solution were taken from the serosal chamber at 20, 40, 60 and 100 min after adding markers. Meanwhile, the same volume of buffer was taken from the mucosal side to keep the volume balance across sides. Fluorescence intensity of FD-4 in the medium was measured at excitation wavelength of 485 nm and emission wavelength of 538 nm using a fluorescence plate reader (Thermo Scientific, Marietta, OH, USA). HRP was measured according to the method described previously [38]. In short, HRP activity was measured by adding a 'start solution' (50 mL of 0.2 mol/L NaH_2PO_4, 1 mL of 0.2 mol/L Na_2HPO_4, 20.4 μL of 30% H_2O_2, 1.7 mL of 1% dianisidine peroxide substrate, made up to 204 mL with water) to the HRP samples and left for 10 min, after which the reaction was terminated by addition of 120 μL of 4% sodium azide and the absorbance was read at 460 nm. The relation between peroxidase concentration and absorbance is linear in the concentration range 10–100 pmol/L. The apparent permeation coefficient (Papp) was calculated as:

Papp (cm/s): (dc/dt)×V/c0/A,Papp (cm/s): (dc/dt)×V/c0/A,

Whereby dc/dt is the change of serosal concentration in the 20- to 100-min period (cm/s); V is the volume of the chamber, c_0 is the initial marker concentration in the mucosal reservoirs and A the area of the exposed intestine in the chambers (cm^2).

Mucosal, liver and blood homogenate extracts and biochemical assays

An acid extract was prepared from 1 g of homogenized (Braun homogenizer at 900 rpm) intestinal mucosa or liver that was placed in 10 mL ice-cold perchloric acid (PCA) 10% solution and centrifuged at 15,000 × g for 15 min at 4 °C. The resulting acid extract (0.5 mL) was transferred to tubes containing 50 μL γ-glu-glu internal standard solution. Samples were snap frozen in liquid nitrogen and stored at -80 °C until analysis of GSH and glutathione disulfide (GSSG). The biuret reaction was applied to determine the total protein content. Mucosal GSH and GSSG were measured using a modified high performance liquid chromatography (HPLC) method [39, 40]. The derivation procedure included the reaction of 100 mmol/L iodoacetic acid solution with thiols to form S-carboxymethyl derivatives followed by chromophore derivation of primary amines with dinitrofluorobenzene (DNFB, 1% (v/v) in ethanol). GSH and GSSG were separated through EC250/4.6 Nucleosil 120-7 NH_2 aminopropyl column (Machery-Nagel, Düren, Germany) protected by the same NH_2 guard column (CC8/4). Chromatographic runs were performed at a flow-rate of 1.5 mL/min, starting at 80% solvent A/20% solvent B for 5 min followed by a 10 min linear gradient to 1% solvent A/99% solvent B and a 10 min isocratic period at 1% solvent A/99% solvent B (solvent A: water-methanol solution (1:4, v/v), solvent B: 0.5 mol/L sodium acetate–64% methanol). The column was re-equilibrated to the initial conditions for 15 min while maintaining the column temperature at 40 °C. The UV detector was set at 365 nm for absorption measurements. GSH and GSSG were identified by retention times of authentic standards. Concentrations were determined by using the internal and external standards and expressed as µmol/g protein. In addition, a phosphate buffered aqueous extract was made by mixing approximately 1 g of homogenized mucosa in 10 mL ice cold 1% Triton-X-100 phosphate buffer solution (pH = 7.0), by using an Ultra-Turrax dispensing machine (IKA-Werke GmbH & Co. KG, Staufen, Germany). The supernatant was transferred to 2 mL tubes, snap frozen and stored at -80 °C until analysis. Supernatants were used for the determination of GSH-Px activity and malondialdehyde (MDA; expressed as nmol/g protein) concentration. Assessment of GSH-Px activity (expressed as U/g protein) in EDTA plasma and mucosa was determined spectrophotometrically [41]. The thiobarbituric acid reactive substances (TBARS) method was used to measure MDA concentration in EDTA plasma, liver and mucosa extracts [42].

Statistical analysis

Linear mixed models were fitted to assess the influence of birth weight category (NBW/LBW), feeding (artificially reared/sow-reared) and days postnatal (as a categorical variable) on the quantitative outcome variables. To model the dependence between observations within the same litter, random intercept terms for litter were added to the model. Depending on the research question, separate analyses were carried out in subgroups (e.g. only in sow-reared piglets) or time points were analyzed separately. In the subgroup analyses where no random effect terms was needed, a multiple linear regression model was fitted. Post hoc tests to compare mean values between the different time points (days postnatal) were carried out using Tukey's honestly significant difference. Models were fitted using the Mixed Model procedure of the JMP Pro11 software (SAS Institute, Cary, NC, USA). Significance for the fixed effects was tested using an F-test with Kenward-Roger correction. Data are expressed as means and their standard errors (S.E.), and $P < 0.05$ was considered significant.

RESULTS

Body weight of piglets

Average body weight of LBW and NBW piglets sampled at birth (d0) was 0.77 ± 0.07 and 1.29 ± 0.08 kg, respectively (Figure 1). The body weight of both birth weight categories did not significantly change during the first 3 days. Afterwards, in both sow- and artificially reared piglets the body weight gradually increased ($P < 0.001$). At each point in time, LBW sow-reared piglets showed lower body weights compared to their NBW sow-reared littermates ($P < 0.01$). A similar observation was seen in artificially reared piglets at d5 ($P < 0.001$) and d19 ($P < 0.05$). Body weights of LBW and NBW artificially reared piglets were significantly higher, compared to respectively LBW and NBW sow-reared piglets at d19 ($P < 0.05$).

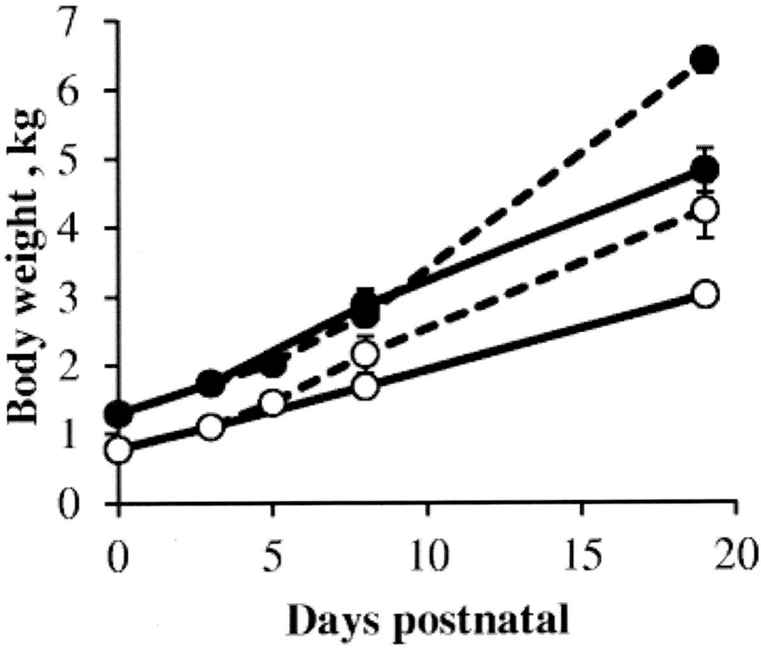

Figure 1. Body weights (kg) of LBW (open circle) and NBW (*closed circle*) from sow- (*full line*) and artificially reared (*dashed line*) piglets during the suckling period. Values are means ± SE ($n = 6$)

Histo-morphological measurements in the proximal and distal small intestine

The histo-morphological parameters were stable between d0 and d3 (Figure 2). After d3, villus height significantly decreased ($P < 0.05$; Figure 2a), whereas villus width ($P < 0.05$; Figure 2b) and crypt depth significantly increased ($P < 0.05$; Figure 2c). Furthermore, the transfer of piglets to a nursery caused villus atrophy in the proximal and distal ($P < 0.05$) SI from d3 to d5. Villi were significantly wider (at d19: $P < 0.001$) and crypts deeper (at d8 and d19: $P < 0.01$) in artificially reared piglets compared to sow-reared piglets. In the distal SI of NBW sow-reared piglets villus heights were consistently higher compared to those of their LBW littermates ($P < 0.05$). In contrast, villus heights in the distal SI of NBW artificially reared piglets were on average 193 µm lower at d19 compared to their LBW littermates ($P < 0.05$).

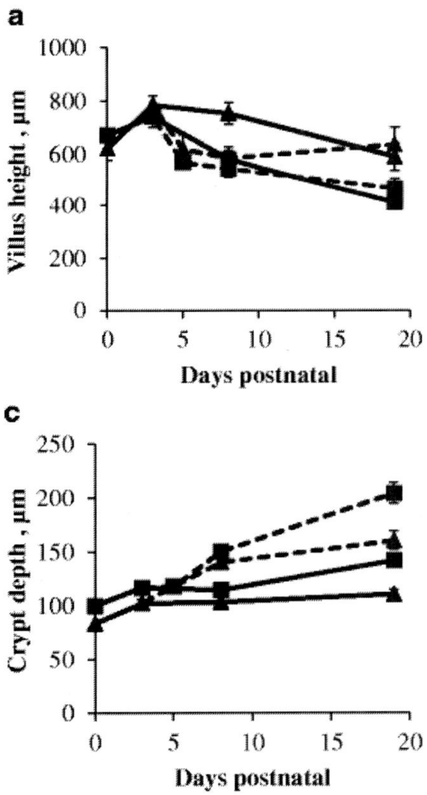

Figure 2. Villus height (µm) (**A**), villus width (µm) (**B**) and crypt depth (µm) (**C**) in proximal (*square*) and distal (*triangle*) SI of sow- (*full line*) and artificially reared (*dashed line*) piglets expressed as µm. Values are means ± SE ($n = 12$ as LBW and NBW piglets were pooled together)

PCNA and caspase-3 protein expression in the proximal and distal small intestine

PCNA protein expression remained at the same level during the suckling period in the proximal and distal SI of sow-reared piglets (Figure 3a). Transferring the piglets at d3 to a nursery significantly reduced PCNA protein expression in the proximal SI at d5 ($P < 0.05$) as compared to d3, whereas in the distal SI its level remained unaffected. PCNA protein expression in the proximal and distal SI of artificially reared piglets was significantly lower compared to sow-reared piglets at d8 ($P < 0.05$) and at d19 in the proximal SI ($P < 0.001$) (Figure 3a).

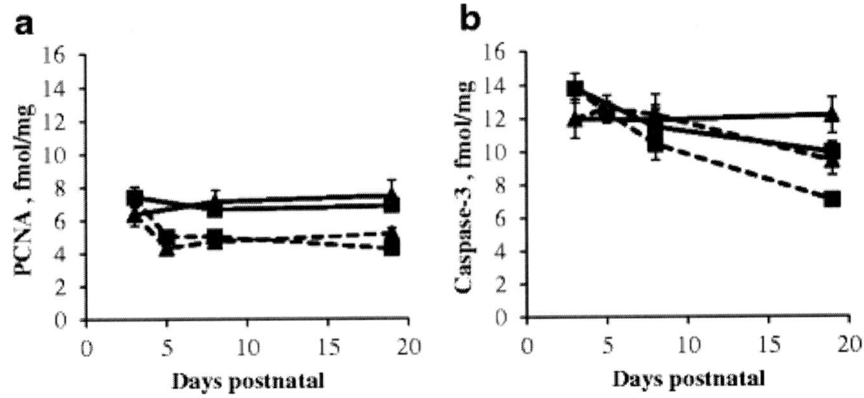

Figure 3. Relative protein expression of PCNA (fmol/mg) (**A**) and caspase-3 (fmol/mg) (**B**) in the proximal (*square*) and distal (*triangle*) SI of sow- (*full line*) and artificially reared (*dashed line*) piglets. Values are means ± SE ($n = 12$ as LBW and NBW piglets were pooled together)

Caspase-3 protein expression significantly decreased in the proximal SI of sow-reared ($P < 0.05$) and artificially reared piglets ($P < 0.001$) piglets from d3 to d19 (Figure 3b). In contrast, in the distal SI of sow- and artificially reared piglets, no age-related differences were observed. Caspase-3 protein expression was significantly lower in the proximal ($P < 0.05$) and distal ($P < 0.05$) SI of artificially reared piglets compared to sow-reared piglets at d19.

Occludin and claudin-3 protein expression in the proximal and distal small intestine

Occludin expression in the proximal SI of sow-reared piglets decreased from d3 to d19 ($P < 0.001$), whereas in the distal SI its level remained unchanged (Figure 4a). Occludin expression dropped significantly in the proximal ($P < 0.05$) but not in the distal SI of artificially reared piglets, from d3 to d5, but returned to its initial levels afterwards. The expression of occludin in the SI was comparable for sow- and artificially reared piglets.

Claudin-3 expression in the proximal and in the distal SI of artificially reared piglets dropped significantly from d3 to d5 ($P < 0.01$; Figure 4b). Claudin-3 expression increased from d5 to d19 and was significantly higher in the SI of artificially reared piglets compared to sow-reared piglets at d19 ($P < 0.01$). Claudin-3 expression in the distal SI of LBW sow-reared piglets was on average 33% higher compared to NBW sow-reared piglets at d3, d8 and d19 ($P < 0.01$).

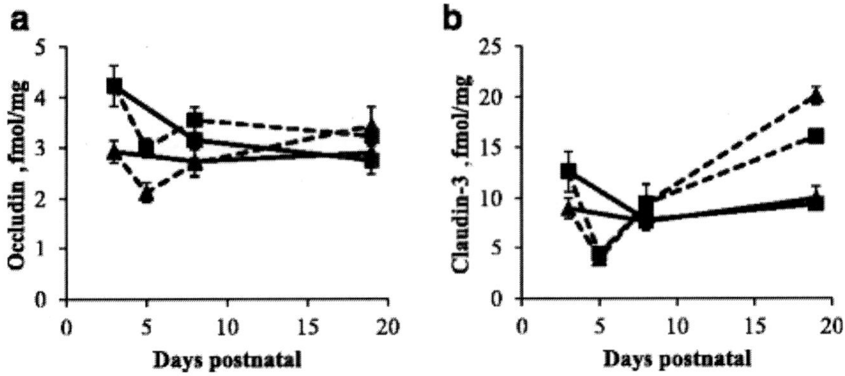

Figure 4. Relative protein expression of occludin (fmol/mg) (**A**) and claudin-3 (fmol/mg) (**B**) in the proximal (*square*) and distal (*triangle*) SI of sow- (*full line*) and artificially reared (*dashed line*) piglets. Values are means ± SE ($n = 12$ as LBW and NBW piglets were pooled together)

Ex vivo permeability in the proximal and distal small intestine

FD-4 and HRP permeability in the proximal and distal SI remained stable in sow-reared piglets (Figure 5a and b). FD-4 permeability in the proximal SI significantly increased in 5-day-old NBW artificially reared piglets when transferred to a nursery at d3 ($P < 0.05$; Figure 5a). Given the observation that the Papp of FD-4 was already high at 3 days of age in the proximal SI of LBW piglets, no changes could be observed after the start of artificial rearing. In the distal SI, both FD-4 ($P < 0.05$; Figure 5a) and HRP ($P < 0.01$; Figure 5b) permeabilities were significantly increased in both 5-day-old LBW and NBW artificially reared piglets, compared to d3. LBW sow-reared piglets consistently showed significantly higher FD-4 (proximal: on average 0.5×10^{-6} cm/s higher; distal: on average 0.2×10^{-6} cm/s higher) and HRP (proximal: on average 0.09×10^{-6} cm/s higher; distal: on average 0.06×10^{-6}cm/s higher) permeability when compared to their NBW littermates ($P < 0.05$). No such differences were observed in artificially reared piglets.

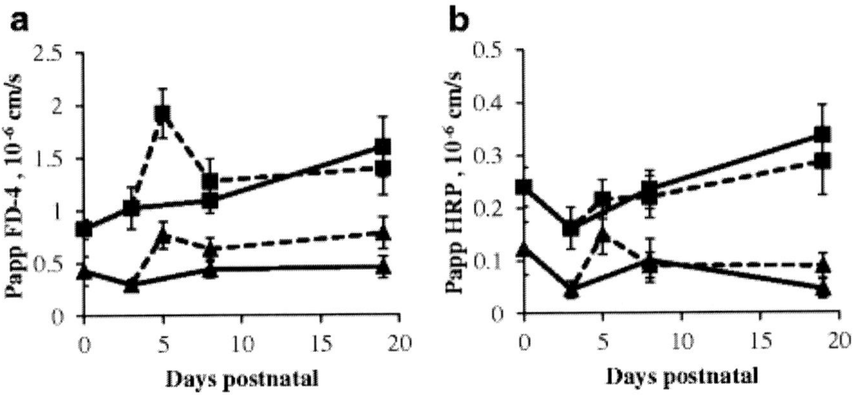

Figure 5. Intestinal permeability indicated by the Papp of FD-4 (10^{-6} cm/s) (**A**) and HRP (10^{-6} cm/s) (**B**) in the proximal (*square*) and distal SI (*triangle*) of sow- (*full line*) and artificially reared (*dashed line*) piglets. Values are means ± SE ($n = 6$)

Mucosal redox state represented by GSH concentration, GSSG concentration, GSSG/GSH ratio, GSH-Px activity and MDA concentration

GSH concentration was constant from birth until d8. In the distal SI mucosa GSH concentrations remained stable SI, whereas in the proximal SI it decreased significantly after d8 ($P < 0.01$) of sow-reared piglets (Figure 6a). The concentration of GSSG and the GSSG/GSH ratio did not show any age-related changes in sow-reared piglets ($P > 0.05$; Figure 6b and c). The activity of GSH-Px in the proximal and distal SI mucosae of sow-reared piglets increased from birth until d3 ($P < 0.05$). After d8, GSH-Px activity decreased significantly ($P < 0.01$), reaching a similar activity at d19 as observed at birth (Figure 6d).

In the proximal intestine of sow-reared pigs, the concentration of MDA abruptly dropped after birth ($P < 0.001$; Figure 6e) and remained stable from d3 until d19, whereas in the distal intestine the decrease was more spread out in time ($P < 0.01$). Transferring piglets to a nursery decreased the concentration of GSH, whereas the GSSG concentration and GSSG/GSH ratio were increased in the SI of artificially reared piglets at d5. In the proximal SI of artificially reared piglets GSH concentration peaked, while GSSG concentration and GSSG/GSH ratio showed a minimum at d8 ($P < 0.05$). However by d19, these redox parameters returned to the values noted at birth ($P < 0.05$). GSH-Px activity showed a minimum at d5 ($P < 0.05$), but recovered from d8 to d19 ($P < 0.01$) in the SI of artificially reared piglets.

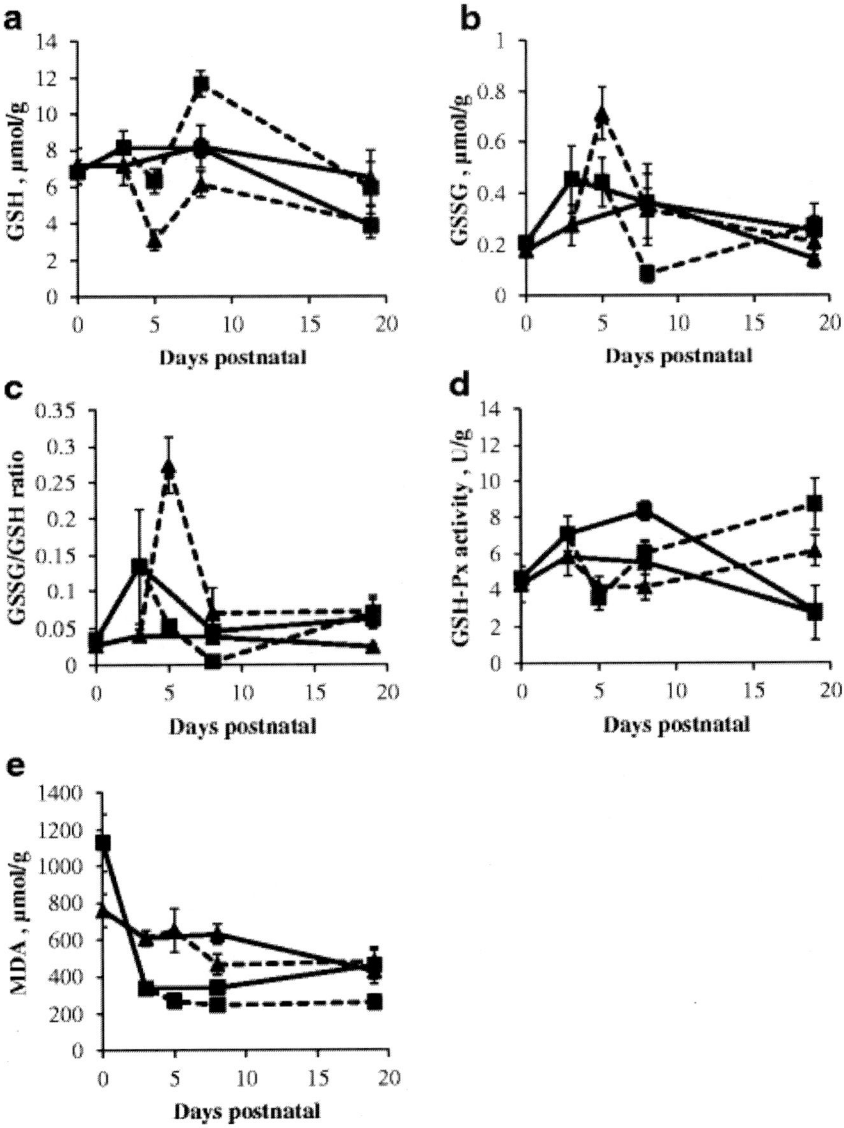

Figure 6. Mucosal GSH concentration (µmol/g) (**A**), GSSG concentration (µmol/g) (**B**), GSSG/GSH ratio (**C**) and GSH-Px activity (U/g) (**D**), MDA concentration (µmol/g) (**E**) of the proximal (*square*) and distal SI (*triangle*) in sow- (*full line*) and artificially reared (*dashed line*) piglets. Values are means ± SE (*n* = 12 as LBW and NBW piglets were pooled together)

In the proximal SI, GSH concentration was significantly higher ($P < 0.05$), while GSSG concentration ($P < 0.01$) and consequently the GSSG/GSH ratio ($P = 0.001$) was significantly lower in artificially reared piglets compared to sow-reared piglets at d8. GSH-Px activity was significantly lower at d8 ($P < 0.001$) and significantly higher at d19 ($P < 0.001$) in both regions of the SI of artificially reared piglets compared to sow-reared piglets. GSH-Px activity in the SI of LBW sow-reared piglets was significantly higher than their NBW littermates (proximal: on average 1.38 U/g higher; distal: on average 0.64 U/g higher) ($P < 0.001$). A similar observation for both birth weight categories was made in the distal SI of artificially reared piglets (on average 1.24 higher, $P < 0.01$).

At d8, MDA concentration in the proximal SI of sow-reared piglets was significantly higher compared to artificially reared piglets ($P < 0.01$).

Liver redox state represented by GSH concentration, GSSG concentration, GSSG/GSH ratio, GSH-Px activity and MDA concentration

Liver GSH concentration, GSSG/GSH ratio and GSH-Px activity in sow-reared piglets did not significantly change (Figure 7a, c and d). GSSG concentration decreased significantly from d0 to d3 ($P < 0.05$; Figure 7b) in sow-reared piglets and returned to concentrations seen at birth afterwards (Figure 7b). MDA concentration showed a significant increase after d3 in both LBW and NBW piglets ($P < 0.05$; Figure 7e).

When piglets were introduced to a milk replacer, GSH ($P < 0.05$) and GSSG ($P < 0.01$) concentrations significantly decreased whereas GSSG/GSH ratio ($P < 0.05$) and MDA concentration ($P < 0.001$) significantly increased from d3 to d5. Meanwhile, GSH-Px activity remained unchanged. All redox parameters in the liver were stable from d5 to d8 in artificially reared piglets, except GSSG/GSH ratio that showed a significant decrease ($P < 0.01$). Towards d19, GSH concentration ($P < 0.01$) and GSH-Px activity ($P < 0.05$) significantly increased while MDA concentration ($P < 0.01$) significantly decreased in artificially reared piglets.

At d8, a lower GSH concentration and a higher MDA were observed in artificially reared piglets compared to sow-reared piglets ($P < 0.001$). At d19, higher GSH-Px activities were observed in artificially reared piglets compared to sow-reared piglets ($P < 0.01$).

Systemic redox state represented by GSH concentration, GSSG concentration, and GSSG/GSH ratio, GSH-Px activity and MDA concentration

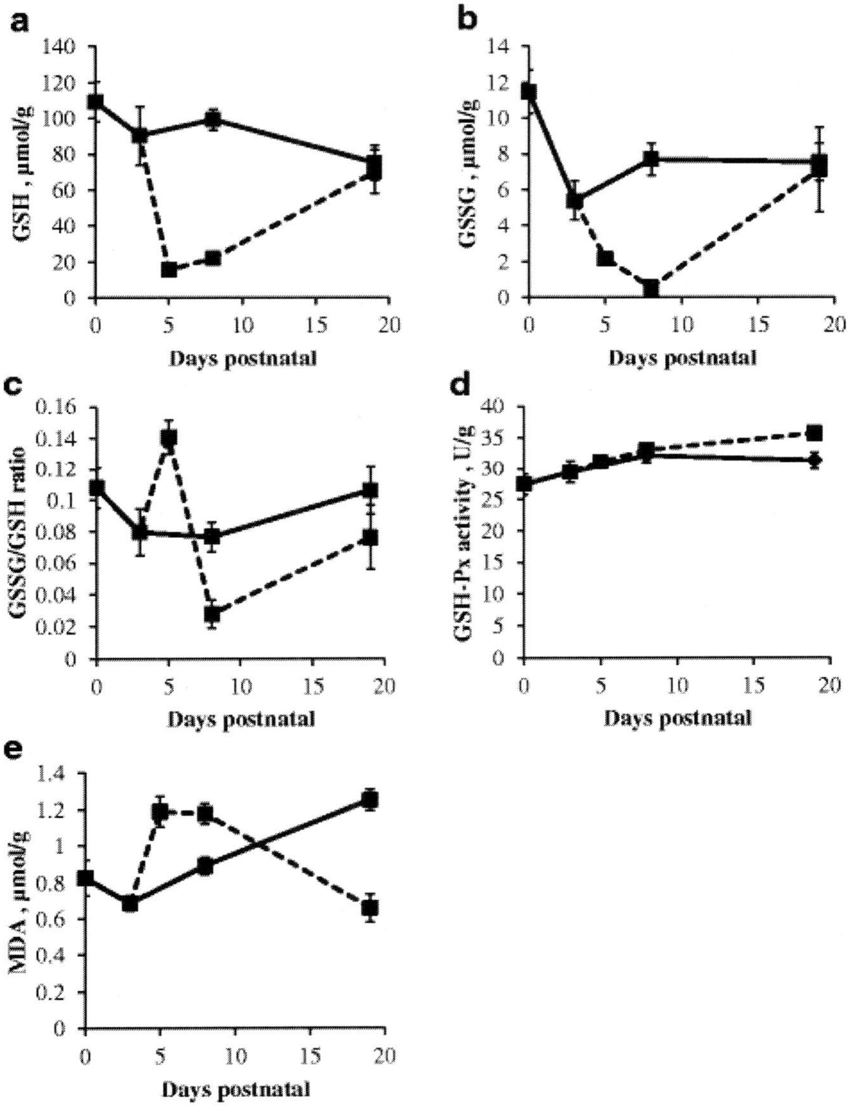

Figure 7. Liver GSH concentration (µmol/g) (**A**) and GSSG concentration (µmol/g) (**B**), GSSG/GSH ratio (**C**), GSH-Px activity (U/g) (**D**) and MDA (µmol/g) (**E**) in sow- (*full line*) and artificially reared (*dashed line*) piglets. Values are means ± SE ($n = 12$ as LBW and NBW piglets were pooled together)

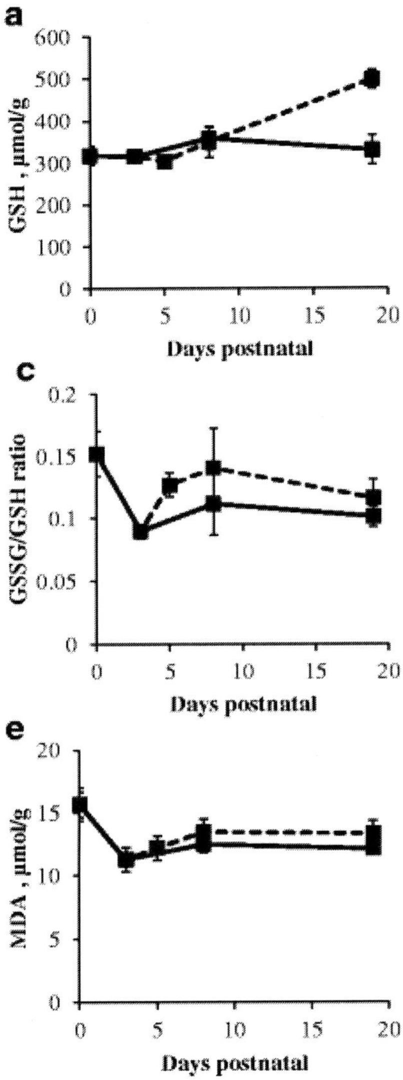

Figure 8. Systemic GSH concentration (µmol/g) (**A**), GSSG concentration (µmol/g) (**B**), GSSG/GSH ratio (**C**), GSH-Px activity (U/g) (**D**) and MDA concentration (µmol/g) (**E**) in sow- (*full line*) and artificially reared (*dashed line*) piglets. Values are means ± SE ($n = 12$ as LBW and NBW piglets were pooled together)

The GSH-Px activity gradually, significantly increased during the investigated time frame in both sow- and artificially reared piglets ($P < 0.01$; Figure 8d). GSH concentration in erythrocytes significantly increased in artificially reared piglets ($P < 0.01$) from d5 to d19, but remained constant in sow-reared piglets from d0 to d19 (Figure 8a). Plasma MDA concentration decreased 6.77 µmol/g during the first 3 days of life in LBW piglets ($P < 0.01$; Figure 8e). Plasma GSH-Px activity was significantly lower in artificially reared piglets compared to sow-reared piglets at d8 and d19 ($P < 0.05$). At d19, GSH-Px activity was still lower when LBW piglets were artificially reared ($P < 0.05$). Erythrocyte GSSG concentrations ($P < 0.01$; Figure 8b) and GSSG/GSH ratio ($P < 0.05$; Figure 8c) were significantly increased at d5 when artificially reared piglets were transferred to a nursery at d3 (Figure 8b and c). The concentration of GSH ($P < 0.01$) and GSSG ($P < 0.05$) significantly increased in artificially reared piglets from d5 to d19. At d19, GSH concentration and GSSG concentration were significantly higher in artificially reared when compared to sow-reared piglets ($P < 0.05$). GSH concentration was significantly lower in LBW sow-reared piglets when compared to their NBW littermates. ($P < 0.05$).

DISCUSSION

Given the need for an alternative rearing strategy that lowers the challenges that LBW and supernumerary piglets face during the suckling period, we aimed to investigate the responses to full artificial rearing. Our data demonstrated that artificial rearing beneficially affected piglet performance, notwithstanding impairing effects on small intestinal architecture, permeability and redox state in both LBW and NBW piglets.

Artificial rearing influences piglet performance

This study documents the implications of full artificial rearing of LBW and supernumerary piglets. Under standard rearing conditions these pigs are at high risk of succumbing due to insufficient nutrient intake, increased disease susceptibility, and physiological deficits (e.g. lower energy reserves) [1]. Our study demonstrated that transitioning piglets to a nursery with ad libitum access to a milk replacer led to significantly higher body weights of LBW and NBW piglets compared to sow-reared piglets at d19. The experiment was terminated on the same day as when the conventionally reared piglets were weaned on the farm with a 3-week batch system. For this specific farm, the average age at weaning is 19.6 d. Furthermore, milk production of the sow strongly decreases towards the end of the suckling phase. Around d 18–19, sow milk starts to be very limiting and becomes hard to compare to the ad libitum access for the artificially reared piglets [43]. Using a weigh-suckle-weigh technique, De Vos et al. [26] showed that piglets with ad libitum access to milk replacer have a higher relative energy intake compared to sow-reared piglets [28]. Our findings confirm this and other previous research where LBW piglets receiving an energy rich diet—comparable with our LBW piglets fed a milk replacer ad libitum—presented a comparable body weight gain as NBW piglets receiving a lower energy intake—comparable with our NBW piglets fed by the sow [44]. In addition, the milk replacer used in our study contained spray-dried plasma which could have contributed further to the higher weight gain in the artificially reared group. Ermer et al. [45] showed that spray-dried porcine plasma increased feed intake. Thus next to ad libitum access to feed, diet composition cannot be neglected.

Artificial rearing influences small intestinal architecture

Small intestinal morphology is one of the major indicators reflecting gut health in pigs [46]. However, caution should be taken when evaluating morphology alone as a measure of gut health. For example, Enterotoxigenic *Escherichia coli*, the major causal agent of neonatal diarrhea, may occur without histological changes in the intestine [47]. Notwithstanding, stereological analysis of small intestinal morphology will provide the most accurate estimation of the intestinal absorptive surface area [48], a proxy of the surface can be calculated using villus height and villus width [47]. In our study, the mucosal surface area of the proximal small intestine at d19 was markedly larger in artificially reared piglets than sow-reared piglets. In this regard, feeding a milk replacer shows promise as the increased mucosal surface suggests a higher ability to absorb nutrients. Moreover, previous research showed an increased activity of maltase and sucrase when piglets are fed a milk replacer [49, 50, 51]. Thus, artificial rearing seems to improve the digestive capacity at the level of the small intestine.

The deepening of the crypts could be a response to promote mucus secretion rather than lead to enterocyte maturation and proliferation and thus an increase in PCNA expression. Previous research showed that breast-fed infants showed a delay in the mucin degradation when compared to artificially reared infants [52]. Phillips [53] showed that crypt goblet cells have the ability to restitute the mucus layer and showed a decrease in the percentage of villus epithelial volume occupied by mucin secretory granules.

Transferring piglets to a nursery at d3 exposed them to stressful effects caused by psychological, environmental or nutritional factors similar to those encountered during the conventional weaning process [54]. However, it is difficult to unravel the separate contributions of these factors. Previous studies showed significant villus atrophy at d4 and deeper crypts at d7 in piglets that were separated from the sow and still fed sow's milk compared to unweaned piglets [55, 56]. Similarly, our study showed transient villus atrophy in piglets transferred to a nursery at d3. This villus height reduction is analogue to the intestinal morphological changes as a result of inadequate food intake immediately after conventional weaning [56, 57, 58, 59]. In contrast to conventional weaning, villus length is rapidly restored. This could be related to the inclusion of spray-dried plasma in the milk replacer since this is known to increase villus height [8, 60, 61, 62, 63].

Artificial rearing affects small intestinal tight junction protein expression

We hypothesized that artificial rearing influences small intestinal physiology. The intestinal epithelium plays a critical role in the transport of nutrients and macromolecules. At the same time, it has to provide an effective barrier to harmful macromolecules and microorganisms [64]. Epithelial cells constitute a dynamic barrier where large molecules can be transported by transcytosis and this can be measured by HRP [65]. Tight junctions (TJs) are essential components of the physical intercellular barrier and their presence and functionality changes under different physiological and pathological conditions [66, 67]. Well-formed TJs are characterized by low solute permeability which can be determined by measuring FD-4 permeability [65]. The family of junctional adhesion molecules, the claudin and occludin families, are structural transmembrane TJ components that have the potential to mediate cell–cell adhesion [66, 68]. Within TJs, claudins are the main determinants of the selective pore properties [68, 69], while the role of occludin in barrier functioning is more diverse [70, 71]. In mice, the expression of claudin-3 is promoted during the first 3 weeks of life concomitant with the establishment of the intestinal microbiota [72, 73]. Our study showed that claudin-3 expression rose particularly in artificially reared piglets. Possibly the different microbiota fingerprint [74] and the absence of milk born IgA's [34] can be held responsible. Claudin-3 is known to be a "tightening" claudin [75]. This could explain why permeability is seemingly unaffected by the observed increase in claudin-3. However, at the start of artificial rearing, claudin-3 as well as occludin protein levels transiently dropped. This drop is reflected in a concomitantly increased permeability for FD-4 and HRP. Previous studies showed lower abundances of occludin mRNA and protein, claudin-3 protein, and increased lactulose permeability after weaning [50, 76]. Thus artificial rearing induced a similar response as seen after conventional weaning.

Artificial rearing resulted in a redox imbalance

Previously, we investigated the link between oxidative stress, intestinal integrity, and permeability in intestinal epithelial cells in vitro [12] and in vivo during normal suckling [21]. Vergauwen et al. [12] and others showed a redistribution of TJ proteins during times of imposing reactive species and could relate these responses to a compromised permeability [12, 77]. The current study demonstrated that transfer of piglets to a nursery resulted in oxidative stress.

Glutathione (GSH) is an important regulator of the redox status within intestinal epithelial cells [13]. The liver is the major site of GSH biosynthesis and exports GSH via the bile to the proximal SI [78]. Thus GSH originating from the liver supports mucosal GSH by decreasing lipid peroxidation and maintaining the GSSG/GSH redox homeostasis in the proximal intestine [11, 79, 80]. In our study, artificial rearing resulted in the liver in a decreased GSH content, an increased GSSG/GSH ratio and MDA concentration. After this first phase, the redox parameters returned to their initial values. GSH-Px activity in the liver seemed unaffected during the transition period, while GSH-Px activity transiently dropped in the proximal and distal SI when piglets were introduced to a nursery. Previously, conventional weaning caused a drop in GSH-Px activity but GSH-Px activity increased afterwards as part of a feedback mechanism [31]. Our study shows that artificial rearing increases oxidative stress in the SI and as a result the GSH-Px activity increased and was higher than the activity seen in sow-reared piglets. Furthermore, LBW piglets showed a higher GSH-Px activity compared to their NBW littermates. It is clear that the antioxidant capacity of the liver helps protecting the proximal intestine by secreting GSH into the lumen of proximal SI [78, 79]. This could explain the increased mucosal GSH concentration of the proximal SI and the concomitant massive decrease of the liver GSH concentration. On the other hand, a remarkable drop of the GSH concentration in the distal SI was observed from d3 until d5. Consequently, this resulted in a high concentration of MDA and GSSG, resulting in a higher GSSG/GSH ratio in the distal SI. Furthermore, the GSSG/GSH ratio increased systemically upon transfer to a nursery. Degroote et al. [31] already showed that conventional weaning increased the GSSG/GSH ratio.

Our research presents a window of opportunity for antioxidant supplementation to protect piglets from redox imbalance due to artificial rearing. NBW artificially reared piglets were more susceptible to redox imbalance and loss of intestinal integrity upon transferal to a nursery (from d3 to d5). Taken together, these findings favor the transfer of both LBW and NBW piglets to a nursery as a solution for LBW and supernumerary piglets.

Further research is necessary to elucidate how these artificially reared LBW and NBW piglets will respond to the introduction to solid food.

CONCLUSIONS

In conclusion, we demonstrated that artificial rearing influences morphological and functional parameters in the small intestine, liver and blood in a way similar to what is seen after conventional weaning. Nevertheless, growth performance of artificially reared piglets was positively influenced. In addition, artificially reared piglets rapidly recovered from redox imbalances and restored intestinal permeability within a couple of days. Further research is needed to explore the possibility to supplement LBW and NBW piglets with antioxidants prior to initiating artificial rearing. Thus, artificial rearing is a valuable alternative to raise LBW or supernumerary piglets.

ACKNOWLEDGMENTS

We thank K. Huybrechts, G. Vrolix, K. Jennes, M. De Reys, S. Coolsaet, A. Ovyn and T. Van der Eecken for their technical assistance.

Funding

This work was supported by a grant from the government agency for Innovation by Science and Technology (IWT-LO 100856).

Availability of data and materials

The datasets supporting the conclusions of this article are included within the article.

Authors' contributions

Author contributions: CVG, JM and SDM concept and design of research; HV, JDG, WW, CVG and JM performed experiments; HV and JDG analyzed the data; HV and EF design and implementation of the statistical model; HV, JDG, CVG and JM interpreted the results of the experiments; HV and CVG prepared the figures; HV and CVG drafted manuscript; HV, JDG, CVG, JM, SDS, WW, CC, SVC and SP edited and revised the manuscript; CVG and JM approved the final version of the manuscript. All authors read and approved the final manuscript.

Competing interests

The authors declare that they have no competing interests.

Consent for publication

Not applicable.

Ethics approval

All husbandry and experimental procedures were approved by the Ethical Committee for Animal Testing (ECD) of the University of Antwerp (Belgium) (EC number: 2015-02) in accordance with the European Commission directive for the humane care and use of animals in research (2010/63/EU).

REFERENCES

1. Lay Jr DC, Matteri RL, Carrol JA, Fangman TJ, Safranski TJ. Preweaning survival in swine. J Anim Sci. 2002;80(E. Suppl. 1):E74–86.
2. Tuchscherer M, Puppe B, Tuchscherer A, Tiemann U. Early identification of neonates at risk: traits of newborn piglets with respect to survival. Theriogenology. 2000;54(3):371–88.
3. Milligan BN, Dewey CE, de Grau AF. Neonatal-piglet weight variation and its relation to pre-weaning mortality and weight gain on commercial farms. Prev Vet Med. 2002;56(2):119–27.
4. Friel JK, Diehl-Jones B, Cockell KA, Chiu A, Rabanni R, Davies SS, et al. Evidence of oxidative stress in relation to feeding type during early life in premature infants. Pediatr Res. 2011;69(2):160–4.
5. Ahola T, Levonen AL, Fellman V, Lapatto R. Thiol metabolism in preterm infants during the first week of life. Scand J Clin Lab Invest. 2004;64(7):649–58.

6. Rook D, Te Braake FW, Schierbeek H, Longini M, Buonocore G, Van Goudoever JB. Glutathione synthesis rates in early postnatal life. Pediatr Res. 2010;67(4):407–11.
7. Yin J, Ren W, Liu G, Duan J, Yang G, Wu L, et al. Birth oxidative stress and the development of an antioxidant system in newborn piglets. Free Radic Res. 2013;47(12):1027–35.
8. Tran H, Bundy JW, Li YS, Carney-Hinkle EE, Miller PS, Burkey TE. Effects of spray-dried porcine plasma on growth performance, immune response, total antioxidant capacity, and gut morphology of nursery pigs. J Anim Sci. 2014;92(10):4494–504.
9. Michiels J, De Vos M, Missotten J, Ovyn A, De Smet S, Van Ginneken C. Maturation of digestive function is retarded and plasma antioxidant capacity lowered in fully weaned low birth weight piglets. Br J Nutr. 2013;109(1):65–75.
10. Krueger R, Derno M, Goers S, Metzler-Zebeli BU, Nuernberg G, Martens K, et al. Higher body fatness in intrauterine growth retarded juvenile pigs is associated with lower fat and higher carbohydrate oxidation during ad libitum and restricted feeding. Eur J Nutr. 2014;53(2):583–97.
11. Aw TY. Intestinal glutathione: determinant of mucosal peroxide transport, metabolism, and oxidative susceptibility. Toxicol Appl Pharmacol. 2005;204(3):320–8.
12. Vergauwen H, Tambuyzer B, Jennes K, Degroote J, Wang W, De Smet S, et al. Trolox and ascorbic acid reduce direct and indirect oxidative stress in the IPEC-J2 cells, an in vitro model for the porcine gastrointestinal tract. PLoS One. 2015;10(3):e0120485.
13. Kelly FJ. Glutathione content of the small intestine: regulation and function. Br J Nutr. 1993;69(2):589–96
14. Carrasco-Pozo C, Morales P, Gotteland M. Polyphenols Protect the Epithelial Barrier Function of Caco-2 Cells Exposed to Indomethacin through the Modulation of Occludin and Zonula Occludens-1 Expression. Journal of agricultural and food chemistry. 2013.
15. Bhattacharyya A, Chattopadhyay R, Mitra S, Crowe SE. Oxidative stress: an essential factor in the pathogenesis of gastrointestinal mucosal diseases. Physiol Rev. 2014;94(2):329–54.
16. Wang J, Chen L, Li D, Yin Y, Wang X, Li P, et al. Intrauterine growth restriction affects the proteomes of the small intestine, liver, and skeletal muscle in newborn pigs. J Nutr. 2008;138(1):60–6.
17. Wang X, Wu W, Lin G, Li D, Wu G, Wang J. Temporal proteomic analysis reveals continuous impairment of intestinal development in neonatal piglets with intrauterine growth restriction. J Proteome Res. 2010;9(2):924–35.
18. Mickiewicz M, Zabielski R, Grenier B, Le Normand L, Savary G, Holst JJ, et al. Structural and functional development of small intestine in intrauterine growth retarded porcine offspring born to gilts fed diets with differing protein ratios throughout pregnancy. J Physiol Pharmacol. 2012;63(3):225–39.
19. Ferenc K, Pietrzak P, Godlewski MM, Piwowarski J, Kilianczyk R, Guilloteau P, et al. Intrauterine growth retarded piglet as a model for humans--studies on the perinatal development of the gut structure and function. Reprod Biol. 2014;14(1):51–60.
20. D'Inca R, Gras-Le Guen C, Che L, Sangild PT, Le Huerou-Luron I. Intrauterine growth restriction delays feeding-induced gut adaptation in term newborn pigs. Neonatology. 2011;99(3):208–16.
21. Wang W, Degroote J, Van Ginneken C, Van Poucke M, Vergauwen H, Dam TM et al. Intrauterine growth restriction in neonatal piglets affects small intestinal mucosal permeability and mRNA expression of redox-sensitive genes. FASEB journal : official publication of the Federation of American Societies for Experimental Biology. 2015.
22. Huygelen V, De Vos M, Willemen S, Tambuyzer B, Casteleyn C, Knapen D, et al. Increased intestinal barrier function in the small intestine of formula-fed neonatal piglets. J Anim Sci. 2012;90 Suppl 4:315–7.

23. Huygelen V, De Vos M, Willemen S, Fransen E, Casteleyn C, Van Cruchten S, et al. Age-related differences in mucosal barrier function and morphology of the small intestine in low and normal birth weight piglets. J Anim Sci. 2014;92(8):3398–406.
24. Willemen S, Che L, De Vos M, Huygelen V, Tambuyzer B, Casteleyn C, et al. Perinatal growth restriction is not related to higher intestinal distribution and increased serum levels of 5-hydroxytryptamin in piglets. J Anim Sci. 2012;90 Suppl 4:305–7.
25. Ferrari CV, Sbardella PE, Bernardi ML, Coutinho ML, Vaz Jr IS, Wentz I, et al. Effect of birth weight and colostrum intake on mortality and performance of piglets after cross-fostering in sows of different parities. Prev Vet Med. 2014;114(3-4):259–66.
26. De Vos M, Che L, Huygelen V, Willemen S, Michiels J, Van Cruchten S, et al. Nutritional interventions to prevent and rear low-birthweight piglets. J Anim Physiol Anim Nutr. 2014;98(4):609–19.
27. Donovan TS, Dritz SS. Effect of split nursing on variation in pig growth from birth to weaning. J Am Vet Med Assoc. 2000;217(1):79–81.
28. De Vos M, Huygelen V, Willemen S, Fransen E, Casteleyn C, Van Cruchten S, et al. Artificial rearing of piglets: Effects on small intestinal morphology and digestion capacity. Livest Sci. 2014;159:165–73.
29. Wedig J, Christian MS, Hoberman A, Diener RM, Thomas-Wedig R. A study to develop methodology for feeding 24-hour-old neonatal swine for 3 weeks. Int J Toxicol. 2002;21(5):361–70.
30. Fiorotto ML, Reeds PJ, Cunningham JJ, Pond WG. A semiautomatic device for feeding liquid milk-replacer diets to infant pigs. J Anim Sci. 1993;71(1):78–85.
31. Degroote J, Michiels J, Claeys E, Ovyn A, De Smet S. Changes in the pig small intestinal mucosal glutathione kinetics after weaning. J Anim Sci. 2012;90 Suppl 4:359–61.
32. Zhu LH, Zhao KL, Chen XL, Xu JX. Impact of weaning and an antioxidant blend on intestinal barrier function and antioxidant status in pigs. J Anim Sci. 2012;90(8):2581–9.
33. Sauerwein H, Schmitz S, Hiss S. The acute phase protein haptoglobin and its relation to oxidative status in piglets undergoing weaning-induced stress. Redox Report. 2005;10(6):295–302.
34. Klobasa F, Werhahn E, Butler JE. Composition of sow milk during lactation. J Anim Sci. 1987;64(5):1458–66.
35. Csapo J, Martin TG, Csapo-Kiss ZS, Hazas Z. Protein, fats, vitamin and mineral concentration in porcine colostrum and milk from parturition to 60 days. Int Dairy J. 1996;6(8-9):881–902.
36. Ontsouka CE, Bruckmaier RM, Blum JW. Fractionized milk composition during removal of colostrum and mature milk. J Dairy Sci. 2003;86(6):2005–11.
37. Buesa RJ, Peshkov MV. How much formalin is enough to fix tissues? Ann Diagn Pathol. 2012;16(3):202–9.
38. McKie AT, Zammit PS, Naftalin RJ. Comparison of cattle and sheep colonic permeabilities to horseradish peroxidase and hamster scrapie prion protein in vitro. Gut. 1999;45(6):879–88.
39. Yoshida T. Determination of reduced and oxidized glutathione in erythrocytes by high-performance liquid chromatography with ultraviolet absorbance detection. J Chromatogr B Biomed Appl. 1996;678(2):157–64.
40. Reed DJ, Babson JR, Beatty PW, Brodie AE, Ellis WW, Potter DW. High-performance liquid chromatography analysis of nanomole levels of glutathione, glutathione disulfide, and related thiols and disulfides. Anal Biochem. 1980;106(1):55–62.
41. Hernandez P, Zomeno L, Arino B, Blasco A. Antioxidant, lipolytic and proteolytic enzyme activities in pork meat from different genotypes. Meat Sci. 2004;66(3):525–9.
42. Grotto D, Santa Maria LD, Boeira S, Valentini J, Charao MF, Moro AM, et al. Rapid quantification of malondialdehyde in plasma by high performance liquid chromatography-visible detection. J Pharm Biomed Anal. 2007;43(2):619–24.

43. Aguinaga MA, Gomez-Carballar F, Nieto R, Aguilera JF. Production and composition of Iberian sow's milk and use of milk nutrients by the suckling Iberian piglet. Animal. 2011;5(9):1390–7.
44. Han F, Hu L, Xuan Y, Ding X, Luo Y, Bai S, et al. Effects of high nutrient intake on the growth performance, intestinal morphology and immune function of neonatal intra-uterine growth-retarded pigs. Br J Nutr. 2013;110(10):1819–27.
45. Ermer PM, Miller PS, Lewis AJ. Diet preference and meal patterns of weanling pigs offered diets containing either spray-dried porcine plasma or dried skim milk. J Anim Sci. 1994;72(6):1548–54.
46. Caspary WF. Physiology and pathophysiology of intestinal absorption. Am J Clin Nutr. 1992;55(1 Suppl):299s–308s.
47. Egberts HJ, de Groot EC, van Dijk JE, Vellenga L, Mouwen JM. Tight junctional structure and permeability of porcine jejunum after enterotoxic Escherichia coli infection. Res Vet Sci. 1993;55(1):10–4.
48. Van Ginneken C, Van Meir F, Weyns A. Stereologic characteristics of pig small intestine during normal development. Dig Dis Sci. 2002;47(4):868–78.
49. Cera KR, Mahan DC, Reinhart GA. Effect of weaning, week postweaning and diet composition on pancreatic and small intestinal luminal lipase response in young swine. Journal of animal science. 1990;68(2).
50. Wang J, Zeng L, Tan B, Li G, Huang B, Xiong X, et al. Developmental changes in intercellular junctions and Kv channels in the intestine of piglets during the suckling and post-weaning periods. J Anim Sci Biotechnol. 2016;7:4.
51. Huygelen V, De Vos M, Prims S, Vergauwen H, Fransen E, Casteleyn C, et al. Birth weight has no influence on the morphology, digestive capacity and motility of the small intestine in suckling pigs. Livest Sci. 2015;182:129–36
52. Midtvedt AC, Carlstedt-Duke B, Midtvedt T. Establishment of a mucin-degrading intestinal microflora during the first two years of human life. J Pediatr Gastroenterol Nutr. 1994;18(3):321–6.
53. Phillips TE. Both crypt and villus intestinal goblet cells secrete mucin in response to cholinergic stimulation. Am J Physiol. 1992;262(2 Pt 1):G327–31.
54. Funderburke DW, Seerley RW. The effects of postweaning stressors on pig weight change, blood, liver and digestive tract characteristics. J Anim Sci. 1990;68(1):155–62.
55. van Beers-Schreurs HM, Nabuurs MJ, Vellenga L, Kalsbeek-van der Valk HJ, Wensing T, Breukink HJ. Weaning and the weanling diet influence the villous height and crypt depth in the small intestine of pigs and alter the concentrations of short-chain fatty acids in the large intestine and blood. J Nutr. 1998;128(6):947–53
56. Kelly D, Smyth JA, McCracken KJ. Digestive development of the early-weaned pig. 1. Effect of continuous nutrient supply on the development of the digestive tract and on changes in digestive enzyme activity during the first week post-weaning. Br J Nutr. 1991;65(2):169–80.
57. Kelly D, Smyth JA, McCracken KJ. Digestive development of the early-weaned pig. 2. Effect of level of food intake on digestive enzyme activity during the immediate post-weaning period. Br J Nutr. 1991;65(2):181–8.
58. Spreeuwenberg MA, Verdonk JM, Gaskins HR, Verstegen MW. Small intestine epithelial barrier function is compromised in pigs with low feed intake at weaning. J Nutr. 2001;131(5):1520–7.
59. Hedemann MS, Hojsgaard S, Jensen BB. Small intestinal morphology and activity of intestinal peptidases in piglets around weaning. J Anim Physiol Anim Nutr. 2003;87(1-2):32–41.
60. Pierce JL, Cromwell GL, Lindemann MD, Russell LE, Weaver EM. Effects of spray-dried animal plasma and immunoglobulins on performance of early weaned pigs. J Anim Sci. 2005;83(12):2876–85.
61. Touchette KJ, Allee GL, Matteri RL, Dyer CJ, Carroll JA. Effect of spray-dried plasma and Escherichia coli on intestinal morphology and the hypothalamic–pituitary–adrenal (HPA) axis of the weaned pig. J Anim Sci. 1999;77(Supplement 1):56.

62. Jiang R, Chang X, Stoll B, Fan MZ, Arthington J, Weaver E, et al. Dietary plasma protein reduces small intestinal growth and lamina propria cell density in early weaned pigs. J Nutr. 2000;130(1):21–6.
63. van Dijk AJ, Niewold TA, Margry RJ, van den Hoven SG, Nabuurs MJ, Stockhofe-Zurwieden N, et al. Small intestinal morphology in weaned piglets fed a diet containing spray-dried porcine plasma. Res Vet Sci. 2001;71(1):17–22.
64. Ramanan D, Cadwell K. Intrinsic Defense Mechanisms of the Intestinal Epithelium. Cell host & microbe. 2016.
65. Boudry G. The Ussing chamber technique to evaluate alternatives to in-feed antibiotics for young pigs. Animal Res. 2005;54:219–30.
66. Schneeberger EE, Lynch RD. The tight junction: a multifunctional complex. Am J Physiol Cell Physiol. 2004;286(6):C1213–28.
67. Lu Z, Ding L, Lu Q, Chen Y-H. Claudins in intestines: distribution and functional significance in health and diseases. Tissue Barriers. 2013;1(3):e24978.
68. Anderson JM, Van Itallie CM, Fanning AS. Setting up a selective barrier at the apical junction complex. Curr Opin Cell Biol. 2004;16(2):140–5.
69. Furuse M, Tsukita S. Claudins in occluding junctions of humans and flies. Trends Cell Biol. 2006;16(4):181–8.
70. Yu AS, McCarthy KM, Francis SA, McCormack JM, Lai J, Rogers RA, et al. Knockdown of occludin expression leads to diverse phenotypic alterations in epithelial cells. Am J Physiol Cell Physiol. 2005;288(6):C1231–41.
71. Raleigh DR, Boe DM, Yu D, Weber CR, Marchiando AM, Bradford EM, et al. Occludin S408 phosphorylation regulates tight junction protein interactions and barrier function. J Cell Biol. 2011;193(3):565–82.
72. Patel RM, Myers LS, Kurundkar AR, Maheshwari A, Nusrat A, Lin PW. Probiotic bacteria induce maturation of intestinal claudin 3 expression and barrier function. Am J Pathol. 2012;180(2):626–35.
73. Holmes JL, Van Itallie CM, Rasmussen JE, Anderson JM. Claudin profiling in the mouse during postnatal intestinal development and along the gastrointestinal tract reveals complex expression patterns. Gene Expression Patterns. 2006;6(6):581–8.
74. Mackie RI, Sghir A, Gaskins HR. Developmental microbial ecology of the neonatal gastrointestinal tract. Am J Clin Nutr. 1999;69(5):1035s–45s.
75. Milatz S, Krug SM, Rosenthal R, Gunzel D, Muller D, Schulzke JD, et al. Claudin-3 acts as a sealing component of the tight junction for ions of either charge and uncharged solutes. Biochim Biophys Acta. 2010;1798(11):2048–57.
76. Wang H, Zhang C, Wu G, Sun Y, Wang B, He B, et al. Glutamine enhances tight junction protein expression and modulates corticotropin-releasing factor signaling in the jejunum of weanling piglets. J Nutr. 2015;145(1):25–31.
77. Fukui A, Naito Y, Handa O, Kugai M, Tsuji T, Yoriki H et al. Acetyl salicylic acid induces damage to intestinal epithelial cells by oxidation-related modifications of ZO-1. American journal of physiology Gastrointestinal and liver physiology. 2012.
78. Lee TK, Li L, Ballatori N. Hepatic glutathione and glutathione S-conjugate transport mechanisms. Yale J Biol Med. 1997;70(4):287–300.
79. Aw TY. Biliary glutathione promotes the mucosal metabolism of luminal peroxidized lipids by rat small intestine in vivo. J Clin Investig. 1994;94(3):1218–25.
80. Ballatori N, Truong AT. Relation between biliary glutathione excretion and bile acid-independent bile flow. Am J Physiol. 1989;256(1 Pt 1):G22–30.

CHAPTER 9

Association of Growth Rate with Hormone Levels and Myogenic Gene Expression Profile in Broilers

Yingping Xiao[1] , Choufei Wu[2] , Kaifeng Li[1] , Guohong Gui[1] , Guolong Zhang[3] and Hua Yang[1]

[1] *Institute of Quality and Standards for Agro-products, Zhejiang Academy of Agricultural Sciences, Hangzhou 310021, China.*

[2] *College of Life Sciences, Huzhou University, Huzhou 313000, China.*

[3] *Department of Animal Science, Oklahoma State University, Stillwater, Oklahoma 74078, USA*

ABSTRACT

Background

The growth rate often varies among individual broilers of the same breed under a common management condition. To investigate whether a variation in the growth rate is associated with a difference in hormone levels and myogenic gene expression profile in broilers, a feeding trial was conducted with 10,000 newly hatched Ross 308 chicks in a commercial production facility under standard management. At 38 d of age, 30 fast-, 30 medium-, and 30 slow-growing broilers were selected among 600 healthy male individuals. The levels of insulin-like growth factor-1 (IGF-1), triiodothyronine (T3), thyroxine (T4), and growth hormone in the serum or breast muscle were assayed by ELISA or RIA kits, and the expression levels of several representative pro- and anti-myogenic genes in the breast muscle were also measured by real-time PCR.

Results

Results showed that both absolute and relative weights of the breast muscle were in linear positive correlations with the body weight of broilers ($P < 0.001$). Fast-growing broilers had higher concentrations of IGF-1 than slow-growing broilers ($P < 0.05$) in both the serum and breast muscle. The serum concentration of T3 was significantly higher in fast-growing birds than in slow-growing birds ($P < 0.05$). However, no difference was observed in growth hormone or T4 concentration among three groups of birds. Additionally, a decreased expression of an anti-myogenic gene (*myostatin*) and increased expressions of pro-myogenic genes such as *myogenic differentiation factor 1, myogenin, muscle regulatory factor 4, myogenic factor 5, IGF-1,* and *myocyte enhancer factor 2B, C,* and *D* were observed in fast-growing broilers ($P < 0.05$), relative to slow-growing broilers.

Conclusions

Collectively, these findings suggested that the growth rate is linked to the hormone and myogenic gene expression levels in broiler chickens. Some of these parameters such as serum concentrations of IGF-1 and T3 could be employed to breed for enhanced growth.

Keywords: Breast muscle Broiler Growth performance Hormone Myogenic gene expression

BACKGROUND

Muscle growth, also known as myogenesis, is a complicated but precisely regulated developmental process. Myogenic regulatory factors (MRF) and myocyte enhancer factor 2 (MEF2) proteins are key transcription factors that are positively involved in the regulation of skeletal muscle development [1]. The MRF family of transcription factors is comprised of a group of basic helix-loop-helic proteins such as myogenic differentiation factor 1 (MyoD1), myogenic factor 5 (Myf5), myogenin, and MRF4, whereas the MEF2 family consists of four members denoted as MEF2A, B, C, and D. On the other hand, myostatin (MSTN), a member of the transforming growth factor β (TGF-β) superfamily secreted from skeletal muscle, acts as a potent negative regulator of muscle differentiation and growth [2]. MSTN modifies the muscle fiber-type composition by regulating MyoD and MEF2 expression [3]. Inhibition of MSTN causes myofibre hypertrophy [4], while

MSTN over-expression decreases the skeletal muscle mass and fiber size [5].

Additionally, a number of hormones are known to impact on animal growth. The major hormones required to support normal growth in chickens are growth hormone (GH), 3,5,3′-triiodothyronine (T3), thyroxine (T4), and insulin-like growth factor-1 (IGF-1) [6]. IGF-1 have been shown to stimulate the growth of the skeletal muscle by enhancing the rate of protein synthesis and therefore, the concentration of IGF-1 is often positively correlated with the body weight (BW) in broiler chickens [6, 7, 8].

It is well known that the growth rate of individual broilers of the same breed exhibits a normal Gaussian distribution under a common management condition. However, it remains elusive about whether there is a difference between fast- and slow-growing broilers in growth-related hormone and myogenic gene expression levels. The objective of the present study was to investigate the association of the growth rate with the concentrations of growth-related hormones in the circulation as well as both pro- and anti-myogenic gene expression levels in the breast muscle of broilers.

METHODS

Animals

A flock of 10,000 day-of-hatch Ross 308 broiler chicks were raised in a commercial production facility under standard management, with a 24-h photoperiod and 32-34 °C in the first two days, followed by a 16-h photoperiod and a reduction by 2-3 °C per week to a final temperature of 20 °C. The broilers were allowed ad libitum access to water and commercial mash feed. At 38 d of age, 600 healthy male broilers were randomly picked and weighed, from which 30 broilers of the highest, medium, and lowest BW were selected as H, M, and L groups, respectively.

Sampling

Ten birds were randomly selected from each group. Blood (4 mL each) was collected from the wing vein and centrifuged at 3,000 × *g* and 4 °C for 10 min to obtain serum, which was stored at -80 °C for further analysis. Chickens were then killed by cervical dislocation. The entire breast muscle (including pectoralis major and minor) was weighed after removal. Approximately 3 g of each muscle sample was immediately snap frozen in liquid nitrogen and stored at -80 °C for further analysis as described below.

Tissue preparation and hormone concentration measurement

The breast muscle samples were homogenized (1:19, w/v) in chilled physiological saline. After centrifuging at 4 °C and 5000 × *g* for 10 min, the supernatants were collected for subsequent assay for various hormones. The IGF-1 concentrations in the serum and muscle were measured with a commercial chicken-specific ELISA kit following the manufacturer's protocols (Jiancheng Bioengineering, Nanjing, China). Serum concentrations of GH, T3 and T4 were measured with the respective commercial RIA kits (Beijing North Institute of Biological Technology, China).

Extraction and quantification of total DNA and proteins

The breast muscle was homogenized in chilled physiological saline (1:10, w/v). DNA and proteins were extracted from the muscle homogenates as described [9]. The concentrations of DNA were measured using a NanoDrop1000 Spectrophotometer (NanoDrop Products, Wilmington, DE), and the protein concentrations were measured with a protein assay kit (Jiancheng Bioengineering, Nanjing, China). The percentages of proteins and nucleic acids in each breast muscle sample were calculated.

RNA extraction and quantitative real-time RT-PCR

Total RNA was extracted from the breast muscle using RNAiso (Takara Bio, Dalian, Liaoning, China). The first-strand cDNA was synthesized using the SuperScript® III Reverse Transcriptase with random primers and an RNase inhibitor (Invitrogen, Carlsbad, CA, USA) as per the manufacturer's instructions. Gene-specific primers for *IGF-1*, *MSTN*, *MyoD1*, *myogenin*, *MRF4*, *Myf5*, *MEF2A*, *MEF2B*, *MEF2C*, and *MEF2D* were designed using Primer Premier 6.0 (Premier, Ontario, Canada) (Table 1). PCR was performed on the ABI 7500 Real-Time PCR Detection System (Applied Biosystems, Foster City, CA, USA) using SYBR Premix Ex Taq II Kit (Takara Bio) and 40 cycles of 95 °C for 15 s and 60 °C for 30 s. All measurements were performed in triplicate. The fold difference was calculated using the ΔΔCt method [10, 11] using the geometric means of 18S rRNA and *glyceraldehyde 3-phosphate dehydrogenase* (*GAPDH*) mRNA for data normalization.

Table 1. Primers Used for Real-Time PCR

Gene	GenBank Accession Number	Primer Sequence (5' to 3')	Product Size, bp
IGF-1	NM_001004384	GAGCTGGTTGATGCTCTTCAGTT	148
		CCAGCCTCCTCAGGTCACAACT	
MSTN	NM_001001461	CGCTACCCGCTGACAGTGGAT	132
		CAGGTGAGTGTGCGGGTATTTCT	
MyoD1	NM_204214	CCGACGGCATGATGGAGTACA	131
		GTCGAGGCTGGAAACAACAGAA	
Myogenin	D90157	GGAGGCTGAAGAAGGTGAACGA	127
		CTCTGCAGGCGCTCGATGTACT	
MRF4	D10599	CAGGCTGGATCAGCAGGACAA	106
		GCCGCAGGTGCTCAGGAAGT	
Myf5	NM_001030363	CAGAGACTCCCCAAAGTGGAGAT	106
		GTCCCGGCAGGTGATAGTAGTTC	
MEF2A	NM_204864	CGGAGGACAGATTCAGCAAACTA	109
		GACACTGGAACCGTAACCGACAT	
MEF2B	XM_430389	CACGCCATCAGCATCAAGTCA	156
		GGGGTAGCCCTTGGAGTAGTCAT	
MEF2C	XM_001231661	GCCGTCTGCCCTCAGTCAACT	137
		GGGTGGTGGTACGGTCTCTAGGA	
MEF2D	NM_001031600	GTGTCTCCCAAGCGACTCACTCT	109
		GTGTTGTATGCGGTCGGCAT	
GAPDH	NM_204305	GCCACACAGAAGACGGTGGAT	86
		GTGGACGCTGGGATGATGTTCT	
18S	AF173612	CCGGACACGGACAGGATTGACA	94
		CAGACAAATCGCTCCACCAACTAAG	

Statistical analysis

All data were presented as means ± standard error of the mean (SEM). Statistics were performed using one-way ANOVA using the SPSS software, version 16.0 (Chicago, IL, USA). The differences were considered to be significant, if $P < 0.05$.

RESULTS

Body weight and the breast muscle weight

The BW of 600 male Ross 308 broilers that were randomly picked exhibited a normal Gaussian distribution ranging from 1,250 to 3,330 g with an average of 2,213 g (Figure 1). Thirty heaviest (H), 30 medium (M), and 30 lightest (L) broilers were further selected. The average BW was 2,784, 2,221, and 1,493 g for the H, M, and L groups, respectively, and the H group weighed nearly twice as much as the L group (Figure 2). As expected, the absolute breast muscle weight was also obviously different ($P < 0.001$) among three groups, with the H group doubling the weight of the L group (Figure 3a). However, the differences among the relative breast muscle weight (breast muscle weight: BW ratio) in the H and M groups were significantly higher than that in the L group ($P = 0.026$ between H and L; $P = 0.045$ between M and L) (Figure 3b).

Concentrations of total DNA, RNA, and proteins in the breast muscle

Total DNA concentration in the breast muscle of the H group was 17% ($P = 0.042$) and 18% ($P = 0.045$) lower than that of the M and L groups, whereas no difference was observed between the M and L groups (Table 2). On the other hand, the total RNA concentration between the H and M groups showed no difference ($P = 0.884$), but was approximately 42% and 39% higher than that in the L group, respectively (Table 2). No difference was seen with the total protein concentration in the breast muscle among three groups of broilers ($P = 0.661$) (Table 2). The protein:DNA ratio was positively correlated with the BW ($P = 0.049$), and the RNA:protein ratio showed the same trend ($P = 0.086$). The RNA:DNA ratio also exhibited an obvious correlation with the BW among the H, M, and L groups ($P = 0.006$), with the H group being significantly higher than the M group and the M group significantly higher than the L group (Table 2).

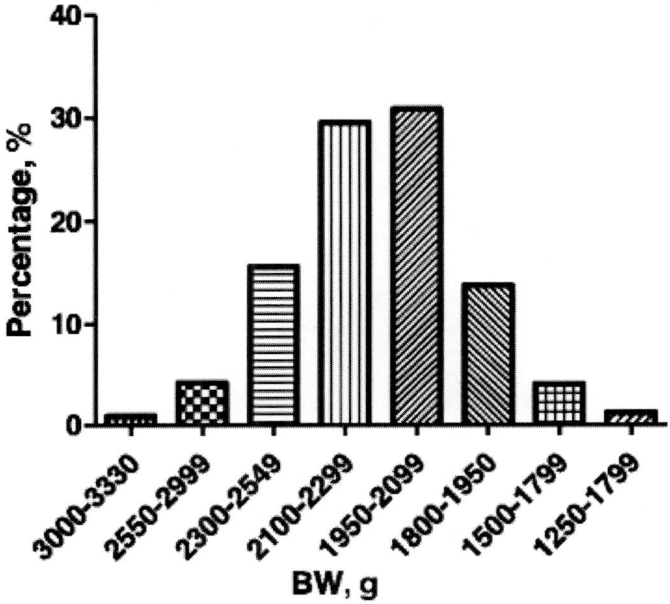

Figure 1. Body weight (BW) distribution of broilers. The data were based on 600, 38-day-old male broilers that were randomly picked and weighed from a group of 10,000 Ross 308 chickens raised under standard management in a single commercial house

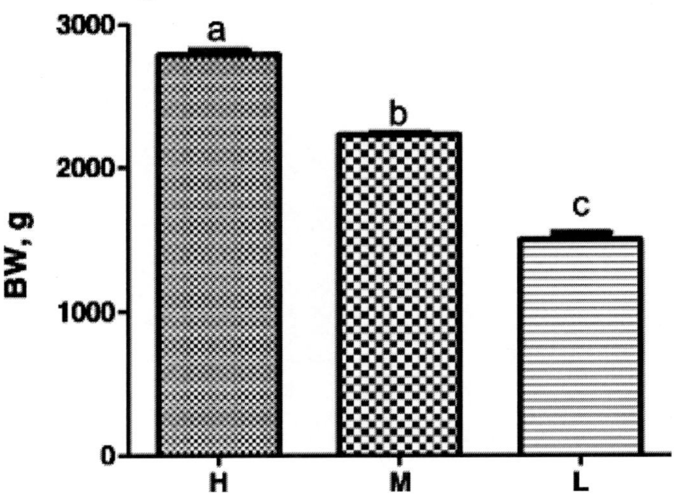

Figure 2. Average body weight (BW) of three groups of broilers. Each bar represents mean ± standard error (SE) of thirty, 38-day-old male broilers per group. Means with no common letter differ significantly ($P < 0.05$). H = fast-growing broilers (heavy BW); M = medium-growing broilers (average BW); L = slow-growing broilers (low BW)

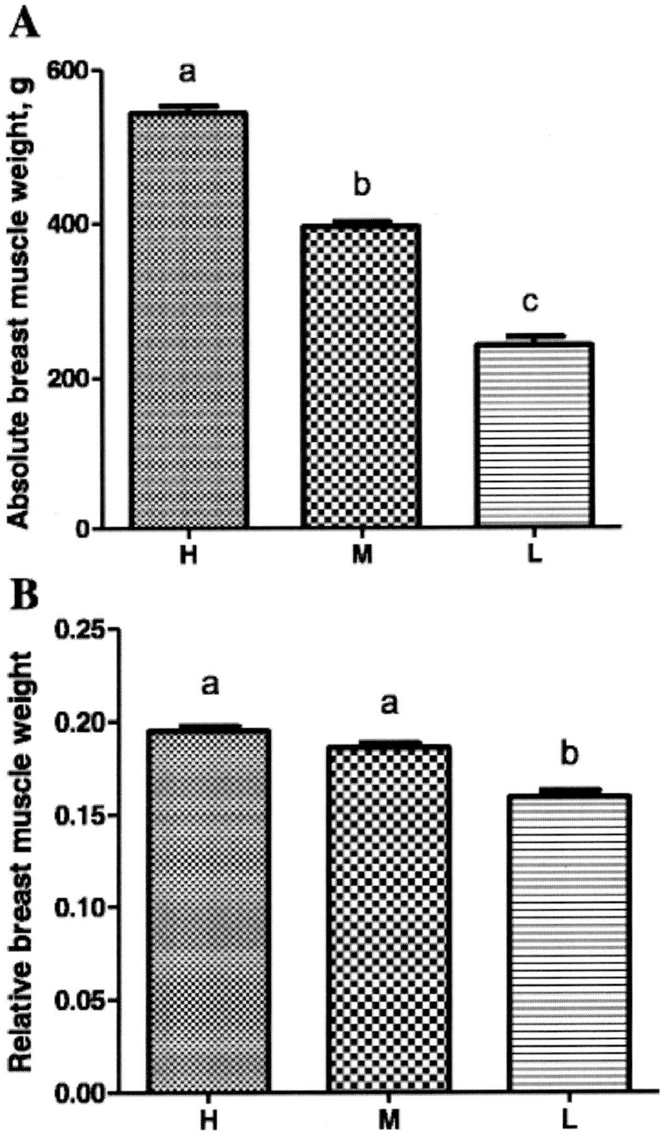

Figure 3. Absolute (**A**) and relative breast muscle weight (**B**) of three groups of broilers. The relative breast muscle weight is the ratio of absolute breast muscle weight to BW. Each bar represents mean ± standard error (SE) of thirty, 38-day-old male broilers per group. Means with no common letter differ significantly ($P < 0.05$). H = fast-growing broilers (heavy BW); M = medium-growing broilers (average BW); L = slow-growing broilers (low BW)

Table 2. The concentrations of DNA, RNA, and proteins in the breast muscle of broilers ($n = 10$)

Item	H	M	L	SEM	*P*-value
DNA, mg/g	1.68^{b}	2.02^{a}	2.06^{a}	0.07	0.044
RNA, mg/g	3.04^{a}	2.97^{a}	2.14^{b}	0.11	0.036
Protein, mg/g	86.32	88.65	84.97	2.54	0.661
Protein:DNA	49.52^{a}	43.87^{ab}	39.71^{b}	1.61	0.049
RNA:Protein	0.035	0.033	0.025	0.003	0.086
RNA:DNA	1.81^{a}	1.47^{b}	1.06^{c}	0.08	0.006

$^{a-c}$Means in the same row with different superscript letters differ significantly ($P < 0.05$)
H = fast-growing broilers (heavy BW group); M = medium-growing broilers (average BW group); L = slow-growing broilers (low BW group)

Concentrations of hormones in the serum and breast muscle

The serum concentration of IGF-1 showed an obvious linear relationship with the BW (Figure 4a). The H group had approximately 18% ($P = 0.044$) and 33% ($P = 0.027$) higher IGF-1 concentrations than the M and L groups, respectively. The serum concentration of T3 in the L group was approximately 50% ($P = 0.017$) and 47% ($P = 0.029$) lower than those in the H and M groups, respectively (Figure 4b). However, no significant difference was observed in the serum concentrations of GH and T4 among the H, M, and L groups (Figure 4c and d). Similar to the serum concentration, the IGF-1 concentration in the breast muscle was also linearly correlated with the BW (Figure 5). Comparing with the L group broilers, the IGF-1 concentration in the breast muscle showed a 18% increase ($P = 0.013$) in the H group.

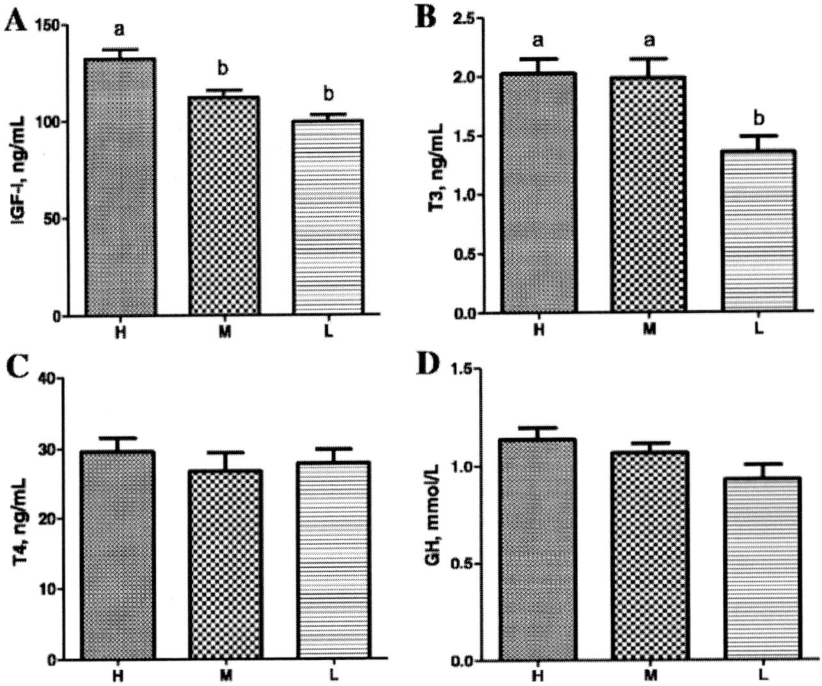

Figure 4. The serum concentrations of fours hormones including IGF-1 (**A**), T3 (**B**), T4 (**C**), and GH (**D**) in broilers. All hormones were measured by chicken-specific ELISA on 38-day-old male broilers. Each bar represents mean ± standard error (SE) of ten replicates. Means with no common letter differ significantly ($P < 0.05$). H = fast-growing broilers (heavy BW); M = medium-growing broilers (average BW); L = slow-growing broilers (low BW). IGF-1, insulin-like growth factor-1; T3, triiodothyronine; T4, thyroxine; GH, growth hormone

Expression levels of myogenic genes in the breast muscle

To study the correlation between the expression levels of major myogenic genes and the BW, total RNA was isolated from the breast muscle and subjected to reverse transcription and real-time PCR analysis of the genes for *MyoD1, Myogenin, MRF4, Myf5, MEF2A, MEF2B, MEF2C, MEF2D, IGF-1,* and *MSTN*. As shown in Table 3, the expression levels of all pro-myogenic genes but *MEF2A* showed a significant decrease ($P < 0.05$) in the L group relative to the H group. The pro-myogenic genes, namely *MEF5C*, showed a strong linear correlation ($P < 0.05$) with the BW, with the H group giving the highest expression and the L group the lowest expression. No difference

($P > 0.05$) was observed with *MyoD1, MRF4* or *MEF2B* between the H and M group, whereas *Myogenin* and *MEF2D* showed no difference between the M and L group. As for the expression of *MSTN*, an anti-myogenic gene, the H and M group showed no statistical difference, but both groups were significantly lower than the L group (Table 3). Interestingly, the *IGF-1* mRNA expression level was also significantly reduced ($P = 0.026$), relative to the H or M group (Table 3).

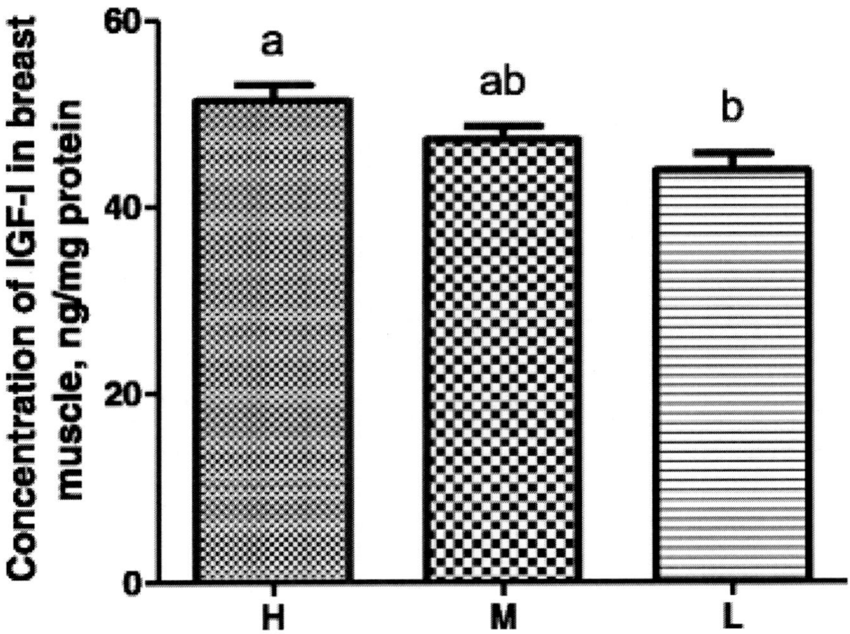

Figure 5. The concentrations of IGF-1 in the breast muscle of broilers. IGF-1 (insulin-like growth factor-1) was measured by chicken-specific ELISA on 38-day-old male broilers. Each bar represents mean ± standard error (SE) of ten replicates. Means with no common letter differ significantly ($P < 0.05$). H = fast-growing broilers (heavy BW); M = medium-growing broilers (average BW); L = slow-growing broilers (low BW)

Table 3. The relative mRNA levels in the breast muscle of broilers ($n = 10$)

Item	H	M	L	SEM	P-value
MyoD1	0.84^{ab}	1.00^{a}	0.66^{b}	0.065	0.017
Myogenin	2.94^{a}	1.00^{b}	1.29^{b}	0.111	<0.001
MRF4	1.22^{a}	1.00^{a}	0.78^{b}	0.078	0.047
Myf5	0.85^{a}	1.00^{a}	0.61^{b}	0.076	0.044
MEF2A	1.23	1.00	1.14	0.080	0.110
MEF2B	1.06^{a}	1.00^{a}	0.60^{b}	0.042	<0.001
MEF2C	1.22^{a}	1.00^{b}	0.74^{c}	0.062	0.024
MEF2D	1.32^{a}	1.00^{b}	0.89^{b}	0.052	0.015
MSTN	1.15^{b}	1.00^{b}	1.82^{a}	0.101	0.041
IGF-1	0.94^{a}	1.00^{a}	0.51^{b}	0.060	0.031

$^{a-c}$Means in the same row with different superscript letters differ significantly ($P < 0.05$)
H = fast-growing broilers (heavy BW group); M = medium-growing broilers (average BW group); L = slow-growing broilers (low BW group)

DISCUSSION

Genetic differences among the chicken breeds of different growth rates have been extensively studied and a number of genes and quantitative trait loci have been reported in controlling the growth rate of chickens [12, 13, 14, 15]. However, little is known regarding genetic and hormonal variations among individual broilers of the same breed under a common management. This study demonstrated that there is a large variation in the growth rate among individual birds raised under industrial standard management practices, with BW ranging from 1,250 g to 3,330 g. To elucidate the mechanism of the BW variations in a cohort of broiler chickens, the serum concentrations of GH, T3, T4, and IGF-1 were measured in the present study. The result showed that the T3 serum concentration was higher in heavier broilers than lighter ones, with no difference seen with the T4 concentration. This is consistent with an earlier report that the plasma concentration of T3, but not T4, was reduced in the sex-linked dwarf chickens relative to the normal control breed [16]. Broilers with heavier BW had a higher serum concentration of IGF-1 in our study, which is in agreement with an earlier observation that a reduction in the growth rate was associated with a decrease in circulating concentrations of IGF-1 in chickens [6]. Thus, higher levels of T3 and IGF-1 in the circulation accelerate growth in broilers and can be good predictors of faster growth.

Muscle growth is known to be accelerated by activation of the IGF-1 signaling pathway [8, 17], which is consistent with our observation that IGF-1 in the breast muscle were higher at both protein and mRNA levels in fast-growing than slow-growing broilers, which is probably due to increased methylation in the promoter region of the *IGF-1 receptor* gene in slow-growing broilers [18].

Besides *IGF-1*, altered expressions of multiple myogenic genes was also observed between heavy and light broilers in this study. Pro-myogenic genes such as *MyoD1* and *Myf5* are important in the formation of skeletal muscle [19, 20]. Other pro-myogenic genes such as *myogenin* and *MRF4* are directly involved in the differentiation of myotubes [21]. Consistently, we observed a significant increase in the expression of *MyoD1, Myf5, myogenin*, and *MRF4* in fast-growing birds relative to slow-growing ones. Similarly, three members of the pro-myogenic MEF2 family, namely *MEF2B, MEF2C* and *MEF2D*, were also significantly up-regulated in fast-growing broilers. On the other hand, the anti-myogenic *MSTN* mRNA levels are significantly lower in heavy broilers than light ones.

The skeletal muscle is a heterogeneous tissue composed of individual muscle fibers, diversified in size, shape and contractile protein content with different morphology, metabolism, and physiology [22]. Muscle mass can be affected by alterations in satellite cell activities, which dictate mature muscle size [22]. In the present study, fast-growing broilers had a greater breast muscle yield than slow-growing ones, reminiscent of previous studies [8, 23]. It is well known that DNA contents, protein:DNA and RNA:DNA ratios reflect cell population, cell size, and cellular metabolic activities, respectively [24]. An increase in the amount of total cellular RNA is likely due to enhanced synthetic activities [9]. In this study, increased protein:DNA and RNA:DNA ratios and total RNA content, accompanied with a decrease in the DNA content, in the breast muscle of fast-growing broilers suggested that heavier birds has less muscle cells but higher cellular activities. The results coincide with previous studies showing that heavier chickens had greater muscle mass with more satellite cells with enhanced proliferative activities [23, 25].

Hatching weight was shown to be a primary predictor of the market weight in chicks. The average BW of broilers at 42 d of age with heavy (48.3 g) and light hatching weight (41.7 g) became 2,368 g and 2,116 g, respectively [8]. Similarly, broilers on d 41 hatching at 53.1 g was about 1.1-fold heavier that those hatching at 43.5 g [23]. It is important to note that, although the breast muscle weight differs significantly between heavy and light broilers, the abdominal fat content shows no difference at marketing [23], suggesting that the genes involved in adipogenesis may not play a major role in the big variation in BW seen in a commercial broiler production facility.

In this study, we observed a significant difference in both mRNA and protein levels of IGF-1 in the breast muscle and circulation between fast- and slow-growing chickens on d 38. In fact, such a difference in the IGF-1 protein concentration is also evident on day of hatch between the broilers with heavy hatching weight and the ones with light hatching weight [23]. Consistently, IGF-1 levels are positively associated with growth rate in broiler strains divergently selected for high or low growth potential [26, 27], suggesting that IGF-1 can be a good predictor for growth [28]. However, the heritability of IGF-1 is only 0.1 [29] and therefore, its potential might still be limited as a useful parameter for future broiler breeding for enhanced performance. Nevertheless, to our knowledge this is the first study to reveal a positive correlation between IGF-1 and growth rate in a large population of broilers growing under standard commercial management.

It is noted that the serum concentration of T3 was also observed to be significantly different between fast- and slow-growing broilers on d 38 in our study. It will be important to examine a possible difference in its concentration on d 0 in order to explore the growth predictive value of T3.

CONCLUSIONS

A variation in the growth rate among individual broilers of the same breed is caused by an alteration in the growth-related hormone and expression of the genes involved in myogenesis. Serum concentration of IGF-1 and T3 is positively correlated with the growth rate of broilers. Most pro-myogenic genes such as *MyoD1, Myogenin, MRF4, Myf5, MEF2B, MEF2C,* and *MEF2D* are also associated positively with the growth, while *MSTN*, an anti-myogenic gene, shows a negative association. Some of these parameters may have potential to be further explored for the breeding purpose.

ACKNOWLEDGEMENTS

The authors gratefully thank all of the staff of the Wanming Poultry Farm (Hangzhou, China) for their assistance in feeding and care of the animals.

We also acknowledge the members of the Institute of Quality and Standards for Agro-products, Zhejiang Academy of Agricultural Sciences for their assistance with sample collections.

Funding

This research was supported by the Special Fund for Public Projects of Zhejiang Province (2016C32073) and International Cooperation Program of Zhejiang Academy of Agricultural Sciences. The funding body has not participated in or interfered with the research.

Availability of data and materials

Not applicable.

Authors' contributions

YX, GZ, and HY conceived the study. YX, CW, KL, and GG participated in sample collection and performed all the assays. YX and HY drafted and GZ revised the manuscript. All authors read and approved the final manuscript.

Competing interests

The authors declare that they have no competing interests.

Consent for publication

Not applicable.

Ethics approval and consent to participate

All procedures involving animals were approved by Zhejiang Academy of Agricultural Sciences Institutional Animal Care and Use Committee.

/zero/1.0/) applies to the data made available in this article, unless otherwise stated.

REFERENCES

1. Perry RL, Rudnick MA. Molecular mechanisms regulating myogenic determination and differentiation. Front Biosci. 2000;5:D750–67.
2. Jia Y, Gao G, Song H, Cai D, Yang X, Zhao R. Low-protein diet fed to crossbred sows during pregnancy and lactation enhances myostatin gene expression through epigenetic regulation in skeletal muscle of weaning piglets. Eur J Nutr. 2016;55:1307–14.
3. Hennebry A, Berry C, Siriett V, O'Callaghan P, Chau L, Watson T, et al. Myostatin regulates fiber-type composition of skeletal muscle by regulating mef2 and myod gene expression. Am J Physiol Cell Physiol. 2009;296:C525–34.
4. Wang Q, McPherron AC. Myostatin inhibition induces muscle fibre hypertrophy prior to satellite cell activation. J Physiol. 2012;590:2151–65.
5. Reisz-Porszasz S, Bhasin S, Artaza JN, Shen R, Sinha-Hikim I, Hogue A, et al. Lower skeletal muscle mass in male transgenic mice with muscle-specific overexpression of myostatin. Am J Physiol Endocrinol Metab. 2003;285:E876–88.
6. Scanes CG. Perspectives on the endocrinology of poultry growth and metabolism. Gen Comp Endocrinol. 2009;163:24–32.
7. Boschiero C, Jorge EC, Ninov K, Nones K, do Rosario MF, Coutinho LL, et al. Association of igf1 and kdm5a polymorphisms with performance, fatness and carcass traits in chickens. J Appl Genet. 2013;54:103–12.
8. Wen C, Wu P, Chen Y, Wang T, Zhou Y. Methionine improves the performance and breast muscle growth of broilers with lower hatching weight by altering the expression of genes associated with the insulin-like growth factor-i signalling pathway. Br J Nutr. 2014;111:201–6.
9. Johnson LR, Chandler AM. Rna and DNA of gastric and duodenal mucosa in antrectomized and gastrin-treated rats. Am J Physiol. 1973;224:937–40.
10. Xiao YP, Wu TX, Sun JM, Yang L, Hong QH, Chen AG, et al. Response to dietary l-glutamine supplementation in weaned piglets: A serum metabolomic comparison and hepatic metabolic regulation analysis. J Anim Sci. 2012;90:4421–30.
11. Schmittgen TD, Livak KJ. Analyzing real-time pcr data by the comparative c(t) method. Nat Protoc. 2008;3:1101–8.
12. Zheng Q, Zhang Y, Chen Y, Yang N, Wang XJ, Zhu D. Systematic identification of genes involved in divergent skeletal muscle growth rates of broiler and layer chickens. BMC Genomics. 2009;10:87.
13. Buzala M, Janicki B, Czarnecki R. Consequences of different growth rates in broiler breeder and layer hens on embryogenesis, metabolism and metabolic rate: A review. Poult Sci. 2015;94:728–33.
14. Pauwels J, Coopman F, Cools A, Michiels J, Fremaut D, De Smet S, et al. Selection for growth performance in broiler chickens associates with less diet flexibility. PLoS One. 2015;10, e0127819.
15. Davis RV, Lamont SJ, Rothschild MF, Persia ME, Ashwell CM, Schmidt CJ. Transcriptome analysis of post-hatch breast muscle in legacy and modern broiler chickens reveals enrichment of several regulators of myogenic growth. PLoS One. 2015;10, e0122525.
16. Scanes CG, Marsh J, Decuypere E, Rudas P. Abnormalities in the plasma concentrations of thyroxine, tri-iodothyronine and growth hormone in sex-linked dwarf and autosomal dwarf white leghorn domestic fowl (gallus domesticus). J Endocrinol. 1983;97:127–35.

17. Guernec A, Berri C, Chevalier B, Wacrenier-Cere N, Le Bihan-Duval E, Duclos MJ. Muscle development, insulin-like growth factor-i and myostatin mrna levels in chickens selected for increased breast muscle yield. Growth Horm IGF Res. 2003;13:8–18.
18. Hu Y, Xu H, Li Z, Zheng X, Jia X, Nie Q, et al. Comparison of the genome-wide DNA methylation profiles between fast-growing and slow-growing broilers. PLoS One. 2013;8, e56411.
19. Rudnicki MA, Jaenisch R. The myod family of transcription factors and skeletal myogenesis. Bioessays. 1995;17:203–9.
20. Shin J, McFarland DC, Velleman SG. Heparan sulfate proteoglycans, syndecan-4 and glypican-1, differentially regulate myogenic regulatory transcription factors and paired box 7 expression during turkey satellite cell myogenesis: Implications for muscle growth. Poult Sci. 2012;91:201–7.
21. Comai G, Tajbakhsh S. Molecular and cellular regulation of skeletal myogenesis. Curr Top Dev Biol. 2014;110:1–73.
22. Sobolewska A, Elminowska-Wenda G, Bogucka J, Szpinda M, Walasik K, Bednarczyk M, et al. Myogenesis--possibilities of its stimulation in chickens. Folia Biol (Krakow). 2011;59:85–90.
23. Sklan D, Heifetz S, Halevy O. Heavier chicks at hatch improves marketing body weight by enhancing skeletal muscle growth. Poult Sci. 2003;82:1778–86.
24. Uni Z, Noy Y, Sklan D. Development of the small intestine in heavy and light strain chicks before and after hatching. Br Poult Sci. 1996;37:63–71.
25. Wen C, Chen Y, Wu P, Wang T, Zhou Y. Mstn, mtor and foxo4 are involved in the enhancement of breast muscle growth by methionine in broilers with lower hatching weight. PLoS One. 2014;9, e114236.
26. Scanes CG, Dunnington EA, Buonomo FC, Donoghue DJ, Siegel PB. Plasma concentrations of insulin like growth factors (igf-)i and igf-ii in dwarf and normal chickens of high and low weight selected lines. Growth Dev Aging. 1989;53:151–7.
27. Beccavin C, Chevalier B, Cogburn LA, Simon J, Duclos MJ. Insulin-like growth factors and body growth in chickens divergently selected for high or low growth rate. J Endocrinol. 2001;168:297–306.
28. Duclos MJ. Insulin-like growth factor-i (igf-1) mrna levels and chicken muscle growth. J Physiol Pharmacol. 2005;56 Suppl 3:25–35.
29. Pym RA, Johnson RJ, Etse DB, Eason P. Inheritance of plasma insulin-like growth factor-i and growth rate, food intake, food efficiency and abdominal fatness in chickens. Br Poult Sci. 1991;32:285–93.

CHAPTER 10

Fatty Acid Content in Epididymal Fluid and Spermatozoa During Sperm Maturation in Dogs

Daniel S. Ramos Angrimani[1] , Marcilio Nichi[1] , João Diego A. Losano[1] , Cristina F. Lucio[1] , Gisele A. Lima Veiga[1] , Márcia V. M. Junqueira Franco[2] and Camila I. Vannucchi[1]*

[1] *Department of Animal Reproduction, School of Veterinary Medicine and Animal Science, University of São Paulo, Rua Prof. Orlando Marques de Paiva, 87, São Paulo 05508-270, Brazil.*

[2] *Department of Internal Medicine, Division of Nutrition, University of São Paulo, Av. Bandeirantes, 3900 - 14040-040, Ribeirão Preto, Brazil.*

ABSTRACT

Background

During sperm maturation, there is a reorganization of fatty acids from plasmatic membrane of the spermatozoa, which allows higher membrane integrity and acquisition of sperm motility. However, the fatty acid profile during sperm maturation remains unclear in dogs. Thus, the aim of this study was to identify the fatty acids from the epididymal spermatozoa and plasma during the sperm maturation, and observed changes in the motility and plasmatic membrane parameters. Twenty one adult dogs were used, subsequently to bilateral orchiectomy and epididymal storage, sperm samples were collected from the different segments of the epididymis. Samples were evaluated for conventional microscopy, computer-assisted motility analysis, sperm plasma membrane permeability and the fatty acid analysis (lipids were extracted, transmethylated and analyzed by chromatography).

Results

Caput and corpus sperm showed lower values for the motility variables evaluated and plasmatic membrane integrity, indicating different levels of the fatty acids organization. Saturated, monounsaturated and polyunsaturated fatty acids were in higher concentrations in the spermatozoa from epididymis cauda. Highlighting the presence of caprylic, stearic and docosahexaenoic acids.

Conclusions

These findings demonstrate the influence of the fatty acid profile during sperm maturation, assigning physical and chemical changes in sperm cells, essential for fertilization.

Keywords: Dogs Epididymis Fatty acids Sperm maturation

BACKGROUND

Sperm maturation occurs along the transit through the epididymal regions, which allows for functional and morphological changes in the spermatozoa [1]. After these modifications, the spermatozoa stabilize its plasma membrane through structural changes in lipid profile and, ultimately, acquire progressive motility [2, 3]. In addition, the remodeling events of the sperm plasma membrane are important to provide cellular integrity and fertilization ability [4, 5]. Molecular and biochemical organization are essential in order to prevent sperm damage during storage and transport through both male and female reproductive tract after ejaculation [5].

Sperm plasmatic membrane is composed by a bilayer of phospholipids, with the saturated or unsaturated fatty acids being the functional units [5]. The origin of the plasma membrane lipid content derives from the spermatogonial cell, however, it can be modified during spermatogenesis and sperm maturation [6]. While passing through the first epididymal region (caput), sperm plasmatic membrane is flexible, facilitating lipid remodeling [7]. However, across the other epidydimal segments (corpus and cauda), sperm becomes more stable and able to withstand the damages of the storage area situated at the cauda epididymides [3]. The lipid organization of the spermatozoa during epididymal maturation is mediated by protein and lipid constituents of the epididymal fluid [5, 8, 9]. Hence, sperm plasma membrane changes in accordance with the environment in which sperm is located. Only at the final stage of epididymal maturation, the plasmatic membrane reaches its final composition, which consists, generally, in 70% phospholipids, 25% neutral lipids (mainly cholesterol) and 5% glycolipids [6, 10]. Among the polyunsaturated fatty acids, the docosahexaenoic acid (DHA) represents 50% of the total sperm content [11].

In mammals, the fatty acids composition of the sperm plasmatic membrane is related to physicochemical properties and, consequently, interferes with the functional competence of the sperm cells. DHA is essential for the fluidity of the plasma membrane, and thus for the acquisition of sperm motility and acrosome reaction [12]. Moreover, DHA is related to the physiological occurrence of oxidative stress in sperm cells [13]. The sperm membrane is rich in polyunsaturated fatty acids (PUFAs), which sensitize to the attack of reactive oxygen species. Such lipid peroxidation is related physiologically to several steps of the fertilization process such as sperm hyperactivation, capacitation, acrosome reaction and fertilization [13]. Simultaneously with the aforementioned changes in sperm lipid membrane, sperm motility parameters gradually change along the epididymal transit, from immobility or vibration to progressive movements [2, 14, 15]. During spermiation to epididymal caput, spermatozoa are immobile. On the other hand, progressive motility, speed and linearity is acquired in the epididymal corpus [16]. However, sperm ability to acquire motility depends on changes in plasma membrane composition [17]. Thus, specific studies are required in order to increase the physiological knowledge regarding sperm maturation and to improve the reproductive biotechnologies involved with semen manipulation [18].

Despite the importance of sperm lipid profile, the exact composition of the fatty acids that structure sperm plasma membrane has not yet been explored for epididymal samples in dogs, only in ejaculated semen [19] and seminal plasma [20]. Moreover, the exact mechanism that structures sperm plasma membrane should be investigated in dogs. To our knowledge, the specific modifications in dog sperm and the exact composition of the epididymal fluid have not yet been fully established or elucidated. As the ideal experimental model for humans, the studies concerning sperm maturation in dogs can enable the development of new biotechnologies, such as male contraception, in vitro sperm maturation and treatments associated with sperm maturation [21].

Hence, the aim of this study was to identify the fatty acid profile of epididymal spermatozoa and epididymal fluid during sperm maturation along the epididymides, and also, characterize changes in sperm motility parameters and plasmatic membrane.

METHODS

The present study was approved by the Bioethics Committee of the School of Veterinary Medicine and Animal Science - University of São Paulo (protocol number: 2277/2011). Unless otherwise stated, all chemicals and reagents were purchased from Sigma–Aldrich (St. Louis, MO, USA).

Twenty one mature dogs (aged from 1 to 6 yr) of distinct breeds and body weights were used (Table 1). At macroscopical exam, testis and epididymides were considered normal. All animals were tested against brucellosis, using the *Antigen Rapid Canine Brucella Ab Ttest* (Bioeasy, Minas Gerais, Brazil), detecting the anti-*Brucella canis* IgG antibodies.

After the bilateral orchiectomy, testicles–epididymis were immediately stored at 5 °C and samples were transported to the laboratory. Epididymal storage has been proved safe at 4 °C for no more than 24 h [22]. Therefore, after 18–24 h, we collected the epididymal sperm by individual incisions in the caput, corpus and cauda epididymides with a scalpel blade. Dissection was performed carefully to avoid sectioning of blood vessels. The flowing epididymal fluid and sperm was collected with an automatic pipette, and the total volume of the epididymal sample (fluid + sperm) was resuspended in 300 μL of PBS medium (phosphate buffered saline), according to [23]. The dilution rate varied according to the epididymal size, similar to a previously described protocol [24]. To calculate the epididymal sample (fluid + sperm) dilution factor, tubes containing solely the PBS medium were weighted before (tube + PBS) and after the addition of the epididymal sample (tube + PBS + epididymal sample). Due to the small volume recovered, the samples from the left and right epididymides of the same animal were pooled.

Although the ejaculate sperm analysis has not been performed, the epididymal sperm were within adequate parameters for dogs, according to [25] and [23] assuring that testicular and epidydimal abnormalities were ruled out.

Table 1. Age, weight and breed of the dogs from which testis and epididymides were harvested (n = 21)

	Breed	Weight, kg	Age, yr
Dog 1	Mixed Breed	19.6	1
Dog 2	Labrador Retriever	26.6	3
Dog 3	Mixed Breed	6.4	1
Dog 4	Mixed Breed	7.8	5
Dog 5	Poodle	12.8	5
Dog 6	Mixed Breed	18	3
Dog 7	Mixed Breed	12.6	2
Dog 8	Mixed Breed	7.8	1
Dog 9	Mixed Breed	14.6	2
Dog 10	Mixed Breed	10.5	2
Dog 11	Mixed Breed	12.8	2
Dog 12	Mixed Breed	17.9	1
Dog 13	Mixed Breed	48	2
Dog 14	Labrador Retriever	19.9	6
Dog 15	Mixed Breed	40	3
Dog 16	German Shepherd	11.9	2
Dog 17	Mixed Breed	14.8	2
Dog 18	Shetland Sheepdog	7.6	6
Dog 19	Schnauzer	7.5	6
Dog 20	Mixed Breed	45	3
Dog 21	Mixed Breed	13.7	1

Immediate sperm evaluation

Samples from the epididymal segments (i.e. caput, corpus and cauda) were immediately evaluated for sperm motility and vigor by conventional microscopy (Nikon, Eclipse E200, Japan) at 1,000× magnification and for computer-assisted sperm analysis (CASA; HTM-IVOS Ultimate 12.3; Hamilton Thorne Biosciences, Beverly, MA, USA), according to a previously described protocol [26]. Briefly, 10 µL of each sample was deposited on microscope slides previously warmed at 37 °C and covered by coverslips. Eight fields of view were randomly selected and the following variables were assessed: VAP (average pathway velocity, µm/s), VSL (strait-line velocity, µm/s), VCL (curvilinear velocity, µm/s), ALH (amplitude of lateral head displacement, µm), BCF (beat cross frequency, Hz), STR (straightness -VSL/VAP, %), LIN (Linearity - VSL/VCL,%); MOT (motility, %), PROG (progressive motility, %). Sperm were also divided into four groups based on velocity: rapid (RAP, VAP > 50 µm/s, %), medium (MED, 30 µm/s < VAP < 50 µm/s, %), slow (SLOW, VAP < 30 µm/s or VSL < 15 µm/s, %) and non-moving spermatozoa (STATIC, %).

Samples were also evaluated for plasma membrane permeability using the eosin/nigrosin stain [27]. In brief, 5 µL of semen and 5 µL of the previously prepared stain were placed in a pre-warmed (37 °C) glass slide. The sperm smear was evaluated under light microscopy (Nikon, Eclipse E200, Japan) at 1,000× magnification. We considered intact sperm (membrane integrity) as sperm with no stain; and damaged sperm (membrane lesion) as pink colored cells. Results were analyzed by counting 200 cells and expressed as percentage of stained sperm (%).

Semen processing for fatty acid analysis

Immediately after semen analysis, samples were centrifuged at 800 ×g for 10 min to separate the spermatozoa from the epididymal fluid. The volume of each sample contained an equivalent concentration of 1.5 million spermatozoa, by pooling the sample of four dogs, totalizing five pools of each segment of the epididymides (caput, corpus and cauda), according to previous works [28, 29]. For the epididymal fluid analysis we used a fixed volume of 60 µL. Samples were stored at −20 °C until analysis.

For the fatty acids extraction, we utilized the transesterification method described by Lepage and Roy [30]. Samples were transferred to a glass cuvette, to which 10 µL of sodium chloride and triphenylphosphate (10 mg/mL; as internal standard) was added. Subsequently, 1 mL of the solution was combined to methanol:acetyl chloride, in a 100:5 proportion (3 mL:150 µL), and the cuvette was maintained at 100 °C for 60 min for fatty acids transmethylation [31].

After incubation, the cuvettes were maintained at room temperature for the addition of 1 mL of hexane, which allowed the solubilization of fatty acids, enabling the passage through gas chromatography. Samples were vortexed for 60 s and then centrifuged (640 ×g for 5 min). The supernatant was transferred to a glass jar and dried by N_2 vapor. Sequentially, the sample was suspended in 50 µL of hexane, vortexed for 60 s and 1 µL of this solution was injected into the gas chromatograph (GC-17A®, Shimadzu, Kyoto, Japan).

The fatty acids were classified in accordance with the retention time of a pre-established curve, then sorted into saturated fatty acids (butyric, caproic, caprylic, capric, undecanoic, lauric, tridecanoic, myristic, pentadecanoic, palmitic, heptadecanoic, stearic, arachidic, heneicosanoic, behenic, tricosanoic and lignoceric), monounsaturated fatty acids (myristoleic, pentadecenoic, palmitoleic, elaidic, oleic, eicosenoic, erucic and nervonic) and polyunsaturated fatty acids (linoleic, linolelaidic, gamma-linolenic, alpha-linolenic, eicosadienoic, arachidonic, eicosatrienoic, eicosapentaenoic acid and docosahexaenoic).

Stastical analysis

All data were analyzed using the SAS for Windows (SAS Institute Inc., Cary, NC, USA). The effect of sperm origin (epididymal spermatozoa from caput, corpus and cauda) and the matrix used (epididymal fluid vs. sperm) on fatty acids content was determined using parametric (Student's *t* test for two treatments or LSD test for more than two treatments) and nonparametric (Wilcoxon) tests, according to the residue normality (Gaussian distribution) and variance homogeneity of each variable. The different fatty acids were also divided, grouped and analyzed according to the number of unsaturations (i.e. saturated, monounsaturated and polyunsaturated).

A probability value of $P < 0.05$ was considered statistically significant. The results are reported as untransformed means ± SEM. Pearson's correlation was used to calculate the relationship between the variables studied in each variable group. In order to perform the correlation analysis between the concentration of fatty acids and the sperm analysis, data (i.e. sperm motility and plasma membrane integrity) were merged according to the pool of samples adopted for the fatty acids analysis.

RESULTS

Samples from the epididymal cauda presented a higher sperm motility and vigor in comparison to the corpus and caput segments, respectively (Table 2). The same pattern was observed for the percentage of sperm with plasmatic membrane integrity, i.e., cauda sperm had greater integrity compared to the corpus and caput (Table 2). The Computer Assisted Semen Analysis (CASA) showed significant and progressive increase in sperm motility, progressive motility, rapid speed, average path velocity (VAP), straight line velocity (VSL), curvilinear velocity (VCL), straightness (STR) and linearity (LIN), while passaging through epididymal segments (caput, corpus and cauda) (Table 2). The sperm variables related to medium speed, amplitude of lateral head displacement (ALH) and beat cross frequency (BCF) were not different between the corpus and cauda groups, while caput sperm showed lower results (Table 2). The index of spermatozoa with progressive motility in the caput did not differ from the corpus samples, but was lower than the cauda of the epididymides (Table 2). However, the percentage of static spermatozoa from the epididymis caput was greater than corpus and cauda sperm (Table 2).

Table 2. Effect of sperm origin (epididymal cauda, corpus and caput) on subjective sperm motility, vigor, plasma membrane integrity and computer assisted sperm analysis (CASA) in dogs

	Caput	Corpus	Cauda
Sperm motility, %	0.0 ± 0.0^{c}	27.74 ± 3.0^{b}	69.7 ± 4.0^{a}
Sperm vigor (1–5)	0.0 ± 0.0^{c}	2.0 ± 0.1^{b}	2.6 ± 0.1^{a}
Intact plasma membrane, %	40.1 ± 4.7^{c}	74.5 ± 2.3^{b}	92.6 ± 1.1^{a}
% of Motile	0.0 ± 0.0^{c}	34.5 ± 3.5^{b}	71.0 ± 4.4^{a}
% of Progressive	0.0 ± 0.0^{b}	5.5 ± 0.9^{b}	27.8 ± 2.9^{a}
% of Rapid	0.0 ± 0.0^{c}	18.2 ± 2.4^{b}	54.2 ± 4.5^{a}
% of Medium	0.0 ± 0.0^{b}	16.3 ± 1.7^{a}	16.8 ± 2.6^{a}
% of Slow	6.0 ± 2.5	6.0 ± 2.4	10.5 ± 1.4
% of Static, %	94.0 ± 2.5^{a}	54.8 ± 4.4^{b}	23.1 ± 3.6^{c}
VAP, µm/s	0.0 ± 0.0^{c}	66.3 ± 5.5^{b}	94.9 ± 5.6^{a}
VSL, µm/s	0.0 ± 0.0^{c}	37.8 ± 3.7^{b}	67.2 ± 4.6^{a}
VCL, µm/s	0.0 ± 0.0^{c}	142.5 ± 10.3^{b}	178.5 ± 10.0^{a}
ALH, µm/s	0.0 ± 0.0^{b}	7.3 ± 0.6^{a}	7.8 ± 0.3^{a}
BCF, µm/s	0.0 ± 0.0^{b}	32.0 ± 3.1^{a}	28.3 ± 2.2^{a}
Straightness, STR - %	0.0 ± 0.0^{c}	48.9 ± 3.5^{b}	67.4 ± 1.4^{a}
Linearity, LIN - %	0.0 ± 0.0^{c}	26.3 ± 2.4^{b}	38.9 ± 1.5^{a}

Different superscripts in the same line indicate significant differences ($P < 0.05$)

In the corpus segment, there was a positive correlation between sperm straightness and the percentage of sperm with intact plasmatic membrane ($r = 0.49$; $P = 0.03$). In the epididymides cauda, a positive correlation was observed between the percentage of sperm with intact membrane and subjective motility ($r = 0.80$; $P < 0.0001$), CASA motility ($r = 0.63$; $P = 0.003$) and progressive motility ($r = 0.46$; $P = 0.04$).

Regarding the identification of the saturated fatty acids, we observed higher content in the epididymal fluid from epididymal cauda in comparison to the caput, but not different from the epididymal corpus (Figure 1). On the other hand, sperm samples from the cauda had higher concentration of saturated fatty acids compared to epididymal corpus and caput (Figure 1). Regarding the matrix origin, sperm had higher concentration of saturated fatty acids compared to the epididymal fluid in the cauda ($P = 0.0005$) and corpus segments ($P = 0.004$) (Figure 1).

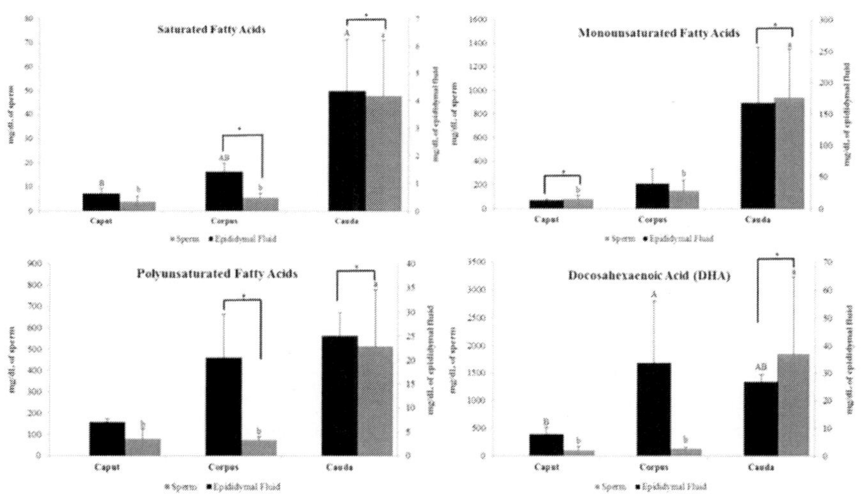

Figure 1. Mean and standard error ($\bar{X} \pm SE$) of saturated ($\times 10^3$ mg/dL), monounsaturated (mg/dL), polyunsaturated fatty acids (mg/dL) and docosahexaenoic acid (DHA; mg/dL) in sperm and epididymal fluid according to the epididymal segment (caput, corpus and cauda) in dogs. ^{a,b}values with different superscript letters differ significantly between segments for epididymal sperm ($P < 0.05$). ^{A,B}values with different superscript letters differ significantly between segments for epididymal fluid ($P < 0.05$). *Indicates statistical difference between epididymal fluid and sperm ($P < 0.05$)

No difference was observed between caput, corpus and cauda for monounsaturated and polyunsaturated fatty acid concentrations in the epididymal fluid (Figure 1). However, sperm from the epididymal cauda presented higher monounsaturated and polyunsaturated fatty acid content compared with the corpus and caput of the epididymis (Figure 1). Moreover, regarding the matrix origin, sperm had higher concentration of monounsaturated and polyunsaturated fatty acids compared to the epididymal fluid in the caput ($P = 0.05$) and cauda ($P = 0.05$) segments, and in the corpus ($P = 0.01$) and cauda epididymides ($P = 0.04$), respectively (Figure 1). For the concentration of docosahexaenoic acid (DHA; Figure 1), we observed statistical difference between the epididymal fluid and sperm in the cauda of epididymis ($P = 0.005$). Moreover, corpus epididymal fluid had greater DHA concentration compared only to the caput, while cauda sperm present the highest content compared to the other epididymal segments (Figure 1). In the fluid and sperm of the epididymal caput, corpus and cauda we observed higher concentration of satured fatty acid in comparison to monounsaturated and polyunsaturated (Table 3).

Table 3. Fatty acid concentration (%) of saturated, monounsaturated and polyunsaturated in sperm and epididymal fluid of each epididymal segment (caput, corpus and cauda) in dogs

	Cauda		Corpus		Caput	
	Edidymal Fluid	Sperm	Edidymal Fluid	Sperm	Edidymal Fluid	Sperm
Satured	95.76%	97.04%	95.99%	96.18%	96.92%	96.05%
Monounsatured	3.68%	1.91%	2.63%	2.57%	2%	2.01%
Polyunsaturated	0.54%	1.04%	1.37%	1.24%	1.07%	1.92%

In the epididymal fluid, the following fatty acids were identified: caprylic, capric, heptadecanoic and nervonic, without statistical difference between caput, corpus and cauda (Table 4). However, the concentration of stearic and pentadecenoic fatty acids were higher in the epididymal cauda in comparison to the corpus and caput, and the linoleic fatty acid had lower concentration in the epididymal caput compared with the corpus and cauda, which did not differ significantly (Table 4).

Regarding the fatty acids content in the epididymal sperm (Table 5), caproic and capric fatty acids were present in all epididymal segments, although without statistical difference (Table 5). Concentrations of caprylic, pentadecenoic and stearic were lower in the caput and corpus in comparison to the epididymal cauda (Table 5). The concentration of linoleic fatty acid was higher in cauda epididymides compared to epididymal corpus, without statistical difference with the epididymal caput.

Table 4. Mean and standard error ($\bar{X} \pm SE$) of the epididymal fluid fatty acids (mg/dL) according to the epididymal segment (caput, corpus and cauda) in dogs

	Caput	Corpus	Cauda
Caprylic	31,35.84 ± 1,640.87	3,467.34 ± 917.79	25,493.90 ± 20,432.01
Capric	25.15 ± 19.05	11.25 ± 1.25	25.36 ± 9.78
Heptadecanoic	8.36 ± 1.36	9.47 ± 0.63	23.15 ± 10.78
Nervonic	15.11 ± 3.31	16.72 ± 1.49	41.52 ± 10.56
Stearic	33.10 ± 5.64^a	68.85 ± 11.35^a	171.19 ± 26.22^b
Pentadecenoic	17.06 ± 4.33^a	24.18 ± 3.45^a	129.21 ± 20.21^b
Linoleic	4.68 ± 0.75^a	10.49 ± 2.50^b	11.03 ± 1.17^b

Different superscripts in the same line indicate significant differences ($P < 0.05$)

Table 5. Mean and standard error ($\bar{X} \pm SE$) of the sperm fatty acids (mg/dL) according to the epididymal segment (caput, corpus and cauda) in dogs

	Caput	Corpus	Cauda
Caproic	21.49 ± 13.06	133.54 ± 83.10	1,486.92 ± 1,040.43
Capric	10.34 ± 1.11	109.02 ± 68.63	261.38 ± 94.15
Caprylic	$13,915.63 \pm 9,278.72^a$	$15,273.23 \pm 5,796.17^a$	$166,899.87 \pm 92,944.15^b$
Stearic	137.76 ± 43.45^a	159.02 ± 12.25^a	$1,385.99 \pm 517.63^b$
Pentadecenoic	92.66 ± 18.70^a	75.73 ± 11.47^a	939.84 ± 415.22^b
Linoleic	72.34 ± 61.94^{AB}	40.06 ± 5.71^a	222.23 ± 77.36^b

Different superscripts in the same line indicate significant differences ($P < 0.05$)

DISCUSSION

In the present study, we evaluated sperm motility, plasmatic membrane integrity and fatty acid profile of spermatozoa harvested from the caput, corpus and cauda of the epididymides in dogs. The lipid analysis was performed in the epididymal sperm and fluid as well.

We observed higher concentration of saturated, monounsaturated and polyunsaturated fatty acids in the sperm harvested from the epididymal cauda in comparison with the corpus and caput. This result can be attributed to the necessary acquisition of lipids during the epididymal maturation, which will further act during sperm capacitation and oocyte fertilization in the female tract [32]. Moreover, we observed higher concentration of specific fatty acids. For saturated fatty acids, the caprylic and stearic fatty acids were more prominent. The former fatty acid has antimicrobial and protective functions in the sperm membrane [33], while stearic fatty acid is proved to regulate bovine sperm function, providing energy metabolism and sperm motility [34]. Therefore, increased concentrations of caprylic fatty acid in the spermatozoa of the epididymal cauda may be responsible for the greater integrity of the plasmatic membrane, also observed through our sperm membrane integrity stain analysis.

There was a prevalence of satured fatty acids in all epididymal segments in both spermatozoa and fluid, similar to results obtained by Martinez-Soto, Landeras [11] in men. It is possible to infer that the high percentage of satured fatty acids can minimize the eventual sperm damage during the long term storage in the epididymis, since polyunsaturated fatty acids are more easily oxidized by reactive oxygen species (ROS) [11, 35]. However, this result must be further confirmed by future studies on oxidative stress challenge of stored epididymal sperm.

In the present experiment, the pentadecenoic fatty acid was the predominant monounsaturated fatty acid in the spermatozoa, whereas the docosahexaenoic (DHA), the main polyunsaturated fatty acid. Among the polyunsaturated fatty acids, DHA represents 50% of the total content of human spermatozoa [11]. In boars, DHA content increases during sperm epididymal maturation [36]. Unlike other cells, mature sperm presents a high quantity of polyunsaturated fatty acids at plasma membrane level [37], providing sperm integrity [38] and progressive motility [39], in addition to being considered necessary for the fluidity of the plasma membrane. Therefore, it is possible to affirm that an increased sperm DHA concentration is essential for the final steps of epididymal maturation in dogs, because it is directly involved in events required for fertilization (i.e. motility and plasmatic membrane integrity). Additionally, there was a lack of flagellum movement in the caput, onset of motility in the epididymal corpus and higher percentage of spermatozoa with progressive flagellar movement in the epididymal cauda. Sperm maturation requires changes in the lipid structure of sperm plasma membrane, aiming an efficient transduction of ATP from mitochondria, which is essential for sperm motility acquisition [40]. In the present work, a positive correlation between sperm straightness and progressive motility and plasmatic membrane integrity in the epididymal corpus and cauda, respectively, were verified. Therefore, in dogs, the observed increase in sperm motility in the epididymal caput and corpus is presumably related to plasma membrane integrity, which can be dependent on the incorporation of polyunsaturated fatty acids to the sperm plasma membrane at the end of epididymal maturation.

Regarding the total fatty acids content (saturated, monounsaturated and polyunsaturated), we observed differences in sperm lipid profile throughout epididymal maturation. Modifications on the primary lipid structure of the sperm plasmatic membrane are related to adhesion of proteins and fatty acids of the epididymal lumen [1, 5, 17]. Thus, modifications during sperm maturation in the epididymides occur according to the protein and lipid profile of the epididymal fluid [41]. In fact, in our experiment, a higher concentration of fatty acids, such as stearic and pentadecenoic fatty acids, was observed in the epididymal cauda fluid, where sperm reaches its highest degree of maturation. Such fatty acids were observed previously in ejaculated semen in dogs [20, 42]. In addition, we observed increase in saturated fatty acids between the epididymal corpus and cauda, but not for DHA concentration. Therefore, it can be inferred that during the sperm transit between corpus and cauda epididymides, DHA from the epididymal lumen was incorporated into sperm cells, thus consuming the DHA concentration of the epididymal fluid and increasing the lipid concentration of the sperm plasmatic membrane. For these reasons, in dogs, fatty acids of the epididymal fluid may act as signaling and regulators of the lipid secretion by the epididymal cells at caput and corpus level, as they are later incorporated into the spermatozoa at the final stages of sperm maturation.

CONCLUSIONS

In conclusion, caput and corpus sperm are immature and have low fertilizing potential, as decreased sperm motility and plasma membrane integrity were verified at such stages of sperm maturation in dogs. However, cauda sperm acquire motility and membrane integrity simultaneously to changes in sperm plasma membrane lipid content through epididymal transit. Moreover, sperm lipid profile derives from incorporation of fatty acids from the epididymal fluid, thus, modulating the biochemical content of the sperm plasma membrane.

Our results indicate that the use of spermatozoa from epididymal cauda can be a viable alternative for reproductive biotechnologies in canids, since cauda sperm shows similarities with ejaculated semen in regards to motility and membrane integrity. Furthermore, this study provides information for future researches on contraception, such as the use of inhibitors of the enzymes that are responsible for fatty acid biosynthesis (e.g., stearoyl-CoA desaturase), promoting lipid disruption of sperm from epididymal cauda. On the other hand, the present data also contribute to gamete manipulation techniques, e.g., in vitro sperm maturation, highlighting the importance of fatty acids supplementation (e.g., DHA, caprylic and stearic fatty acids) for immature canine spermatozoa. Furthermore, there is a need for further studies using a larger number of animals in order to improve our comprehension on the lipid modulation during the epididymal maturation in dogs.

ACKNOWLEDGEMENTS

Not applicable.

Funding

This research was financially supported by FAPESP 2011/15689-9.

AUTHORS' CONTRIBUTIONS

Dr. A, N, and V designed the experimental study. Dr. A, L, L and V performed sample collection and analyzed sperm. Dr. J-F, A and N performed the fatty acid analysis. Dr. A, N and V drafted the manuscript, whereas Dr. V provided financial support, completed critical revision and approval of the article. All authors read and approved the final manuscript.

COMPETING INTERESTS

The authors declare that they have no competing interests.

REFERENCES

1. Fouchecourt S, Metayer S, Locatelli A, Dacheux F, Dacheux JL. Stallion epididymal fluid proteome: qualitative and quantitative characterization; secretion and dynamic changes of major proteins. Biol Reprod. 2000;62(6):1790–803.
2. Jervis KM, Robaire B. Dynamic changes in gene expression along the rat epididymis. Biol Reprod. 2001;65(3):696–703.
3. Amann RP, Hammerstedt RH, Veeramachaneni DN. The epididymis and sperm maturation: a perspective. Reprod Fertil Dev. 1993;5(4):361–81.
4. Petruszak JA, Nehme CL, Bartles JR. Endoproteolytic cleavage in the extracellular domain of the integral plasma membrane protein CE9 precedes its redistribution from the posterior to the anterior tail of the rat spermatozoon during epididymal maturation. J Cell Biol. 1991;114(5):917–27. PubMed PMID: 1714914, PubMed Central PMCID: PMC2289119,
5. Parks JE, Hammerstedt RH. Development changes occurring in the lipids of ram epididymal spermatozoa plasma membrane. Biol Reprod. 1985;32(3):653–68.
6. Jones R. Plasma membrane structure and remodelling during sperm maturation in the epididymis. J Reprod Fertil Suppl. 1998;53:73–84.

7. Christova Y, James PS, Cooper TG, Jones R. Lipid diffusion in the plasma membrane of mouse spermatozoa: changes during epididymal maturation, effects of pH, osmotic pressure, and knockout of the c-ros gene. J Androl. 2002;23(3):384–92.
8. Lenzi A, Picardo M, Gandini L, Dondero F. Lipids of the sperm plasma membrane: from polyunsaturated fatty acids considered as markers of sperm function to possible scavenger therapy. Hum Reprod Update. 1996;2(3):246–56.
9. Belleannee C, Belghazi M, Labas V, Teixeira-Gomes AP, Gatti JL, Dacheux JL, et al. Purification and identification of sperm surface proteins and changes during epididymal maturation. Proteomics. 2011;11(10):1952–64.
10. Flesch FM, Gadella BM. Dynamics of the mammalian sperm plasma membrane in the process of fertilization. Biochim Biophys Acta. 2000;1469(3):197–235.
11. Martinez-Soto JC, Landeras J, Gadea J. Spermatozoa and seminal plasma fatty acids as predictors of cryopreservation success. Andrology. 2013;1(3):365–75.
12. Rooke JA, Shao CC, Speake BK. Effects of feeding tuna oil on the lipid composition of pig spermatozoa and in vitro characteristics of semen. Reproduction. 2001;121(2):315–22.
13. Lenzi A, Gandini L, Lombardo F, Picardo M, Maresca V, Panfili E, et al. Polyunsaturated fatty acids of germ cell membranes, glutathione and blutathione-dependent enzyme-PHGPx: from basic to clinic. Contraception. 2002;65(4):301–4.
14. Devi LG, Shivaji S. Computerized analysis of the motility parameters of hamster spermatozoa during maturation. Mol Reprod Dev. 1994;38(1):94–106.
15. Martinez-Pastor F, Diaz-Corujo AR, Anel E, Herraez P, Anel L, de Paz P. Post mortem time and season alter subpopulation characteristics of Iberian red deer epididymal sperm. Theriogenology. 2005;64(4):958–74.
16. Varesi S, Vernocchi V, Faustini M, Luvoni GC. Morphological and acrosomal changes of canine spermatozoa during epididymal transit. Acta Vet Scand. 2013;55:17.
17. Acott TS, Katz DF, Hoskins DD. Movement characteristics of bovine epididymal spermatozoa: effects of forward motility protein and epididymal maturation. Biol Reprod. 1983;29(2):389–99. Epub 1983/09/01.
18. de Souza FF, Barreto CS, Lopes MD. Characteristics of seminal plasma proteins and their correlation with canine semen analysis. Theriogenology. 2007;68(1):100–6.
19. Darin-Bennett A, Poulos A, White IG. The phospholipids and phospholipid-bound fatty acids and aldehydes of dog and fowl spermatozoa. J Reprod Fertil. 1974;41(2):471–4. Epub 1974/12/01.
20. Diaz R, Inostroza K, Risopatron J, Sanchez R, Sepulveda N. Identification of fatty acids in canine seminal plasma. Andrologia. 2014;46(2):194–7.
21. Ma L, Yu H, Ni Z, Hu S, Ma W, Chu C, et al. Spink13, an epididymis-specific gene of the kazal-type serine protease inhibitor (SPINK) family, is essential for the acrosomal integrity and male fertility. J Biol Chem. 2013;288(14):10154–65. PMCID: PMC3617258,
22. Hori T, Uehara Y, Kawakami E, Tsutsui T. Influence of the time between removal and cooling of the canine epididymis on post-thaw caudal epididymal sperm quality. J Vet Med Sci. 2009;71(6):811–5.
23. Angrimani DS, Losano JD, Lucio CF, Veiga GA, Pereda MC, Nichi M, et al. Role of residual cytoplasm on oxidative status during sperm maturation in dogs. Anim Reprod Sci. 2014;151(3–4):256–61.
24. Kaabi M, Paz P, Alvarez M, Anel E, Boixo JC, Rouissi H, et al. Effect of epididymis handling conditions on the quality of ram spermatozoa recovered post-mortem. Theriogenology. 2003;60(7):1249–59. Epub 2003/09/27.
25. Angrimani DS, Lucio CF, Veiga GA, Silva LC, Regazzi FM, Nichi M, et al. Sperm maturation in dogs: sperm profile and enzymatic antioxidant status in ejaculated and epididymal spermatozoa. Andrologia. 2014;46(7):814–9.
26. Iguer-ouada M, Verstegen JP. Evaluation of the "Hamilton thorn computer-based automated system" for dog semen analysis. Theriogenology. 2001;55(3):733–49.

27. Lagergren CG. On the eosin-nigrosin stain and some other methods for the appraisal of sperm vitality with special reference to practical application. Ann Ostet Ginecol. 1953;75(6):998–1005.
28. Intasqui P, Camargo M, Antoniassi MP, Cedenho AP, Carvalho VM, Cardozo KH, et al. Association between the seminal plasma proteome and sperm functional traits. Fertil Steril. 2016;105(3):617–28.
29. Intasqui P, Antoniassi MP, Camargo M, Nichi M, Carvalho VM, Cardozo KH, et al. Differences in the seminal plasma proteome are associated with oxidative stress levels in men with normal semen parameters. Fertil Steril. 2015;104(2):292–301.
30. Lepage G, Roy CC. Improved recovery of fatty acid through direct transesterification without prior extraction or purification. J Lipid Res. 1984;25(12):1391–6.
31. Liu KS. Preparation of fatty-acid methyl esters for Gas-chromatographic analysis of lipids in biological-materials. J Am Oil Chem Soc. 1994;71(11):1179–87.
32. Keber R, Rozman D, Horvat S. Sterols in spermatogenesis and sperm maturation. J Lipid Res. 2013;54(1):20–33.
33. Lin YQ, Li JY, Wang HY, Liu J, Zhang CL, Wang WT, et al. Cloning and identification of a novel sperm binding protein, HEL-75, with antibacterial activity and expressed in the human epididymis. Hum Reprod. 2008;23(9):2086–94.
34. Ortega Ferrusola C, Gonzalez Fernandez L, Morrell JM, Salazar Sandoval C, Macias Garcia B, Rodriguez-Martinez H, et al. Lipid peroxidation, assessed with BODIPY-C11, increases after cryopreservation of stallion spermatozoa, is stallion-dependent and is related to apoptotic-like changes. Reproduction. 2009;138(1):55–63.
35. Irvine DS, Twigg JP, Gordon EL, Fulton N, Milne PA, Aitken RJ. DNA integrity in human spermatozoa: relationships with semen quality. J Androl. 2000;21(1):33–44.
36. Jones RE, Plymate SR. Evidence for the regulation of fatty acid utilization in human sperm by docosahexaenoic acid. Biol Reprod. 1988;39(1):76–80.
37. Lenzi A, Picardo M, Gandini L, Lombardo F, Terminali O, Passi S, et al. Glutathione treatment of dyspermia: effect on the lipoperoxidation process. Hum Reprod. 1994;9(11):2044–50.
38. Lenzi A, Gandini L, Maresca V, Rago R, Sgro P, Dondero F, et al. Fatty acid composition of spermatozoa and immature germ cells. Mol Hum Reprod. 2000;6(3):226–31.
39. Nissen HP, Kreysel HW. Polyunsaturated fatty acids in relation to sperm motility. Andrologia. 1983;15(3):264–9.
40. Amaral A, Lourenco B, Marques M, Ramalho-Santos J. Mitochondria functionality and sperm quality. Reproduction. 2013;146(5):R163–74.
41. Guyonnet B, Dacheux F, Dacheux JL, Gatti JL. The epididymal transcriptome and proteome provide some insights into new epididymal regulations. J Androl. 2011;32(6):651–64.
42. delBarco-Trillo J, Roldan ER. Effects of metabolic rate and sperm competition on the fatty-acid composition of mammalian sperm. J Evol Biol. 2014;27(1):55–62.

CHAPTER 11

Adipose Tissue Lipolysis and Remodeling During The Transition Period of Dairy Cows

G. Andres Contreras* , Clarissa Strieder-Barboza and William Raphael

Department of Large Animal Clinical Sciences, College of Veterinary Medicine, Michigan State University, East Lansing, MI 48824, USA

ABSTRACT

Elevated concentrations of plasma fatty acids in transition dairy cows are significantly associated with increased disease susceptibility and poor lactation performance. The main source of plasma fatty acids throughout the transition period is lipolysis from adipose tissue depots. During this time, plasma fatty acids serve as a source of calories mitigating the negative energy balance prompted by copious milk synthesis and limited dry matter intake. Past research has demonstrated that lipolysis in the adipose organ is a complex process that includes not only the activation of lipolytic pathways in response to neural, hormonal, or paracrine stimuli, but also important changes in the structure and cellular distribution of the tissue in a process known as adipose tissue remodeling. This process involves an inflammatory response with immune cell migration, proliferation of the cellular components of the stromal vascular fraction, and changes in the extracellular matrix. This review summarizes current knowledge on lipolysis in dairy cattle, expands on the new field of adipose tissue remodeling, and discusses how these biological processes affect transition cow health and productivity.

Keywords: Adipose tissue macrophages Adipose tissue remodeling Lipolysis Transition dairy cows

BACKGROUND

Adipose tissues, the energy warehouse during the transition period

Adipose tissue (AT) functions as the major body of energy reserve in mammals. During positive energy balance, AT stores energy surplus as fatty acids (FA) incorporated into triacylglycerols (TAG) in a process known as lipogenesis. In de novo lipogenesis, also defined as de novo FA synthesis, FA are derived from carbohydrate and amino acid carbons through acetylCoA. Additionally, AT directly esterifies circulating free FA from dietary or other metabolic origins (e.g. hepatic lipolysis) [1, 2]. The majority of FA incorporated into TAG come from de novo FA synthesis, however, circulating FA may supply up to 50% of FA, especially during early lactation [2, 3]. When energy is limited, the AT releases FA from TAG through lipolysis [4]. Due to its unique lipogenic and lipolytic responses, AT is the only organ capable of extensive growth and reduction at any stage of life [5].

The transition period of dairy cows is defined as the time when the physiological state changes from gestational non-lactating to non-gestational lactating and occurs from 3 wks before through 3 wks after parturition [4]. During the transition period, hormonal changes associated with parturition and the initiation of lactogenesis, including the reduction in progesterone and the surge of prolactin and growth hormone (GH), favor lipolysis over lipogenesis independently of energy balance status [6]. Lipolysis rate is further exacerbated by intense energy requirements, especially glucose, associated with the rapid fetal growth, parturition, the onset of lactation, and reduced dry matter intake, leading to negative energy balance (NEB) [7, 8]. Intense and protracted lipolysis during the transition period leads to AT size reduction and alterations in the secretion pattern of adipokines that favor the use of FA as an energy source for bodily maintenance and glucose for fetal growth and lactation [8, 9]. In addition, lipolysis induces a remodeling process within the adipose organ that is characterized by an inflammatory response, immune cell infiltration, cellular proliferation, and extracellular matrix (ECM) changes [10, 11]. During the transition period, lipolysis-induced AT remodeling coincides with a period of reduced insulin sensitivity in myocytes, hepatocytes, and adipocytes that redirects energy for milk production in the mammary gland [12]. As lactation progresses, bovine adipocytes become more responsive to insulin, resulting in reduced lipolysis rates and enhanced lipogenesis [2]. AT insulin resistance during the transition period ensures a healthy and productive lactation when moderated; however, when intense and protracted, it predisposes cows to inflammatory and metabolic diseases by limiting AT capacity for energy buffering and contributing to chronically increased plasma FA [12, 13]. In this review, we will discuss the different aspects of adipose tissue lipolysis and remodeling during the transition period, and their impact on transition cow health and lactation performance.

LIPOLYSIS

A key characteristic of the transition period is intense lipolysis. In general, lipolysis can be broadly divided into two categories: basal and demand lipolysis [14]. In humans and rodents, the rate of basal lipolysis is determined by adipocyte size and TAG content [10, 15]. Similarly, in dairy cows, basal lipolysis rate is positively associated with adipocyte size and increases steadily throughout lactation [16, 17, 18, 19]. In contrast, demand lipolysis is regulated hormonally in response to energy demands. Independent of the type of lipolysis, TAG within the adipocyte lipid droplet are broken down by the action of three different lipases: adipose triglyceride lipase (ATGL), hormone sensitive lipase (HSL), and monoglyceride lipase (MGL). However, HSL and MGL have a larger role in demand lipolysis than basal lipolysis. The actions of these lipolytic enzymes, explained in subsequent paragraphs, are regulated by different co-activator proteins that also exhibit changes in expression during the transition period.

Lipases

The primary regulator of basal lipolysis in monogastric animals is ATGL (Table 1). This lipase acts exclusively on TAG. ATGL is activated by CGI-58, also known as α/β hydrolase domain-containing protein 5 [20]. Upon lipolytic stimulus and consequent phosphorylation of lipid droplet associated protein perilipin 1 (PLIN1), CGI-58 is released into the cytoplasm and activates ATGL by direct protein-to-protein interaction [20]. In transition dairy cows, ATGL-encoding gene, *PNPLA2*, is downregulated during the last week of gestation and the first week of lactation compared to the dry period and mid lactation [21, 22]. Similar to the *PNPLA2* expression pattern, ATGL abundance remains decreased during the transition period in comparison to the content observed at

4 wks pre-calving [21]. CGI-58 gene expression is also reduced after calving; however, its protein abundance remains unchanged as the transition period progresses [21, 23]. The patterns of gene expression and the protein abundance of ATGL, as well as its coactivator CGI-58, indicate that basal lipolysis is maintained at a steady rate, suggesting that demand lipolysis is the main source of plasma FA during the transition period.

Table 1. Major components of lipolysis pathways in adipose tissue (AT) of dairy cows

Protein(s)	Encoding gene(s)	Activation stimulus	Function(s)	Transition period dynamics
Catecholamines (adrenaline and noradrenaline)		Metabolic stress [125]	- Bind to βAR on the adipocyte cell surface and initiate the lipolytic cascade [33]	- Plasma adrenaline and noradrenaline: - Increase during the dry period and immediately before calving - Decrease slightly after calving until 60 d in lactation [62]
Adipocyte β-adrenergic receptors (βAR), types $β_1R$, $β_2R$, and $β_3R$	*B1AR* *B2AR* *B3AR*	Catecholamine binding	- Activate adenylyl cyclases that convert ATP to cAMP, and induce PKA activation that, in turn, phosphorylates PLIN1 initiating the lipolytic cascade - βAR are increased in abdominal adipose and during increased rate of lipolysis [33]	- βAR increase during first month after parturition and are positively associated with milk production [27] - B2AR decreases immediately after parturition and increases after peak lactation [27]
Perilipin-1	*PLIN1*	PKA	- Protects the lipid droplet from hydrolytic activity of HSL	- Decreases during first 3 wks following parturition compared to other periods of the lactation cycle [21]
α/β hydrolase domain-containing protein 5 (ABHD5/CGI-58)	*ABHD5*	Phosphorylation of PLIN1	- Increases the activity of ATGL and provides increased DAG substrate for PKA-activated HSL	- Decreased gene expression after calving - Protein abundance remains unchanged through the transition period [21]
Adipose tissue triglyceride lipase (ATGL)	*PNPLA2*	ABHD5/CGI-58	- Hydrolysis of TAG [c]	- Decreased gene expression during the last week of gestation and first week of lactation compared to the dry period and mid lactation - Lower protein abundance around parturition compared to the dry period [21, 22]
Hormone sensitive lipase (HSL)	*LIPE*	Activation of PKA	- Hydrolyzes DAG [b], releasing free fatty acid and MAG [a] - Activity varies depending on anatomical location of AT	- Decreased gene expression during the first 3 wks following parturition compared to the dry period [23] - Protein content unchanged through the transition period [21, 123]
Fatty acid-binding protein (aP2/FABP4)	*FABP4*	Association with HSL	- Transports fatty acids to the plasma membrane	- Increased gene expression during induced milk fat depression [126]
Monoglyceride lipase (MGL)	*MGll*	DAG [b] hydrolysis by HSL	- Hydrolyzes MAG [a], releasing glycerol	- Decreased gene expression during the first 3 wks of lactation [27] - Protein content currently unknown in cattle

[a] MAG: Monoglycerol
[b] DAG: Diglycerol
[c] TAG: Triglycerol

HSL is the rate-limiting factor for demand lipolysis ([24, 25], Table 1). This lipase hydrolyses several lipid substrates including TAG, diglycerides, monoglycerides, and cholesterol esters [14]. Upon activation by protein kinase A (PKA), HSL associates with fatty acid binding protein 4 (FABP4) to form a complex that localizes on the lipid droplet. The expression of *LIPE*, the gene encoding HSL, is reduced during the first 3 wks after parturition compared to the dry period [16, 23]. The protein content of HSL remains unchanged throughout the transition period.

However, the rate of phosphorylation at different active sites, and therefore its lipolytic activity, is increased during the first 3 wks after calving [21]. In rodents, HSL requires phosphorylation at Ser^{563}, Ser^{659}, and Ser^{660} by PKA or protein kinase G (PKG) to initiate TAG hydrolysis. Although HSL serine residue activation sites have not been determined in transition cattle, increased phosphorylation at Ser^{563} and Ser^{660} was reported postpartum and during feed restriction protocols [21, 26]. Notably, HSL activity varies depending on anatomical location of adipose tissues. Locher et al. [26] reported higher phosphorylation rates at Ser^{563} and Ser^{660} in retroperitoneal AT from dairy cows at 21 d of lactation compared to subcutaneous depots. In the same study [26], phosphorylation rates, especially those at Ser^{563}, were associated with higher lipolytic activity.

A complete activation of the lipolytic process requires not only the activation of ATGL and HSL, but also the phosphorylation of PLIN1. This protein protects the lipid droplet from the lipolytic activity of HSL. PLIN1 is phosphorylated by PKA to allow HSL interaction with TAG. In dairy cows, PLIN1 phosphorylation is increased during the first 3 wks after parturition, compared to other periods of the lactation cycle [21]. In transition dairy cows, lipolysis is modulated by post-transcriptional and allosteric changes in the HSL hydrolase pathway. Although differences in lipolysis rate are reported depending on the adipose tissues' anatomical site, it is currently unknown if these changes may impact disease susceptibility or lactation efficiency.

The lipolytic pathway is completed by MGL, which acts exclusively on monoglycerides (Table 1). In dairy cows, the gene transcription patterns of *MGll*, which encodes MGL, follow those of *PNPLA2* and *LIPE*, with lower expression during the first 3 wks of lactation [27]. The dynamics of MGL protein expression are currently unknown in cattle, and it is uncertain if changes in its activity may affect lipolysis rates, as HSL is very active on monoglycerides and MGL hydrolyses these lipid species exclusively.

Regulation of lipolysis during the transition period

In rodents and humans, catecholamines and natriuretic peptides are the most important activators of lipolysis in AT [28]. In the same species, both insulin and catecholamines (acting through alpha adrenergic receptors) are negative regulators of TAG hydrolysis (Table 1). In cattle, the role of catecholamines and insulin in lipolysis modulation is well-documented; however, it is unknown if natriuretic peptides significantly modify lipolytic pathways.

Catecholamines bind to β-adrenergic receptors (βAR) on the adipocyte cell surface, activating adenylyl cyclases that convert ATP to cAMP (Table 1). Accumulation of cAMP induces PKA activation, which in turn phosphorylates PLIN1 and initiates the lipolytic cascade [29]. There are 3 major types of βAR: β_1, β_2, and β_3. In dairy cows, the expression of all βAR-encoding genes was initially reported in mammary gland [30], and,more recently, in subcutaneous AT [16, 27] Only β_1 and β_2 were demonstrated to have lipolytic responses in the AT of cows, and stimulation of the latter was directly associated with increases in plasma FA during NEB [31]. Similar to the gene expression of lipolytic enzymes, the transcription of *B2AR* (encoding β_2AR) is downregulated immediately after parturition and rises again only after peak lactation [27]. A classical study by Jaster and Wegner [32] revealed that the activity and responses to adrenergic stimuli of the βAR in the subcutaneous AT was increased during the first month after parturition compared to the dry period. The dynamics of AT βAR response during the dry and lactation periods were completely characterized by McNamara and Hillers [17] who demonstrated that lipolytic responses to epinephrine peak at 1 month postpartum and remain elevated 6 months into lactation compared to 1 month pre calving. In the same study, lipolytic responses to norepinephrine stimulation remained higher during lactation compared to the dry period. Strikingly, the lipolytic response triggered by βAR stimulation in early lactation is increased in proportion to milk production [27]. Although not described in dairy cows, βAR density and the lipolytic activity in adipocytes after adrenergic stimulation may vary depending on the anatomical location of the adipose depot. In sheep, there are more βAR present in omental adipocytes, compared to subcutaneous adipocytes, after 2 wks of lactation [6]. The impact of the higher content of βAR in abdominal AT is reflected in the higher rate of lipolysis from these depots during NEB states related to lactation. In fact, the FA profiles of the plasma NEFA fraction in cows with intense lipolysis, such as those with displaced abomasum, are remarkably similar to the FA composition of visceral adipose depots, especially in the content of saturated and monounsaturated FA [33].

Adipocytes are one of the most highly insulin-responsive cell types [34]. In adipocytes, insulin stimulates glucose transport and lipogenesis, promotes the uptake of FA from systemic circulation, and inhibits lipolysis [34, 35]. Insulin binds and activates the insulin receptor tyrosine kinase, resulting in the phosphorylation of insulin receptor substrates 1 and 2 (IRS-1/IRS-2) [36]. These events are followed by activation of phosphatidylinositol 3-kinase (PI3K), phophoinositide-dependent kinase1 (PDK1), and protein kinase B (AKT), which mediate insulin metabolic and mitogenic effects, including glucose uptake through glucose transporters (e.g. GLUT4) [35]. Insulin suppresses lipolysis through the activation of AKT, which results in the inhibition of downstream protein kinase A

(PKA) and reduces PKA phosphorylation of PLIN1 [37]. In dairy cows, decreased circulating insulin and AKT phosphorylation in the liver and AT stimulate gluconeogenesis and lipolysis during the transition period and early lactation [8, 12, 38].

During the transition period, changes in the secretion patterns of GH, angiopoietin-like 4 (ANGPTL4), and prolactin further modulate lipolysis. GH activates lipolysis and reduces insulin sensitivity in mammalian AT [39]. In lactating dairy cattle, GH administration augments 2 to 6-fold AT response to adrenergic stimulation [40]. In vitro experiments with AT from lactating animals indicate that GH reduces adipocyte sensitivity to anti-lipolytic molecules, such as adenosine, that inhibit the activity of adenylyl cyclase [41]. Thus, compared to non-lactating cows 4 to 8 wks prior to parturition, cows in the transition period have elevated circulating GH which enhances lipolysis by increasing adipocyte responses to adrenergic stimuli and reducing the inhibitory effects on sympathetic activity through adenosine.

ANGPTL4, also known as fasting-induced adipose factor, is an adipokine secreted during NEB that inhibits adipocyte uptake of FA for esterification [42]. ANGPTL4 plays a key role in enhancing lipolysis in adipocytes, especially during catecholamine stimulation, by increasing cAMP and enhancing the phosphorylation of PKA [43]. ANGPTL4 transcription and synthesis is elevated in response to glucocorticoids [43], possibly indicating that this adipokine facilitates lipolytic response during stress periods, such as parturition. In dairy cows, Koltes and Spurlock [44] described the dynamics of *ANGPTLl4* gene expression following lipolysis induction with GH, as well as during NEB states, including the transition period and during feed restriction. *ANGPTL4* transcription was inversely associated with the degree of NEB in all three models of limited energy status. Although ANGPTL4 protein content in AT from transition cows has not been determined, it is expected to reflect *ANGPTL4* transcription, similar to what has been observed in rodents [45]. Thus, in dairy cattle, ANGPTL4 supports lipolysis during NEB stages, including the transition period.

In addition to supporting lactation, prolactin promotes adipocyte lipolysis in vivo. Furthermore, continuous exposure of fat cells to this pituitary hormone have been shown to reduce AT size in rabbits [46]. However, it is unknown if prolactin exerts the same effect in bovine adipocytes, and its effect on modulating lipolysis intensity during the transition period remains to be elucidated.

ADIPOKINES

The role of AT is now recognized as a major regulator of systemic metabolism which extends beyond energy buffering [47]. This function is accomplished by secreting specialized proteins that exert autocrine, paracrine, and endocrine functions. These proteins are termed adipokines and are produced by the cellular components of AT, such as adipocytes and cells of the stromal vascular fraction (SVF), including immune, vascular, and adipocyte progenitor cells. Despite the ever-expanding list of adipokines, which now accounts for over 300 secretory products [48], few have been studied in dairy cattle (Table 2). Among these signaling molecules, adiponectin and leptin are almost exclusively secreted by AT; while others, such as resistin and retinol binding protein 4, are also produced in the liver. Similarly, interleukin-6 (IL-6) and tumor necrosis factor alpha (TNFα) are not only produced by adipocytes, but also resident immune cells in AT (e.g. macrophages, lymphocytes, polymorphonuclear cells). Recent research provides evidence that the dynamics of adipokine secretion during the transition period drive the homeorhetic potential of dairy cows by redirecting glucose to the mammary gland, increasing FA flow to the liver, and modulating energy intake [49].

Table 2. Adipokine expression dynamics and roles in modulating adipose tissue (AT) lipolysis in transition dairy cows

Protein	Encoding gene	Function(s)	Transition period dynamics
Angiopoietin-like 4 (ANGPTL4)	*ANGPTL4*	- Inhibits LPL activity and adipocyte fatty acid (FA) uptake [42] - Increases cAMP production and enhances PKA phosphorylation through the integration of catecholamine and corticoid signaling [43] - Regulated by PPARs [127]	- Gene expression inversely associated with NEB intensity during transition period [44]
Adiponectin (Acrp30)	*ADIPOQ*	- Improves insulin sensitivity and promotes lipogenesis in adipocytes [9] - Enhances FA β-oxidation in myocytes and hepatocytes [9] - Promotes anti-inflammatory and resolving phenotypes in immune cells [54]	- Decreased gene expression immediately after calving compared to dry period - Plasma concentration peaks between 40–70 d of lactation [53, 54] - Decreased AT expression of its receptors (adipoR1 and adipoR2) during the first 3 wks after calving compared to dry period, increasing steadily to peak around 100 d of lactation [55, 56]
Leptin	*OB*	- Decreases lipogenesis and increases lipolysis and FA oxidation [59] - Reduction in leptin transcription and secretion during the transition period may promote a rapid return to normal DMI [62]	- Leptinemia peaks during the dry period and decreases during the first week following calving [60, 61] - Increased AT gene expression during the dry period contrasted with minimal transcription immediately following parturition [60]
Resistin	*RETN*	- Increases lipolytic rate and the transcription of ATGL and HSL in AT explants from transition cows [66]	- Increased in plasma after calving, returning to pre-partum concentration by 6 wks into lactation [66, 67] - AT transcription and secretion dynamics are similar to that of plasma resistin [66]
Retinol binding protein 4 (RBP4)	RBP4	- Serves as a carrier of retinol [68] - Impairs glucose uptake by adipocytes and hepatocytes by suppressing insulin signaling pathways [69] - Inhibits adipogenesis [70]	- Decreased circulating RBP4 during the first day after parturition followed by return to pre-calving levels by second week of lactation [71, 72]

Adiponectin, also known as ACRP30, is secreted primarily by adipocytes, but is also expressed by cardiomyocytes and skeletal muscle [50]. Adiponectin improves insulin sensitivity and lipogenesis in adipocytes, and FA β-oxidation in myocytes and hepatocytes [9]. These effects are exerted by the activation of its receptors (adipoR1 and adipoR2), which are expressed in liver, AT, and skeletal muscle [51]. In monogastric animals, circulating adiponectin is present in 3 major oligomeric forms (low-molecular weight (LMW), middle molecular weight (MMW), high molecular weight (HMW) and as globular adiponectin [51]. In dairy cows, circulating adiponectin is composed mainly of high molecular weight complexes and its distribution is not affected by stage of lactation [49, 52]. Circulating adiponectin reaches its nadir immediately after calving, then peaks between 40 and 70 d into lactation [53, 54]. Remarkably, concentrations of circulating adiponectin are inversely associated with plasma FA, the main lipolysis biomarker in dairy cows [54]. Similarly, AT expression of genes encoding adipoR1 and adipoR2 is downregulated during the first 3 weeks after calving and then increases steadily to peak at around 100 DIM [55, 56]. The dynamics of the expression and secretion of adiponectin and its receptors may indicate that this adipokine acts as an autocrine, paracrine, and endocrine modulator of the homeorhetic adaptations of AT during the transition period. However, the extent to which adiponectin affects glucose partitioning to the mammary gland in early lactation is currently unknown.

Leptin was one of the first adipokines characterized in mammals. Unlike adiponectin, leptin is expressed and secreted not only by adipocytes, but also by myocytes and mammary gland and gastric mucosa epithelial cells [57]. In dairy cows and other ruminants, the leptin encoding gene (*OB*) is expressed ubiquitously, including in the rumen, abomasum, duodenum, mammary gland, skeletal muscle, pituitary gland, and AT (reviewed in [58]). Similar to adiponectin, leptin acts in an autocrine, paracrine, and endocrine manner to modulate food intake and energy expenditure. This anorexigenic adipokine reduces lipogenesis and increases lipolysis and FA oxidation [59]. In dairy cows, plasma leptin concentration peaks during the dry period and decreases drastically in the first week after calving [60, 61]. In AT, *OB* expression reflects leptinemia dynamics, with the highest expression in the dry period and minimal transcription immediately following parturition [60]. Reduction in leptin transcription and secretion during the transition period may promote a rapid return to normal DMI; however, it is presently unclear if, in over-conditioned cows, hyperleptinemia enhances the onset of pre-calving lipolysis and losses of AT reserves [62, 63].

Resistin is an adipokine secreted by adipocytes and macrophages [64]. AT transcription and secretion levels of resistin are linked to the development of insulin resistance in humans, as well as in rodent models of obesity and metabolic syndrome [65]. In transition dairy cows, plasma concentration of resistin peaks during the first week after calving and returns to pre-partum levels after 6 wks of lactation [66, 67]. Plasma resistin is positively associated with plasma NEFA and negatively correlated with milk production [66, 67]. AT resistin transcription and secretion dynamics are similar to that of plasma resistin, with higher expression during the first week post-calving compared to other stages of lactation and the dry period [66]. Remarkably, resistin was associated with higher rate of lipolysis, as well as enhanced transcription of ATGL and HSL, in AT explants from transition cows [66].

RBP4 is a lipocalin transport protein secreted by adipocytes and hepatocytes that serves as a carrier of retinol [68]. Like resistin, RBP4 impairs glucose uptake by adipocytes and hepatocytes by suppressing insulin signaling pathways [69]. Enhanced expression of RBP4 inhibits adipogenesis, thus impairing the FA buffering capacity of AT during obesity or periods of high lipolytic rate [70]. In dairy cows, plasma RBP4 declines sharply after parturition and returns to pre-calving levels by the second week of lactation [71, 72]. In the AT of transition cows, *RBP4* transcription was detected and found to be positively associated with TNFα secretion [72].

ADIPOSE TISSUE REMODELING

AT lipolysis not only involves the release of FA, neutral lipids, and glycerol, but also induces a remodeling process of the organ [10, 73]. This remodeling is characterized by an inflammatory response with immune cell migration, proliferation of the cellular components of the SVF, and changes in ECM of AT [5]. Intense research during the past two decades has characterized AT remodeling in human diseases that exhibit high AT lipolysis rates, such as obesity [5, 74], metabolic syndrome [75], and lipodystrophy [76]. More recently, different aspects of AT remodeling in dairy cows during the transition period and metabolic disease were evaluated. We summarize these reports from cattle studies below in the context of the current understanding of adipose tissue remodeling in humans and rodent models.

Immune cell infiltration

Studies in multiple animal species and humans show dynamic populations of immune cells located within the SVF of AT [77, 78, 79]. Cells of the innate and adaptive immune systems, such as macrophages, mast cells, lymphocytes, neutrophils, eosinophils, dendritic cells, and NK, exhibit active AT trafficking during health and disease in rodents and humans [80], as well as in dairy cows during the transition period and late lactation [81, 82]

Macrophages are the most predominant cell type in monogastric and ruminant SVF [77, 79]. Adipose tissue macrophages (ATM) are a key component of the inflammatory response during lipolysis. The specific inflammatory phenotype of ATM has been broadly classified in classical ATM (M1), which have active pro-inflammatory responses, and alternative phenotype ATM (M2), which promote inflammation resolution. In dairy cows, ATM infiltration occurs as a response to intense lipolysis. In cases of displaced abomasum in transition dairy cows, macrophages exhibiting a M1 phenotype accumulate in aggregates within omental and subcutaneous depots [81]. During induced NEB in late-lactation cows, lipolysis induced ATM infiltration into subcutaneous and visceral AT, yet no phenotypic change was observed [82].

The role of ATM during lipolysis is to remove lipolytic products such as FA, diglycerides, and monoglycerides that, in excess, cause lipotoxicity in AT [83]. ATM are also involved in the recruitment of new adipocyte progenitors by secreting chemotactic proteins such as osteopontin [83]. The transcription of this latter protein was shown to be upregulated in AT of lactating dairy goats during feed deprivation [84]. The ATM phenotype is plastic in response to the milieu of adipocyte-derived molecules, such as adipokines and FA [79]. During lipolysis, ATM bind to saturated FA at toll-like receptors (TLR), resulting in the activation of NF-κB transcription factor pathways and inflammatory gene transcription, leading to M1 polarization [85, 86, 87]. ATM secretion of cytokines such as TNFα and interleukins 1β and 6 during lipolysis mediates local and systemic inflammation,6 [79, 88]. Interestingly, macrophage infiltration into AT enhances lipolysis, creating a vicious cycle that connects lipolysis, ATM infiltration, and inflammation [89]. To date, this cycle has not been described in cows, but, if exists, could offer a novel explanation of prolonged, severe lipolysis in the transition period.

ATM exhibiting the M2 phenotype possess a restorative and protective function, in contrast to the pro-inflammatory M1 type [90]. The main immune cell-derived inducers of non-classical macrophages are interleukins 4, 10, and 13 [91]. In human and mouse studies, the alternative macrophage phenotype has been induced by omega-3 FA molecules and pharmacological agents through peroxisome proliferator-activated receptors [92, 93, 94]. Given the intense ATM polarization to M1 during excessive lipolysis in the transition period, the possibility of modulating ATM phenotype in transition dairy cows may offer therapeutic alternatives [81].

Reports on trafficking and inflammatory responses of adipose immune cells, other than ATM, in dairy cows are limited. T-cells were reported to comprise 3–5% of the SVF cells in cows, with or without displaced abomasum, during the transition period and in late lactation [81, 82, 95]. Around 4–7% of SVF cells are B lymphocytes and their numbers are higher in visceral AT than other AT. Interactions among different types of immune cells, adipocytes, and other cells in the SVF of AT modulate metabolic responses during the transition period. Therefore, a better understanding of immune cell population trafficking and inflammatory responses is necessary to develop strategies to modulate AT metabolism, especially lipolysis.

Cell proliferation

AT is a plastic organ that adapts to metabolic challenges by expansion or contraction of adipocyte size and numbers. Furthermore, adipocytes have a constant turnover that requires recruitment of new progenitors regularly [96]. Lipolysis induces the recruitment of new adipocytes in rodent models of obesity and adrenergic signaling [97]. In dairy cows, assessment of cell turnover is difficult because genetic tracing tools are not available for large animals. Markers of cell cycle activation, such as *CCNA2*, *CCNB2*, and *MKI67* have been used. In late-lactation cows, short-term intense lipolysis did not increase the expression of these genes [82]. In dairy heifers during peak lactation, Häussler et al. detected a 25-fold increase in the number of preadipocytes in the retroperitoneal AT [98], compared to heifers in mid lactation (105 DIM), using immunohistochemistry with antibodies against Ki67, a cell proliferation marker, and Pref-1 (preadipocyte factor). Remarkably, visceral AT apoptosis was also increased in early lactation compared to mid lactation, as demonstrated by the Terminal deoxynucleotidyl transferase dUTP nick end labeling (TUNEL) assay that detects DNA fragments as a measure of apoptosis. The implications of a potentially increased rate of cell proliferation in AT during the transition period is currently unknown. Increased preadipocyte proliferation, and therefore adipogenesis, especially in subcutaneous AT, would likely be beneficial to transition cow metabolism, by, for example, improving the AT capacity to buffer excess FA released during lipolysis, as has been shown in humans and rodent models of insulin signaling dysregulation [99]

Extracellular matrix

The ECM of AT gives structural support to its cellular components. Proteoglycans and fibrous proteins, such as collagens, are the main classes of AT ECM proteins, and their functions and distribution were recently reviewed in detail by Mariman and Wang [100]. In human obesity, characterized by excessive rates of basal lipolysis, there is enhanced deposition of collagens I and VI and thrombospondin-1 [101, 102, 103]. The presence of these ECM proteins is also associated with ATM infiltration and polarization to the M1 phenotype [102, 104]. High rates of lipolysis observed in cancer cachexia are also associated with changes in the ECM composition that lead to enhanced inflammatory responses by adipocytes [105, 106]. In late-lactation dairy cows, ECM composition is affected by anatomical location, with higher expression of collagens I and VI in subcutaneous fat compared to omental fat [82]. Reports on the dynamics of ECM remodeling during the transition period and early lactation are limited to gene expression evaluation. Akbar et al., [107] reported higher expression of collagen I in subcutaneous AT of transition cows with metritis, compared to healthy animals. In feed-deprived lactating goats, Faulconnier and colleagues described enhanced transcription of *COL3A1* [84]. A reanalysis of microarray data by Sumner-Thompson et al., [108] looking for enrichment in different ECM-related groupings via the Database for Annotation, Visualization, and Integrated Discovery revealed that *THBS1* expression is upregulated during peak lipolysis at the onset of lactation. Proteome analysis focused on the transition period is necessary to evaluate if lipolytic responses during the transition period induce changes in the composition of ECM that impair AT function.

IMPACT OF AT LIPOLYSIS AND REMODELING

Dysregulated inflammatory responses and oxidative stress are recognized as key components of the metabolic stress syndrome in transition dairy cows [7, 109]. AT remodeling and lipolysis can be considered the third key component, also leading to metabolic stress through two different mechanisms. First, and as demonstrated initially in non-ruminants, lipolysis modulates the phenotype of immune cells [110]. FA and other lipolytic products are potent activators of inflammatory pathways in mononuclear cells, lymphocytes, and polymorphonuclear cells [48, 111]. In dairy cows, FA impair the immune responses of lymphocytes and neutrophils by enhancing the pro-inflammatory response and simultaneously reduce their capacity to clear pathogens [64, 112]. Second, excessive lipolysis is a trigger of oxidative stress. In monogastrics, FA in circulation and in tissues become oxidized by free radicals and rapidly deplete antioxidant reserves, leading to oxidative stress [101]. In dairy cows, specific oxidized products of arachidonic acid, 11-hydroxyeicosatetraenoic acid (11-HETE), and linoleic acid, 9-hydroxyoctadecadienoic acid (9-HODE) and 13-HODE, significantly increase during the transition period coinciding with enhanced lipolytic rate [113]. Modulating the rate and intensity of FA release from AT through nutritional interventions, including supplementation of methyl donors choline and methionine [114], and ω-6 polyunsaturated FA [104, 115, 116], improves immune function, reduces oxidative stress, and enhances adipose insulin sensitivity. These studies further demonstrate the importance of lipolysis as a determinant factor for disease susceptibility in transition dairy cows.

Excessive lipolysis and AT remodeling may also impair lactation performance. Increased concentrations of lipolysis biomarkers NEFA and β-hydroxybutyrate during the transition period are associated with reduced milk production and impaired reproductive function [117]. In monogastrics, AT inflammation, a feature of AT remodeling, is directly linked with the development of AT-specific insulin resistance (IR) [11]. In transition cows a transient state of IR is considered homeorhetic, as it guarantees glucose supply for milk synthesis by limiting glucose use by peripheral tissues and triggering AT lipolysis [12]. However, extended periods of IR during the transition period may increase and prolong AT lipolysis, as observed in women with gestational diabetes and metabolic syndrome patients [118, 119]. During active AT remodeling, the expression and secretion of IL1-β, IL-6, resistin, and TNF-α by human adipocytes and ATM is increased, leading to impaired AT insulin signaling [74, 88]. In transition dairy cows, AT expression of IL-6, resistin, TNF-α, and other bioactive peptides associated with IR is increased compared to other stages of lactation [120, 121, 122]. Recent studies provide evidence for AT-specific IR during early lactation in high-yielding dairy cows. Zachut et. al [38] demonstrated a significant reduction in the phosphorylation of downstream insulin signaling pathways, such as IRS-1 and AKT, in subcutaneous adipose, while the activation of these pathways remained intact in the liver. Remarkably, adipose-specific IR was only observed in cows that had higher rates of lipolysis and lost more body condition during early lactation. Despite this evidence, the role of periparturient AT remodeling in the development of IR is still a matter of debate, as Mann and colleagues did not observe any changes in insulin signaling pathways in over-conditioned cows with excessive lipolytic rates [123]. Furthermore, throughout lactation the degree of IR is dynamic and was shown to increase during late lactation, a period characterized by reduced lipolytic activity [124]. These reports underscore that lipolysis regulation by insulin signaling and AT remodeling is a complex process that requires further characterization.

CONCLUSIONS

The impact of intense and extended periods of AT lipolysis on transition cow disease susceptibility and lactation performance may be directly related to the remodeling process, alterations in the expression of adipokines, and the development of insulin resistance (Figure 1). Certainly, AT biology research focus on human obesity and the diabetes epidemic in western countries has expanded our understanding of the role of lipolysis in metabolic and immune function. However, further research on AT lipolysis and remodeling is required to elucidate specific inflammatory and metabolic pathways that link adipocyte and immune cell function in dairy cattle. Improving our understanding of AT function in transition dairy cattle will lead to the identification of new biomarkers of disease and productivity, which will allow for improved herd health and profitability.

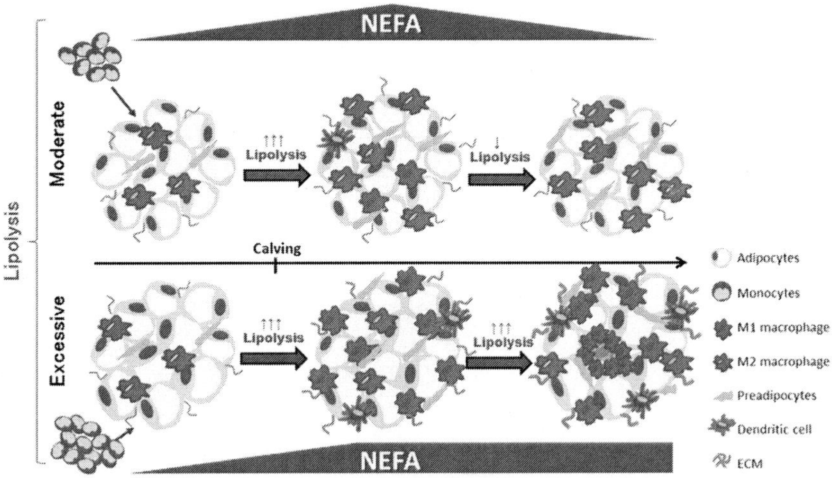

Figure 1. Lipolysis induces adipose tissue (AT) remodeling. This process is characterized by macrophage infiltration and changes in inflammatory phenotype. During moderate lipolysis, macrophage infiltration is limited and involves mainly the M2 phenotype (anti-inflammatory). In contrast, during excessive lipolysis, most infiltrating macrophages are M1 (pro-inflammatory) and enhance lipolysis and reduce adipocyte insulin sensitivity. Excessive lipolysis and AT remodeling increase disease susceptibility and negatively impact lactation performance

ACKNOWLEDGEMENTS

Authors acknowledge Sarah E. Schmidt for assistance with editing the manuscript.

Funding

This project was supported in part by USDA-NIFA (Washington, DC) grants 2014-68004-21972 and 2015-67015-23207, Department of Large Animal Clinical Sciences (East Lansing, MI), the Michigan State University Elwood Kirkpatrick Dairy Science Research Endowment (East Lansing, MI), Michigan Alliance for Animal Agriculture (East Lansing, Michigan), and Michigan Animal Health Foundation.

Availability of data and materials

Data sharing not applicable to this article as no datasets were generated or analyzed in this review article.

Authors' contributions

All authors read and approved the final manuscript. GAC, CSB, and WR reviewed literature available and wrote the manuscript.

Competing interests

The authors declare that they have no competing interest.

Consent for publication

Not applicable.

Ethics approval and consent to participate

Not applicable.

REFERENCES

1. Ameer F, Scandiuzzi L, Hasnain S, Kalbacher H, Zaidi N. De novo lipogenesis in health and disease. Metabolism. 2014;63(7):895–902.
2. McNamara J. Lipid metabolism in adipose tissue during lactation: a model of a metabolic control system. J Nutr. 1994;124(8 Suppl):1383S–91S.
3. McNamara J, Hillers J. Regulation of bovine adipose tissue metabolism during lactation. 1. Lipid synthesis in response to increased milk production and decreased energy intake1. J Dairy Sci. 1986;69(12):3032–41.
4. Contreras GA, Sordillo LM. Lipid mobilization and inflammatory responses during the transition period of dairy cows. Comp Immunol Microbiol Infect Dis. 2011;34(3):281–9.
5. Rutkowski JM, Stern JH, Scherer PE. The cell biology of fat expansion. J Cell Biol. 2015;208(5):501–12.
6. Vernon R, Doris R, Finley E, Houslay M, Kilgour E, Lindsay-Watt S. Effects of lactation on the signal transduction systems regulating lipolysis in sheep subcutaneous and omental adipose tissue. Biochem J. 1995;308:291–6.
7. Sordillo LM, Raphael W. Significance of metabolic stress, lipid mobilization, and inflammation on transition cow disorders. Vet Clin North Am Food Anim Pract. 2013;29(2):267–78.
8. Bell AW, Bauman DE. Adaptations of glucose metabolism during pregnancy and lactation. J Mammary Gland Biol Neoplasia. 1997;2(3):265–78.
9. Stern J, Rutkowski J, Scherer P. Adiponectin, leptin, and fatty acids in the maintenance of metabolic homeostasis through adipose tissue crosstalk. Cell Metab. 2016;23(5):770–84.
10. Kosteli A, Sugaru E, Haemmerle G, Martin JF, Lei J, Zechner R, et al. Weight loss and lipolysis promote a dynamic immune response in murine adipose tissue. J Clin Invest. 2010;120(10):3466–79.
11. Sun K, Kusminski CM, Scherer PE. Adipose tissue remodeling and obesity. J Clin Invest. 2011;121(6):2094–101.
12. De Koster JD, Opsomer G. Insulin resistance in dairy cows. Vet Clin North Am Food Anim Pract. 2013;29(2):299–322.
13. Faulkner A, Pollock HT. Metabolic responses to euglycaemic hyperinsulinaemia in lactating and non-lactating sheep in vivo. J Endocrinol. 1990;124(1):59–66.
14. Lafontan M, Langin D. Lipolysis and lipid mobilization in human adipose tissue. Prog Lipid Res. 2009;48(5):275–97.
15. Magkos F, Fraterrigo G, Yoshino J, Luecking C, Kirbach K, Kelly Shannon C, et al. Effects of moderate and subsequent progressive weight loss on metabolic function and adipose tissue biology in humans with obesity. Cell Metab. 2016;23(4):591–601.
16. Sumner JM, McNamara JP. Expression of lipolytic genes in the adipose tissue of pregnant and lactating Holstein dairy cattle1. J Dairy Sci. 2007;90(11):5237–46.
17. McNamara JP, Hillers JK. Adaptations in lipid metabolism of bovine adipose tissue in lactogenesis and lactation. J Lipid Res. 1986;27(2):150–7.
18. De Koster J, Van den Broeck W, Hulpio L, Claeys E, Van Eetvelde M, Hermans K, et al. Influence of adipocyte size and adipose depot on the in vitro lipolytic activity and insulin sensitivity of adipose tissue in dairy cows at the end of the dry period. J Dairy Sci. 2016;99(3):2319–28.
19. Yang YT, Baldwin RL. Lipolysis in isolated cow adipose cells. J Dairy Sci. 1973;56(3):366–74.
20. Granneman JG, Moore H-PH, Krishnamoorthy R, Rathod M. Perilipin controls lipolysis by regulating the interactions of AB-hydrolase containing 5 (Abhd5) and adipose triglyceride lipase (Atgl). J Biol Chem. 2009;284(50):34538–44.
21. Koltes DA, Spurlock DM. Coordination of lipid droplet-associated proteins during the transition period of Holstein dairy cows. J Dairy Sci. 2011;94(4):1839–48.

22. Ji P, Osorio JS, Drackley JK, Loor JJ. Overfeeding a moderate energy diet prepartum does not impair bovine subcutaneous adipose tissue insulin signal transduction and induces marked changes in peripartal gene network expression. J Dairy Sci. 2012;95(8):4333–51.
23. Khan MJ, Hosseini A, Burrell S, Rocco SM, McNamara JP, Loor JJ. Change in subcutaneous adipose tissue metabolism and gene network expression during the transition period in dairy cows, including differences due to sire genetic merit1. J Dairy Sci. 2013;96(4):2171–82.
24. Miyoshi H, Perfield JW, Obin MS, Greenberg AS. Adipose triglyceride lipase regulates basal lipolysis and lipid droplet size in adipocytes. J Cell Biochem. 2008;105(6):1430–6.
25. Lass A, Zimmermann R, Oberer M, Zechner R. Lipolysis – A highly regulated multi-enzyme complex mediates the catabolism of cellular fat stores. Prog Lipid Res. 2011;50(1):14–27.
26. Locher LF, Meyer N, Weber EM, Rehage J, Meyer U, Dänicke S, et al. Hormone-sensitive lipase protein expression and extent of phosphorylation in subcutaneous and retroperitoneal adipose tissues in the periparturient dairy cow. J Dairy Sci. 2011;94(9):4514–23.
27. Rocco SM, McNamara JP. Regulation of bovine adipose tissue metabolism during lactation. 7. Metabolism and gene expression as a function of genetic merit and dietary energy intake1. J Dairy Sci. 2013;96(5):3108–19.
28. Arner P, Langin D. Lipolysis in lipid turnover, cancer cachexia, and obesity-induced insulin resistance. Trends Endocrinol Metab. 2014;25(5):255–62.
29. Duncan RE, Ahmadian M, Jaworski K, Sarkadi-Nagy E, Sul HS. Regulation of lipolysis in adipocytes. Annu Rev Nutr. 2007;27(1):79–101.
30. Inderwies T, Pfaffl MW, Meyer HHD, Blum JW, Bruckmaier RM. Detection and quantification of mRNA expression of α- and β-adrenergic receptor subtypes in the mammary gland of dairy cows. Domest Anim Endocrinol. 2003;24(2):123–35.
31. Ferlay A, Chilliard Y. Effects of the infusion of non-selective beta-, and selective beta1- or beta2-adrenergic agonists, on body fat mobilisation in underfed or overfed non-pregnant heifers. Reprod Nutr Dev. 1999;39(4):409–21.
32. Jaster EH, Wegner TN. Beta-adrenergic receptor involvement in lipolysis of dairy cattle subcutaneous adipose tissue during dry and lactating state. J Dairy Sci. 1981;64(8):1655–63.
33. Hostens M, Fievez V, Leroy JLMR, Van Ranst J, Vlaeminck B, Opsomer G. The fatty acid profile of subcutaneous and abdominal fat in dairy cows with left displacement of the abomasum. J Dairy Sci. 2012;95(7):3756–65.
34. Kahn BB, Flier JS. Obesity and insulin resistance. J Clin Invest. 2000;106(4):473–81.
35. Laviola L, Perrini S, Cignarelli A, Natalicchio A, Leonardini A, De Stefano F, et al. Insulin signaling in human visceral and subcutaneous adipose tissue in vivo. Diabetes. 2006;55(4):952–61.
36. Kahn SE, Prigeon RL, Schwartz RS, Fujimoto WY, Knopp RH, Brunzell JD, et al. Obesity, body fat distribution, insulin sensitivity and islet β-cell function as explanations for metabolic diversity. J Nutr. 2001;131(2):354S–60S.
37. Choi SM, Tucker DF, Gross DN, Easton RM, DiPilato LM, Dean AS, et al. Insulin regulates adipocyte lipolysis via an akt-independent signaling pathway. Mol Cell Biol. 2010;30(21):5009–20.
38. Zachut M, Honig H, Striem S, Zick Y, Boura-Halfon S, Moallem U. Periparturient dairy cows do not exhibit hepatic insulin resistance, yet adipose-specific insulin resistance occurs in cows prone to high weight loss. J Dairy Sci. 2013;96(6):5656–69.
39. Nellemann B, Vendelbo MH, Nielsen TS, Bak AM, Høgild M, Pedersen SB, et al. Growth hormone-induced insulin resistance in human subjects involves reduced pyruvate dehydrogenase activity. Acta Physiol (Oxf). 2014;210(2):392–402.
40. Houseknecht KL, Dwyer DA, Lanna DPD, Bauman DE. Effect of somatotropin on adipose tissue metabolism: Ontogeny of the enhanced response to adrenergic challenge in the lactating cow. Domest Anim Endocrinol. 1995;12(1):105–13.

41. Houseknecht KL, Bauman DE. Regulation of lipolysis by somatotropin: functional alteration of adrenergic and adenosine signaling in bovine adipose tissue. J Endocrinol. 1997;152(3):465–75.
42. Kersten S. Regulation of lipid metabolism via angiopoietin-like proteins. Biochem Soc Trans. 2005;33(5):1059–62.
43. Gray NE, Lam LN, Yang K, Zhou AY, Koliwad S, Wang J-C. Angiopoietin-like 4 (Angptl4) protein is a physiological mediator of intracellular lipolysis in murine adipocytes. J Biol Chem. 2012;287(11):8444–56.
44. Koltes DA, Spurlock DM. Adipose tissue angiopoietin-like protein 4 messenger RNA changes with altered energy balance in lactating Holstein cows. Domest Anim Endocrinol. 2012;43(4):307–16.
45. Dutton S, Trayhurn P. Regulation of angiopoietin-like protein 4/fasting-induced adipose factor (Angptl4/FIAF) expression in mouse white adipose tissue and 3 T3-L1 adipocytes. Br J Nutr. 2008;100(01):18–26.
46. Fortun-Lamothe L, Langin D, Lafontan M. Influence of prolactin on in vivo and in vitro lipolysis in Rabbits. Comp Biochem Physiol C Pharmacol Toxicol Endocrinol. 1996;115(2):141–7.
47. Ouchi N, Parker JL, Lugus JJ, Walsh K. Adipokines in inflammation and metabolic disease. Nat Rev Immunol. 2011;11(2):85–97.
48. Lehr S, Hartwig S, Sell H. Adipokines: A treasure trove for the discovery of biomarkers for metabolic disorders. Proteomics Clin Appl. 2012;6(1-2):91–101.
49. Giesy SL, Yoon B, Currie WB, Kim JW, Boisclair YR. Adiponectin deficit during the precarious glucose economy of early lactation in dairy cows. Endocrinology. 2012;153(12):5834–44.
50. Jortay J, Senou M, Abou-Samra M, Noel L, Robert A, Many MC, et al. Adiponectin and skeletal muscle: Pathophysiological implications in metabolic stress. Am J Pathol. 2012;181(1):245–56.
51. Kadowaki T, Yamauchi T. Adiponectin and adiponectin receptors. Endocr Rev. 2005;26(3):439–51.
52. Sauerwein H, Häusler S. Endogenous and exogenous factors influencing the concentrations of adiponectin in body fluids and tissues in the bovine. Domest Anim Endocrinol. 2016;56(Supplement):S33–43.
53. Ohtani Y, Takahashi T, Sato K, Ardiyanti A, Song S-H, Sato R, et al. Changes in circulating adiponectin and metabolic hormone concentrations during periparturient and lactation periods in Holstein dairy cows. Anim Sci J. 2012;83(12):788–95.
54. Kabara E, Sordillo LM, Holcombe S, Contreras GA. Adiponectin links adipose tissue function and monocyte inflammatory responses during bovine metabolic stress. Comp Immunol Microbiol Infect Dis. 2014;37(1):49–58.
55. Lemor A, Hosseini A, Sauerwein H, Mielenz M. Transition period-related changes in the abundance of the mRNAs of adiponectin and its receptors, of visfatin, and of fatty acid binding receptors in adipose tissue of high-yielding dairy cows. Domest Anim Endocrinol. 2009;37(1):37–44.
56. Saremi B, Winand S, Friedrichs P, Kinoshita A, Rehage J, Dänicke S, et al. Longitudinal profiling of the tissue-specific expression of genes related with insulin sensitivity in dairy cows during lactation focusing on different fat depots. PLoS One. 2014;9(1):e86211.
57. Cava AL, Matarese G. The weight of leptin in immunity. Nat Rev Immunol. 2004;4(5):371–9.
58. Chilliard Y, Delavaud C, Bonnet M. Leptin expression in ruminants: Nutritional and physiological regulations in relation with energy metabolism. Domest Anim Endocrinol. 2005;29(1):3–22.
59. William Jr WN, Ceddia RB, Curi R. Leptin controls the fate of fatty acids in isolated rat white adipocytes. J Endocrinol. 2002;175(3):735–44.

60. Block S, Butler W, Ehrhardt R, Bell A, Van Amburgh M, Boisclair Y. Decreased concentration of plasma leptin in periparturient dairy cows is caused by negative energy balance. J Endocrinol. 2001;171(2):339–48.
61. Holtenius K, Agenäs S, Delavaud C, Chilliard Y. Effects of feeding intensity during the dry period. 2. Metabolic and hormonal responses. J Dairy Sci. 2003;86(3):883–91.
62. Weber C, Hametner C, Tuchscherer A, Losand B, Kanitz E, Otten W, et al. Variation in fat mobilization during early lactation differently affects feed intake, body condition, and lipid and glucose metabolism in high-yielding dairy cows. J Dairy Sci. 2013;96(1):165–80.
63. Ehrhardt RA, Foskolos A, Giesy SL, Wesolowski SR, Krumm CS, Butler WR, et al. Increased plasma leptin attenuates adaptive metabolism in early lactating dairy cows. J Endocrinol. 2016;229(2):145–57.
64. Schwartz DR, Lazar MA. Human resistin: Found in translation from mouse to man. Trends Endocrinol Metab. 2011;22(7):259–65.PubMedPubMedCentralGoogle Scholar
65. Ding Q, White SP, Ling C, Zhou W. Resistin and cardiovascular disease. Trends Cardiovasc Med. 2011;21(1):20–7.
66. Reverchon M, Ramé C, Cognié J, Briant E, Elis S, Guillaume D, et al. Resistin in dairy cows: plasma concentrations during early lactation, expression and potential role in adipose tissue. PLoS One. 2014;9(3):e93198.
67. Weber M, Locher L, Huber K, Kenéz Á, Rehage J, Tienken R, et al. Longitudinal changes in adipose tissue of dairy cows from late pregnancy to lactation. Part 1: The adipokines apelin and resistin and their relationship to receptors linked with lipolysis. J Dairy Sci. 2016;99(2):1549–59.
68. Cheng J, Song Z-Y, Pu L, Yang H, Zheng J-M, Zhang Z-Y, et al. Retinol binding protein 4 affects the adipogenesis of porcine preadipocytes through insulin signaling pathways. Biochem Cell Biol. 2013;91(4):236–43.
69. Klöting N, Graham TE, Berndt J, Kralisch S, Kovacs P, Wason CJ, et al. Serum retinol-binding protein is more highly expressed in visceral than in subcutaneous adipose tissue and is a marker of intra-abdominal fat mass. Cell Metab. 2007;6(1):79–87.
70. Romacho T, Elsen M, Röhrborn D, Eckel J. Adipose tissue and its role in organ crosstalk. Acta Physiol (Oxf). 2014;210(4):733–53.
71. Abd Eldaim MA, Kamikawa A, Soliman MM, Ahmed MM, Okamatsu-Ogura Y, Terao A, et al. Retinol binding protein 4 in dairy cows: its presence in colostrum and alteration in plasma during fasting, inflammation, and the peripartum period. J Dairy Res. 2010;77(01):27–32.
72. Rezamand P, Watts JS, Hunt KM, Bradford BJ, Mamedova LK, Morey SD. Bovine hepatic and adipose retinol-binding protein gene expression and relationship with tumor necrosis factor-α. J Dairy Sci. 2012;95(12):7097–104.
73. Granneman JG, Li P, Zhu Z, Lu Y. Metabolic and cellular plasticity in white adipose tissue I: effects of β3-adrenergic receptor activation. Am J Physiol Endocrinol Metab. 2005;289(4):E608–16.
74. Martinez-Santibañez G, Nien-Kai Lumeng C. Macrophages and the regulation of adipose tissue remodeling. Annu Rev Nutr. 2014;34(1):57–76.
75. Pope BD, Warren CR, Parker KK, Cowan CA. Microenvironmental control of adipocyte fate and function. Trends Cell Biol. 2016;26(10):745–55.
76. Bindlish S, Presswala LS, Schwartz F. Lipodystrophy: syndrome of severe insulin resistance. Postgrad Med. 2015;127(5):511–6.
77. Ampem G, Azegrouz H, Bacsadi Á, Balogh L, Schmidt S, Thuróczy J et al. Adipose tissue macrophages in non-rodent mammals: a comparative study. Cell Tissue Res. 2016;363:461.
78. Weisberg SP, McCann D, Desai M, Rosenbaum M, Leibel RL, Ferrante AW. Obesity is associated with macrophage accumulation in adipose tissue. J Clin Invest. 2003;112(12):1796–808.

79. Nishimura S, Manabe I, Nagasaki M, Eto K, Yamashita H, Ohsugi M, et al. CD8+ effector T cells contribute to macrophage recruitment and adipose tissue inflammation in obesity. Nat Med. 2009;15(8):914–20.
80. Mathis D. Immunological goings-on in visceral adipose tissue. Cell Metab. 2013;17(6):851–9.
81. Contreras GA, Kabara E, Brester J, Neuder L, Kiupel M. Macrophage infiltration in the omental and subcutaneous adipose tissues of dairy cows with displaced abomasum. J Dairy Sci. 2015;98(9):6176–87.
82. Contreras G, Thelen K, Schmidt S, Strieder-Barboza C, Preseault C, Raphael R, et al. Adipose tissue remodeling in late-lactation dairy cows during feed restriction-induced negative energy balance. J Dairy Sci. 2016;99(12):10009–21.
83. Lee Y-H, Petkova A, Granneman J. Identification of an adipogenic niche for adipose tissue remodeling and restoration. Cell Metab. 2013;18(3):355–67.
84. Faulconnier Y, Chilliard Y, Torbati MBM, Leroux C. The transcriptomic profiles of adipose tissues are modified by feed deprivation in lactating goats. Comp Biochem Physiol Part D Genomics Proteomics. 2011;6(2):139–49.
85. Velloso LA, Folli F, Saad MJ. TLR4 at the crossroads of nutrients, gut microbiota, and metabolic inflammation. Endocr Rev. 2015;36(3):245–71.
86. Grant RW, Stephens JM. Fat in flames: influence of cytokines and pattern recognition receptors on adipocyte lipolysis. Am J Physiol Endocrinol Metab. 2015;309(3):9.
87. Suganami T, Tanimoto-Koyama K, Nishida J, Itoh M, Yuan X, Mizuarai S, et al. Role of the toll-like receptor 4/NF-κB pathway in saturated fatty acid–induced inflammatory changes in the interaction between adipocytes and macrophages. Arterioscler Thromb Vasc Biol. 2007;27(1):84–91.
88. Wisse BE. The inflammatory syndrome: the role of adipose tissue cytokines in metabolic disorders linked to obesity. J Am Soc Nephrol. 2004;15(11):2792–800.
89. Gao D, Madi M, Ding C, Fok M, Steele T, Ford C, et al. Interleukin-1beta mediates macrophage-induced impairment of insulin signaling in human primary adipocytes. Am J Physiol Endocrinol Metab. 2014;307(3):10.
90. Shaul ME, Bennett G, Strissel KJ, Greenberg AS, Obin MS. Dynamic, M2-like remodeling phenotypes of CD11c + adipose tissue macrophages during high-fat diet–induced obesity in mice. Diabetes. 2010;59(5):1171–81.
91. Sica A, Mantovani A. Macrophage plasticity and polarization: in vivo veritas. J Clin Invest. 2012;122(3):787–95.
92. Chang HY, Lee HN, Kim W, Surh YJ. Docosahexaenoic acid induces M2 macrophage polarization through peroxisome proliferator-activated receptor gamma activation. Life Sci. 2015;120:39–47.
93. Chazaud B. Macrophages: Supportive cells for tissue repair and regeneration. Immunobiology. 2014;219(3):172–8.
94. Osborn O, Olefsky JM. The cellular and signaling networks linking the immune system and metabolism in disease. Nat Med. 2012;18(3):363–74.
95. Aylward BAA, Clark M, Barnard A, Jen W, Gittens C, Gressley T, et al. Immune cells populate mesenteric adipose tissues of Holstein Friesian cows. J Dairy Sci. 2015;98(Sppl. 2):292.Google Scholar
96. Spalding KL, Arner E, Westermark PO, Bernard S, Buchholz BA, Bergmann O, et al. Dynamics of fat cell turnover in humans. Nature. 2008;453(7196):783–7.
97. Lee Y-H, Petkova A, Mottillo E, Granneman J. In vivo identification of bipotential adipocyte progenitors recruited by β3-adrenoceptor activation and high-fat feeding. Cell Metab. 2012;15(4):480–91.
98. Häussler S, Germeroth D, Friedauer K, Akter SH, Dänicke S, Sauerwein H. Characterization of the dynamics of fat cell turnover in different bovine adipose tissue depots. Res Vet Sci. 2013;95(3):1142–50.
99. Gustafson B, Hedjazifar S, Gogg S, Hammarstedt A, Smith U. Insulin resistance and impaired adipogenesis. Trends Endocrinol Metab. 2015;26(4):193–200.
100. Mariman ECM, Wang P. Adipocyte extracellular matrix composition, dynamics and role in obesity. Cell Mol Life Sci. 2010;67(8):1277–92.

101. Khan T, Muise ES, Iyengar P, Wang ZV, Chandalia M, Abate N, et al. Metabolic dysregulation and adipose tissue fibrosis: role of collagen VI. Mol Cell Biol. 2009;29(6):1575–91.
102. Divoux A, Tordjman J, Lacasa D, Veyrie N, Hugol D, Aissat A, et al. Fibrosis in human adipose tissue: composition, distribution, and link with lipid metabolism and fat mass loss. Diabetes. 2010;59(11):2817–25.
103. Varma V, Yao-Borengasser A, Bodles AM, Rasouli N, Phanavanh B, Nolen GT, et al. Thrombospondin-1 Is an adipokine associated with obesity, adipose inflammation, and insulin resistance. Diabetes. 2008;57(2):432–9.
104. Guglielmi V, Cardellini M, Cinti F, Corgosinho F, Cardolini I, D'Adamo M, et al. Omental adipose tissue fibrosis and insulin resistance in severe obesity. Nutr Diabetes. 2015;5(8):e175.
105. Dahlman I, Mejhert N, Linder K, Agustsson T, Mutch DM, Kulyte A, et al. Adipose tissue pathways involved in weight loss of cancer cachexia. Br J Cancer. 2010;102(10):1541–8.
106. Tsoli M, Schweiger M, Vanniasinghe AS, Painter A, Zechner R, Clarke S, et al. Depletion of white adipose tissue in cancer cachexia syndrome is associated with inflammatory signaling and disrupted circadian regulation. PLoS One. 2014;9(3):e92966.
107. Akbar H, Cardoso FC, Meier S, Burke C, McDougall S, Mitchell M, et al. Postpartal subclinical endometritis alters transcriptome profiles in liver and adipose tissue of dairy cows. Bioinform Biol Insights. 2014;8:45–63.
108. Sumner-Thomson JM, Vierck JL, McNamara JP. Differential expression of genes in adipose tissue of first-lactation dairy cattle. J Dairy Sci. 2011;94(1):361–9.
109. Bradford BJ, Yuan K, Farney JK, Mamedova LK, Carpenter AJ. Invited review: Inflammation during the transition to lactation: New adventures with an old flame. J Dairy Sci. 2015;98(10):6631–50.
110. Lumeng CN. Innate immune activation in obesity. Mol Aspects Med. 2013;34(1):12–29.
111. Namgaladze D, Brüne B. Macrophage fatty acid oxidation and its roles in macrophage polarization and fatty acid-induced inflammation. Biochim Biophys Acta. 2016;1861(11):1796–807.
112. Scalia D, Lacetera N, Bernabucci U, Demeyere K, Duchateau L, Burvenich C. In vitro effects of nonesterified fatty acids on bovine neutrophils oxidative burst and viability1. J Dairy Sci. 2006;89(1):147–54.
113. Raphael W, Halbert L, Contreras GA, Sordillo LM. Association between polyunsaturated fatty acid-derived oxylipid biosynthesis and leukocyte inflammatory marker expression in periparturient dairy cows. J Dairy Sci. 2014;97(6):3615–25.
114. Sun F, Cao Y, Cai C, Li S, Yu C, Yao J. Regulation of nutritional metabolism in transition dairy cows: energy homeostasis and health in response to post-ruminal choline and methionine. PLoS One. 2016;11(8):e0160659.
115. Pires JAA, Pescara JB, Brickner AE, Silva del Rio N, Cunha AP, Grummer RR. Effects of abomasal infusion of linseed oil on responses to glucose and insulin in Holstein Cows. J Dairy Sci. 2008;91(4):1378–90.
116. Salin S, Taponen J, Elo K, Simpura I, Vanhatalo A, Boston R, et al. Effects of abomasal infusion of tallow or camelina oil on responses to glucose and insulin in dairy cows during late pregnancy. J Dairy Sci. 2012;95(7):3812–25.
117. Ospina PA, Nydam DV, Stokol T, Overton TR. Associations of elevated nonesterified fatty acids and β-hydroxybutyrate concentrations with early lactation reproductive performance and milk production in transition dairy cattle in the northeastern United States. J Dairy Sci. 2010;93(4):1596–603.
118. Hernandez TL, Van Pelt RE, Anderson MA, Reece MS, Reynolds RM, de la Houssaye BA, et al. Women with gestational diabetes mellitus randomized to a higher-complex carbohydrate/low-fat diet manifest lower adipose tissue insulin resistance,

inflammation, glucose, and free fatty acids: a pilot study. Diabetes Care. 2016;39(1):39–42.
119. Capurso C, Capurso A. From excess adiposity to insulin resistance: The role of free fatty acids. Vascul Pharmacol. 2012;57(2–4):91–7.
120. Sadri H, Bruckmaier RM, Rahmani HR, Ghorbani GR, Morel I, Van Dorland HA. Gene expression of tumour necrosis factor and insulin signalling-related factors in subcutaneous adipose tissue during the dry period and in early lactation in dairy cows. J Anim Physiol Anim Nutr (Berl). 2010;94(5):e194–202.
121. Schmitt E, Ballou MA, Correa MN, DePeters EJ, Drackley JK, Loor JJ. Dietary lipid during the transition period to manipulate subcutaneous adipose tissue peroxisome proliferator-activated receptor-γ co-regulator and target gene expression. J Dairy Sci. 2011;94(12):5913–25.
122. Zachut M. Defining the adipose tissue proteome of dairy cows to reveal biomarkers related to peripartum insulin resistance and metabolic status. J Proteome Res. 2015;14(7):2863–71.
123. Mann S, Nydam DV, Abuelo A, Leal Yepes FA, Overton TR, Wakshlag JJ. Insulin signaling, inflammation, and lipolysis in subcutaneous adipose tissue of transition dairy cows either overfed energy during the prepartum period or fed a controlled-energy diet. J Dairy Sci. 2016;99(8):6737–52.
124. Oliveira LH, Nascimento AB, Monteiro Jr PLJ, Guardieiro MM, Wiltbank MC, Sartori R. Development of insulin resistance in dairy cows by 150 days of lactation does not alter oocyte quality in smaller follicles. J Dairy Sci. 2016;99(11):9174–83.
125. Mordak R, Stewart PA. Periparturient stress and immune suppression as a potential cause of retained placenta in highly productive dairy cows: examples of prevention. Acta Vet Scand. 2015;57(1):1–8.
126. Harvatine KJ, Perfield JW, Bauman DE. expression of enzymes and key regulators of lipid synthesis is upregulated in adipose tissue during CLA-induced milk fat depression in dairy cows. J Nutr. 2009;139(5):849–54.
127. Zhu P, Goh Yan Y, Chin Hwee Fang A, Kersten S, Tan Nguan S. Angiopoietin-like 4: a decade of research. Biosci Rep. 2012;32(3):211–9.

CHAPTER 12

Artificial Blood for Dogs

Kana Yamada[1], Kyoko Yokomaku[1], Moeka Kureishi[1], Motofusa Akiyama[1], Kiyohito Kihira[2] & Teruyuki Komatsu[1]

[1]*Department of Applied Chemistry, Faculty of Science and Engineering, Chuo University, 1-13-27 Kasuga, Bunkyo-ku, Tokyo 112-8551, Japan.*

[2]*JEM Utilization Center, Human Spaceflight Technology Directorate, Japan Aerospace Exploration Agency (JAXA), 2-1-1 Sengen, Tsukuba-shi, Ibaraki 305-8505, Japan.*

ABSTRACT

There is no blood bank for pet animals. Consequently, veterinarians themselves must obtain "blood" for transfusion therapy. Among the blood components, serum albumin and red blood cells (RBCs) are particularly important to save lives. This paper reports the synthesis, structure, and properties of artificial blood for the exclusive use of dogs. First, recombinant canine serum albumin (rCSA) was produced using genetic engineering with *Pichia* yeast. The proteins showed identical features to those of the native CSA derived from canine plasma. Furthermore, we ascertained the crystal structure of rCSA at 3.2 Å resolution. Pure rCSA can be used widely for numerous clinical and pharmaceutical applications. Second, hemoglobin wrapped covalently with rCSA, hemoglobin–albumin cluster (Hb-rCSA$_3$), was synthesized as an artificial O_2-carrier for the RBC substitute. This cluster possesses satisfactorily negative surface net charge ($pI = 4.7$), which supports enfolding of the Hb core by rCSA shells. The anti-CSA antibody recognized the rCSA exterior quantitatively. The O_2-binding affinity was high ($P_{50} = 9$ Torr) compared to that of the native Hb. The Hb-rCSA$_3$ cluster is anticipated for use as an alternative material for RBC transfusion, and as an O_2 therapeutic reagent that can be exploited in various veterinary medicine situations.

INTRODUCTION

Japan has become "a pet superpower". Official estimates put the pet population at 19,791,000 (9,917,000 dogs and 9,874,000 cats), which is far greater than the number of young people under the age of 15 (at 16,040,000)[1,2]. In the country, the human birthrate has been gradually falling while the average age of the population has been rapidly climbing. Pet animals, which can enrich human life, are now treated as family members. As one might expect, the demand for pet medical care continues to grow, and the frequency of "blood transfusion" continues to increase. Nevertheless, the shortage of blood products, especially for dogs, is little recognized. Because blood banks for pet animals are not approved in Japan, veterinarians themselves must obtain blood for transfusion therapy of hemorrhagic-shocked or anemic patients. The most difficult problem is how to find a donor. Among blood components, serum albumin and red blood cell (RBC) are particularly important to save lives. The challenges confronting the present study were two-fold. First, a recombinant canine serum albumin (rCSA) was produced using genetic engineering procedures. In the mammalian circulatory system, serum albumin is the most abundant protein (typical concentration of 2–5 g/dL). It plays two major roles of (i) transporting diverse metabolites and drugs and (ii) maintaining colloid osmotic pressure of the bloodstream[3,4]. Valenta *et al.* reported the expression of rCSA using *E. coli*, although the quantity of the protein was insufficient to explore physicochemical features[5]. Some concern also persists about the contamination of toxic lipopolysaccharide by *E. coli* membranes. If it were possible to produce rCSA using *Pichia* yeast as a host cell, then rCSA could be manufactured to prepare a plasma expander for clinical use and a new formulation for pharmaceutical and biochemical applications. We have established the efficient *Pichia pastoris* expression system of rCSA, and defined its physicochemical properties. Moreover, X-ray crystallography has revealed the three-dimensional (3D) structure of the protein at 3.2 Å resolution.

A second challenge is the creation of artificial O_2-carrier as an RBC substitute. Naturally, dogs have different blood types. For example, dog erythrocyte antigen (DEA) classification divides blood types into nine kinds (DEA1.1, 1.2, 1.3, 3, 4, 5, 6, 7, and 8)[6]. The blood of DEA1.1-negative donor can be transfused into any dog. On the one hand, transfusion of the DEA1.1-positive donor's blood into a DEA1.1-negative recipient generates an antibody. Consequently, a second infusion of the DEA1.1-positive type blood will cause an acute hemolytic reaction. To conduct safe blood transfusion, it is necessary to confirm blood compatibility in advance. Of course, veterinarians would not need to find a donor and test blood compatibility before infusion if an RBC substitute for dogs were realized.

Studies of RBC substitutes for human use have continued since the 1980s[7,8,9,10]. The products with the greatest potential are hemoglobin (Hb)-based O_2 carriers[9,10]. In the United States, poly(ethyleneglycol)-conjugated Hb[11,12,13], polymerized Hb[14], intramolecularly cross-linked Hb[15,16] have been developed and examined in clinical trials[7,8,9,10,17,18,19,20]. Most of these artificial O_2-carriers are designed to increase molecular size and to avoid renal excretion. However, transfusion of such heterogeneous molecules has raised the blood pressure in humans to a greater or lesser extent. Two mechanisms for this side effect are presumed to be (i) nitric oxide (NO: an endothelium-derived relaxing factor) scavenging by Hb diffused into the extravascular space and (ii) oversupply of O_2 to vascular walls by facilitated diffusion [8,17,18,19,20,21,22,23,24]. Despite intensive efforts, no substitute has been able to satisfy all requirements for clinical use. For pet animals, polymerized bovine Hb (Oxyglobin®) used to be commercially available for the treatment of anemic dogs in the United States and the United Kingdom[25]. Alayash *et al.* reported that this formulation contained nonpolymerized Hb of a substantial amount (37.8%)[26]. Some side effects have been reported, such as vomiting, anorexia, and pyrexia, as well as discoloration of skin, mucous membranes, urine, and (darkened) feces[27].

We have recently synthesized a covalent core–shell structured protein cluster comprising a bovine Hb in the center and human serum albumins (HSAs) at the periphery, hemoglobin–albumin cluster (Hb-HSA_3), as a unique artificial O_2-carrier[28,29,30,31]. The Hb nucleus is wrapped covalently with HSA. Thereby the surface net charge of the cluster is negative to a marked degree [isoelectric point (*p*I): 5.1]. As a consequence, the Hb-HSA_3 cluster shows long circulation persistence and does not engender an acute increase in blood pressure[31]. This result is ascribed to the electrostatic repulsion between the negatively charged surface of the cluster and glomerular basement membrane around the endothelial cells. To develop this product for canine use, it is necessary to replace the exterior HSA shell to CSA. Injection of the Hb-HSA_3 cluster into a dog might generate antibodies in the blood[32]. The recipient will therefore experience an immunological response from the second infusion. Unfortunately, it is difficult to obtain canine blood (plasma) of sufficient quantity for the large-scale production of the cluster. To resolve this difficulty, we have used genetically engineered rCSA. A new hemoglobin–albumin cluster, Hb-$rCSA_3$, can be regarded as a promising RBC alternate for dogs as well as an O_2-carrying fluid that is useful in various veterinary medicine scenarios. The Hb-$rCSA_3$ cluster has no blood type. It can be transfused into any dog at anytime, anywhere. Herein, we describe the synthesis, structure, and O_2-binding properties of the Hb-$rCSA_3$ cluster. The rCSA and Hb-$rCSA_3$ cluster can contribute greatly to pet medical care.

RESULTS

Expression and Physicochemical Property of rCSA

Expressions of rCSA using *Pichia* yeast have not been reported in the relevant literature. We have now prepared rCSA using *Pichia pastoris* (GS115) in a gram quantity. *Pichia* yeast secretes the target protein into culture medium so that the accumulated protein can be harvested easily by centrifugation, followed by general chromatographic technique. Results showed that the ideal purification was achieved using a combination of affinity chromatography and anion exchange chromatography.

The obtained rCSA was compared carefully to native CSA derived from canine plasma by physicochemical measurements. SDS-PAGE of rCSA exhibited a single band (Figure 1A and S1A). Its mobility was identical to that observed for CSA. Native-PAGE also demonstrated the same result (Figure 1B and S1B). Size-exclusion chromatography (SEC) measurements of rCSA showed a sharp and single peak that has equivalent mobility for native CSA (Figure 1C). The rCSA purity was ascertained as 99.99%. The isoelectric focusing pattern depicted one band in both rCSA and CSA with a *p*I value of 4.5 (vide infra), which was significantly lower than that of HSA (*p*I = 5.0). Although the numbers of acidic amino acids in CSA and HSA are exactly the same (36 Asp, 62 Glu), CSA possesses fewer basic amino acids (12 His, 56 Lys, 23 Arg) relative to HSA (16 His, 59 Lys, 24 Arg)[5,33]. Consequently, the total charges of the rCSA molecule become negative. MALDI-TOF mass spectrum of rCSA demonstrated a molecular-related ion peak at 65580.224 Da, which was within the experimental range of the simulated mass from the sequence (65709.459 Da) and the observed mass of native CSA (65401.503 Da) (Figure 1D). The UV-visible absorption spectrum of phosphate-buffered saline (PBS) solution (pH 7.4) of rCSA showed a single broad band at 280 nm (data not shown). The molar absorption coefficient (ε_{280}) of rCSA was calculated as 4.50×10^4 M^{-1} cm^{-1}, which is almost identical to the value of CSA. The high ε_{280}(CSA) compared to ε_{280}(HSA) (3.50×10^4 M^{-1} cm^{-1})[3]is attributable to the existence of many Tyr residues (21 Tyr in CSA, 18 Tyr in HSA). The CD spectra of rCSA and CSA in PBS solution were indistinguishable (Figure 1E), suggesting that their peptide backbone conformations are equivalent. All these physicochemical results imply that rCSA is structurally identical to the native CSA.

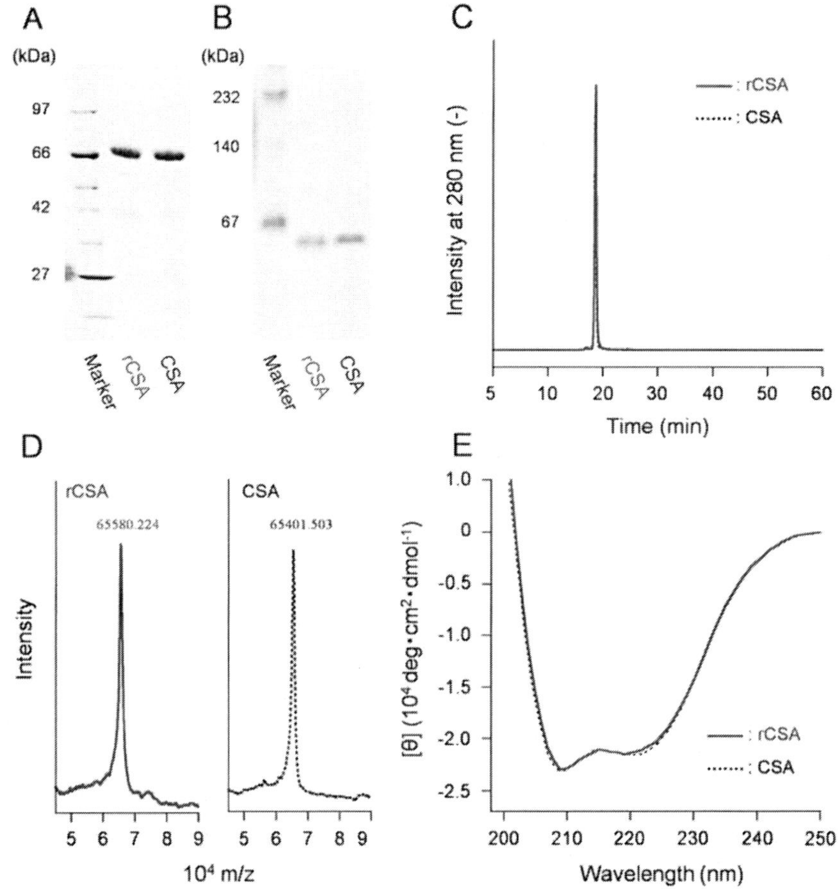

Figure 1: Physicochemical data of rCSA and native CSA. (**A**) SDS-PAGE, (**B**) Native-PAGE, (**C**) SEC profiles, (**D**) MALDI-TOF mass spectra, and (**E**) CD spectra of rCSA and CSA. All results demonstrated clearly that these proteins possess identical molecular weight and structure. Full-length gel images of (**A,B**) are presented in Supplementary Figure S1(A,B).

Crystal Structure of rCSA

The crystal structure of HSA at 2.8 Å resolution was first reported by Carter *et al.* in 199234. Subsequently numerous structures not only of unliganded HSA, but also HSA complexed with specific ligands have been published by several research groups35,36,37,38. More recently, Chruszcz and Minor *et al.* have dissolved crystal structures of bovine serum albumin, equine serum albumin, and leporine serum albumin39. Nonetheless, the crystallographic coordinate of CSA has never been deposited in PDB data. The primary sequence of CSA indicates that the protein is a single polypeptide with 584 amino acids (Figure S2)5,33. Pairwise sequence homology of CSA with respect to HSA is 79.8%. Our results show that the crystals of rCSA belong to the monoclinic space group P 1 21 1 (Table S1), and that the overall structure closely resembles those of HSA and other mammalian albumin (Figure 2A and S3)34,35,36,37,38,39. The rCSA consists of three topologically identical domains (I, residues 1–196; II, residues 197–383; and III, residues 384–584), each of which is divided into two subdomains (A and B). They are composed, respectively, of six α-helices (h1–h6) and four α-helices (h7–h10) (Figure 2B). Despite the symmetrical repeating morphology of domains I–III, their asymmetric spatial alignments form a unique heart-shape structure. As described above, rCSA has fewer basic amino acid residues than HSA has, and also more non-polar amino acid residues. The relative acidic property of rCSA was more clearly visible by its electrostatic potential representation than HSA was (Figure S4).

There are total of 35 Cys residues in rCSA. The Cys-34 in subdomain IA is the only Cys having a free sulfhydryl group which does not participate in a 17 disulfide bridges. In the crystal structure of HSA (PDB ID: 1E78)36, it is apparent that the S atom of Cys-34 is surrounded by a flexible loop (Val-77–Tyr-84; sequence VATLRETY) that prevents the sulfhydryl group from coupling with an external thiol-containing compound (glutathione or cysteine). In terms of the general structure, the environment around the Cys-34 of rCSA has similar features to those of HSA. However, three differences are apparent in the amino acid sequence of the loop in rCSA; VASLRDKY containing shorter Ser, Asp, and positively charged Lys. We inferred that the space around the S atom of Cys-34 in rCSA is slightly flexible and more polar than that of HSA. This finding is consistent with the fact that the mercapto-ratio (free sulfhydryl group ratio) of the Cys-34 residue in rCSA (ca. 30–40%) was lower than the value of HSA (ca. 50–60%).

Synthesis of Hb-rCSA$_3$ Cluster

Succinimidyl-4-(*N*-maleimidomethyl)-cyclohexane-1-carboxylate (SMCC) was used as a cross-linker between the Cys-34 group of rCSA and the surface Lys groups of Hb (Figure 3). First, SMCC was reacted with carbonyl Hb to produce maleimide-activated Hb. Site-specific binding to the free Cys-34 of rCSA and the maleimide-activated Hb yielded a core–shell cluster of Hb at the center and rCSA at the periphery. Later, SEC measurements of the reaction mixture depicted three peaks at the high molecular weight region (Figure 4A, elution time 13–17 min). Native-PAGE also exhibited three new bands above rCSA (Figure 4B and S5A). We collected all the large molecular products using gel filtration chromatography (GFC) and excluded the unreacted rCSA (Figure 4A). Based on the total protein assay and Hb assay, the average rCSA/Hb ratio of the product was ascertained as 3.0 ± 0.2. This protein cluster comprises a mixture of Hb-rCSA$_2$, Hb-rCSA$_3$ (major), and Hb-rCSA$_4$, which is indicated as Hb-rCSA$_{\mathit{3}}$ with italicized subscript 3.

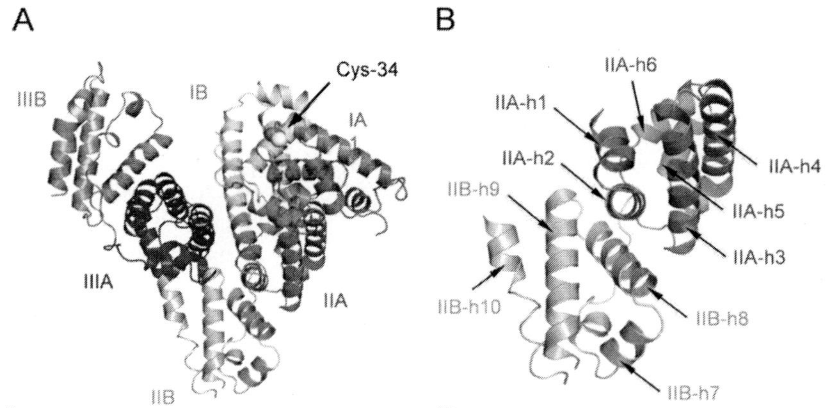

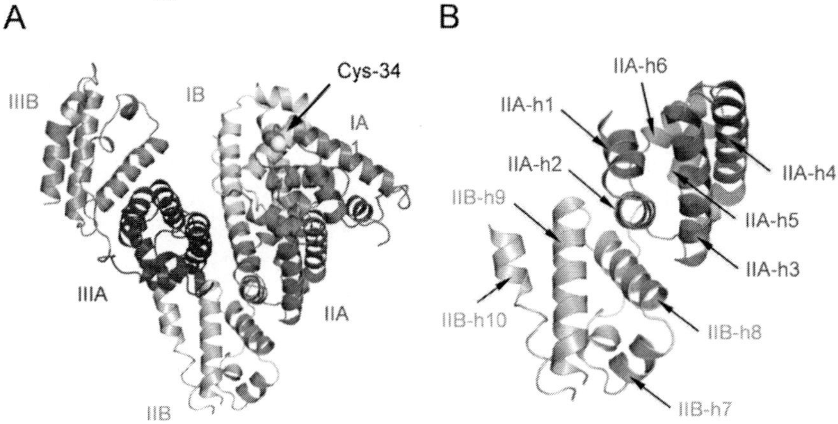

Figure 2: Crystal structure of rCSA. (A) The rCSA comprises three morphologically same domains I–III, each of which is divided into two subdomains: A and B. The Cys-34 in subdomain IA is an only Cys residue with sulfhydryl group. (**B**) Spatial alignments of α-helices in domain II. Subdomain IIA (forest) contains six helices h1–h6; subdomain IIB (pale green) contains four helices h7–h10.

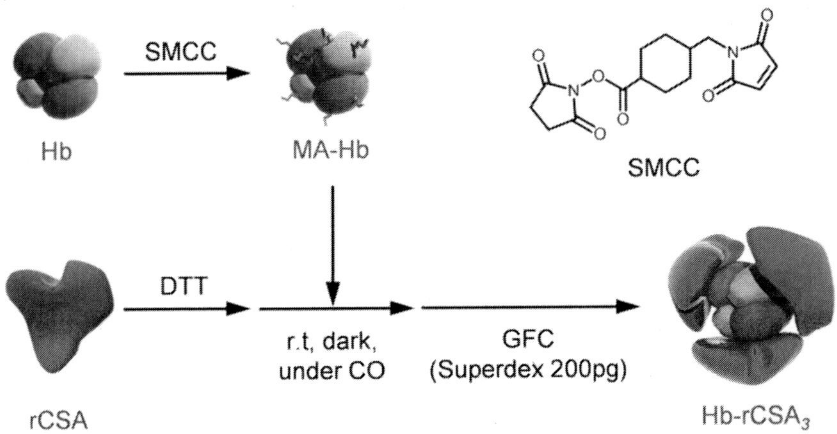

Figure 3: Schematic illustration of the synthetic route of Hb-rCSA$_3$ cluster. The Cys-34 of rCSA and the surface Lys group of Hb were connected covalently with a cross-linking agent (SMCC). The product is a mixture of Hb-rCSA$_m$ (m = 2, 3, 4). The total yield based on the Hb concentration was 75%.

The CD spectrum of the Hb-rCSA$_3$ cluster coincided accurately with the sum of the Hb spectrum and a three-fold amplified rCSA spectrum (Figure 4C). This result proved that the secondary structure of individual protein unit was not perturbed by the coupling and that the rCSA/Hb composition ratio is 3/1 (mol/mol). The isoelectric point of Hb-rCSA$_3$cluster (*p*I: 4.7) was extremely close to the value of rCSA (*p*I: 4.5) (Figure 4D and S5B), which supports enfolding of the Hb core by rCSAs. The surface net charges of this cluster are satisfactorily negative to circulate in the bloodstream for a long period. It is also noteworthy that the molecular features of Hb-rCSA$_3$ cluster are exactly the same as those of the Hb-CSA$_3$ cluster, which was prepared using native CSA.

Immunogenicity

We analyzed immunochemical property of rCSA and Hb-rCSA$_3$ cluster as an index of safety for administration into dogs. First, the immunological reactivity of Hb-rCSA$_3$ cluster against anti-Hb antibody was examined. To perform simple and high sensitive detection of Hb, the immuno-chromatographic quick occult blood test kit was used. For this experiment, we synthesized a cluster bearing a human Hb (HbA) core, HbA-rCSA$_3$. The HbA-antibody immune complexes migrated to the test area and were caught by immobilized rabbit anti-HbA antibody, generating sandwich complexes, which showed a red test-line signal (Figure S6). In case of HbA-$rCSA_3$ cluster, the test line was almost invisible. This means that the central

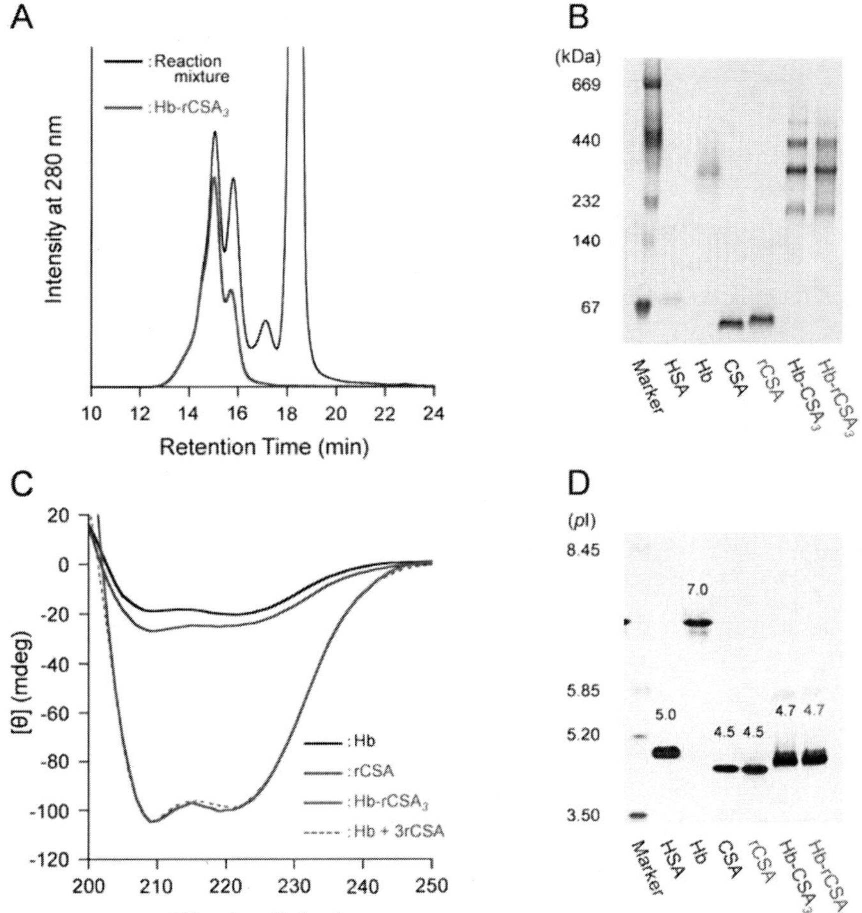

Figure 4: Physicochemical characterizations of Hb-$rCSA_3$ cluster. (**A**) SEC profiles of reaction mixture of maleimide-activated Hb and rCSA (black line) and isolated Hb-$rCSA_3$ cluster (red line). (**B**) Native-PAGE of Hb-$rCSA_3$ cluster and related proteins. (**C**) CD spectra of Hb-$rCSA_3$ cluster ([Hb] = 0.2 μM), Hb (0.2 μM), and HSA (0.2 μM) in PBS solution (pH 7.4) at 25 °C. The sum of Hb + 3rCSA spectrum coincided perfectly with that of Hb-$rCSA_3$. (**D**) IEF patterns of Hb-$rCSA_3$ cluster and related proteins. Full-length gel images of (**B,D**) are presented in Supplementary Figure S5(A,B).

HbA is covered by rCSAs and the anti-HbA antibody cannot recognize the epitope of the HbA molecule. Next, the immunological reactivity of Hb-rCSA$_3$ cluster against anti-CSA antibody was evaluated using a latex agglutination immuno-assay. The absorption intensity of the reactant with rCSA or Hb-rCSA$_3$ cluster demonstrated linear concentration dependence which coincided well with the calibration line of CSA (Figure S7). All these results indicate that Hb-rCSA$_3$ cluster does not bind anti-Hb antibody, but react with anti-rCSA antibody as same as CSA and rCSA.

O_2 Affinity and O_2-Complex Stability

The visible absorption spectra of Hb-rCSA$_3$ cluster in PBS solution (pH 7.4) equilibrated with N_2, O_2, and CO atmosphere (deoxy, oxy, and carbonyl forms) were fundamentally the same as those observed for Hb-HSA$_3$, Hb-CSA$_3$, and native Hb (Figure 5A)[29,40]. The absorption maxima of these clusters showed good agreement, indicating that the wrapping of Hb with different albumins caused no alternation of electronic states of the prosthetic heme groups (Table S2).

The P_{50} of Hb-rCSA$_3$ cluster was found to be 9 Torr (37 °C), whereas native Hb showed P_{50} of 23 Torr in the same experimental conditions (Figure 5B, Table 1)[28,29]. The n value also decreased from 2.6 to 1.6. Covalent wrapping of Hb with rCSA caused P_{50} to decrease (O_2 affinity to increase) 2.6-fold compared to naked Hb. Equivalent P_{50} and n reductions were also observed in Hb-CSA$_3$ and Hb-HSA$_3$ clusters ($P_{50} = 9$ Torr) (Table 1). The increase of O_2 affinity and decrease in cooperativity of Hb-rCSA$_3$ cluster was interpreted by the two structural alternations. First, the sulfhydryl group of Cys-93(β) in Hb was capped by the binding of free SMCC maleimide. Modification of Cys-93(β), which is a neighbor to the proximal histidine [His-92(β)] of the heme group[41], is known to enhance O_2 affinity[12,13,42]. Furthermore, it attenuates the possible motion of the $\alpha_1\beta_1/\alpha_2\beta_2$ interface of Hb[42]. Our detailed inspection revealed that the number of cysteinyl thiols per Hb decreased from 2.0 to 0.2 after the SMCC reaction, suggesting that two Cys-93(β) of Hb were reacted by SMCC maleimide. Fortunately, there is no Lys residue which can bind the opposite-terminal succinimide group of SMCC on Cys-93(β).

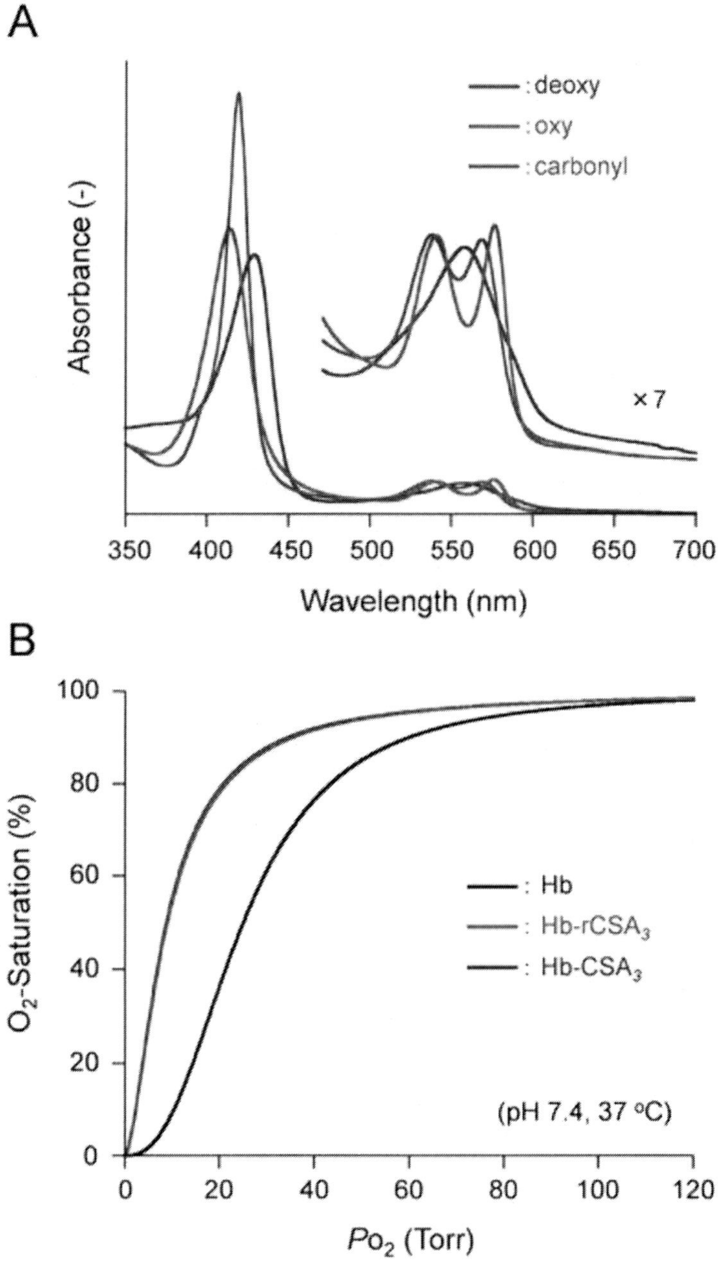

Figure 5: O_2-binding to Hb-rCSA$_3$ cluster. (A) UV-vis absorption spectra of Hb-rCSA$_3$ cluster in deoxy, oxy, and carbonyl forms in PBS solution (pH 7.4) at 25 °C. (**B**) O_2-equilibrium curves of Hb-rCSA$_3$ and Hb-CSA$_3$ clusters in PBS solution (pH 7.4) at 37 °C.

Table 1: O_2-binding parameters of Hb-rCSA$_3$and Hb-CSA$_3$ clusters in PBS solution (pH 7.4) at 37 °C.

Hemoproteins	P_{50} (Torr)	n (−)	k_{ox} (h^{-1})
Hb†	23	2.6	0.037
Hb-rCSA$_3$	9	1.6	0.044
Hb-CSA$_3$	9	1.6	0.044
Hb-HSA$_3$†	9	1.5	0.035

A second reason is the chemical modification of surface amino groups of Lys residues on Hb by SMCC. Such alternation is requisite for formation of the cluster, but modifications of Lys groups on Hb generally engender measureable changes of the O_2 affinity and Hill coefficient11,43,44. Particularly, Lys-82(β) plays a crucially important role in controlling the quaternary structure deformation from the *Relaxed* state to a *Tense* state. We reported previously that Lys-82(β) is a binding partner of Cys-34 of HSA based on the 3D reconstruction of Hb-HSA$_3$28. Therefore, it can be concluded that (i) the capping of Cys-93(β) elevates the O_2 affinity and (ii) modification of surface Lys groups interferes with the Hb quaternary transition, resulting in the decline of O_2-binding cooperativity.

Stability of the oxygenated Hb-rCSA$_3$ cluster in PBS solution (pH 7.4) was examined using an autoxidation rate constant (k_{ox}) of the Hb nuclei at 37 °C. The k_{ox} value of Hb-rCSA$_3$ cluster was ascertained as 0.044 h^{-1}, which was identical to the value of Hb-CSA$_3$ cluster and similar to those of native Hb and Hb-HSA$_3$29. The oxy Hb core in the cluster maintains good stability even after covalent binding with rCSA.

DISCUSSION

The rCSA and Hb-rCSA$_3$ cluster shown in this work were designed for artificial blood products for exclusive use in the medical treatment of dogs. The rCSA has been prepared in a gram quantity using *Pichia* yeast expression. The protein showed identical features to those observed for native CSA. We first ascertained the crystal structure of rCSA at resolution of 3.2 Å, which revealed that the protein consists of three similar helical domains with 17 disulfide bonds. Although our purification protocol did not include a fatty acid removal process, which is typically a powder-charcoal treatment conducted under acidic conditions, the overall structure of rCSA resembled those of the defatted HSA and other mammalian albumin36,39.

We used anion exchange chromatography at the last step of the purification. Thereby, anionic ligands in the protein might be eliminated, yielding a defatted rCSA.

Otagiri *et al.* examined the species differences of serum albumin with respect to drug binding properties, and demonstrated that warfarin (WF) and phenylbutazone (PBZ) cannot bind to CSA45. Commonly used drugs of many kinds with electronegative groups generally associate to the two primary binding sites of HSA, sites 1 and 2 located respectively in subdomains IIA and IIIA3. Both WF and PBZ are well-known site 1 binding drugs. However, ibuprofen, which is a typical site 2 binding drug, associates strongly to CSA with the same binding constant for HSA ($K = 3.15 \times 10^6$ M^{-1})45. Otagiri *et al.* reported that this difference might be attributable to a slight difference of the site 1 microenvironments of CSA and HSA. Curry *et al.* later reported comprehensive structural insight into HSA-drug complexes based on their numerous crystal structure data38. In the HSA-PBZ complex (defatted, PDB ID: 2BXC), the PBZ is accommodated in the center of the site 1 pocket with a single hydrogen bond interaction with the hydroxyl group of Tyr-150. Moreover, weak interaction through water molecules might exist with Lys-199 and Arg-222, which are located 4–6 Å from the edge of PBZ. The superposition image of rCSA and HSA-PBZ shows clearly that the architectures of their subdomains IIA (drug site 1) and IIIA (drug site 2) are identical (Figure S3A). It is remarkable that the side chains of Tyr-150, Lys-199, and Arg-222 are located at the same positions (Figure S3B). The reason why the PBZ does not bind to CSA could not be found from results obtained using the present resolution of our crystal structure analysis.

The only Cys with a free sulfhydryl group in rCSA was found to be Cys-34, which is responsible for the covalent binding to the surface Lys groups of Hb to create the core–shell structured Hb-rCSA$_3$ cluster. The coupling of maleimide-activated Hb with rCSA advanced smoothly in the same way as the reaction with HSA or native CSA. Bovine Hb also possesses a free Cys at the 93 position of the β subunit41. Therefore, the maleimide-activated Hb might produce undesired Hb polymers. Nevertheless, such polymerization did not occur because the SMCC arm (10 Å spacer length) on the Hb is too short to make an intermolecular linkage with different Hb. The Cys-93(β) locates in hydrophobic cavities of Hb, at least 15 Å inside from the molecular surface41. Furthermore, the sulfhydryl groups of Cys-93(β) in Hb were mostly masked by the maleimide-terminal of excess SMCC. We inferred that the SMCC is an excellent cross-linker to synthesize a structurally defined core–shell Hb-rCSA$_3$ cluster. This coupling procedure is likely to be applicable to preparations of different smart clusters composed of various combinations of an enzyme interior and serum albumin exterior.

The negative surface net charge of the Hb-rCSA$_3$ cluster (*p*I: 4.7) proves the enfolding of Hb by rCSA. It can be favorable (i) to circulate in the bloodstream for a long time as an artificial O_2-carrier and (ii) to avoid an undesirable vasopressor response by NO depletion. The immunochemical reactivity of Hb-rCSA$_3$ cluster was markedly low against anti-Hb antibody, but was sufficiently high against anti-CSA antibody. We reasoned that the Hb center becomes a stealth protein by wrapping with HSAs, whereas the antigen epitopes of rCSA units are preserved after conjugation with Hb. These results suggest the immunological safety of Hb-rCSA$_3$ cluster *in vivo*.

The O_2 affinity of Hb-rCSA$_3$ cluster became higher (P_{50} = 9 Torr) than the native Hb because the Cys-93(β) residues in Hb were blocked by the maleimide terminal of free SMCC. This high O_2-affinity can be beneficial for an RBC alternate. Winslow *et al.* reported that the Hb-based O_2-carrier with a low O_2-affinity releases excessive O_2 in the arterioles, causing autoregulatory vasoconstriction[18,46]. Intaglietta *et al.* demonstrated that RBCs having lower P_{50} (10 Torr) provide improvement of microvascular function in a hemorrhagic-shocked hamster model[47]. The high O_2-affinity (lower P_{50}) might be helpful to reduce arteriole O_2-transport, and to eliminate cardiovascular side effects.

We reported previously the tissue distribution and metabolism of Hb-HSA$_3$ cluster in rats[31,48]. The cluster and HSA were similarly distributed in the major organs at 1 h after injection. However, the accumulated quantity of the cluster was greater than that of HSA in the liver after 24 h. Rennen *et al.* demonstrated that the large-size proteins are cleared with predominant uptake by the liver[49]. Furthermore, we found that Hb-HSA$_3$ cluster were metabolized in the body and excreted into urine, which is the same excretion pathway of HSA[48]. The Hb-rCSA$_3$ cluster is likely to show high hepatic distribution after administration into dogs and to be excreted into urine.

We concluded that the Hb-rCSA$_3$ cluster possesses three positive features for RBC substitute: (i) negative surface net charge, (ii) moderately high O_2-affinity, and (iii) identical immunogenicity with CSA. Both rCSA and Hb-rCSA$_3$ cluster can be of great medical importance as a potential plasma expander and O_2-carrying fluid for veterinary blood transfusion in various clinical situations.

METHODS

Materials and Apparatus

Bovine Hb was purified from fresh bovine blood purchased from Tokyo Shibaura Zouki Co. Ltd. Purification procedures of native CSA from canine plasma are presented in Supplementary Materials. All other reagents were purchased from commercial sources as special grades and were used without further purification. Deionized water (18.2 MΩ·cm) was prepared using water purification systems (Elix UV and Milli Q Reference; Millipore Corp.). SDS-PAGE and Native-PAGE were performed using an electrophoresis power supply (EPS 301; GE Healthcare UK Ltd.) and 5–20% polyacrylamide precast gradient gel (Hi-QRAS gel N; Kanto Chemical Co. Inc.). Isoelectric focusing (IEF) was conducted using an electrophoresis power supply (EPS 601; GE Healthcare UK Ltd.) and IEF protein gels (Novex pH 3–10; Thermo Fisher Scientific Inc.). The UV-visible absorption spectra were recorded using a UV-visible spectrophotometer (8543; Agilent Technologies Inc.) connected with a temperature control unit (89090A; Agilent Technologies Inc.). Circulation dichroism (CD) spectra were recorded using a spectrophotometer (J-820; Jasco Corp.). Mass spectra were obtained using a MALDI–TOF mass spectrometer (autoflex equipped with a pulsed N_2 laser (337 nm); Bruker Daltonics K.K.). As the matrix, 0.1% sinapic acid, 0.05% trifluoroacetic acid, and 50% acetonitrile solution were used.

Expression and Purification of rCSA

Complementary (c)DNA was synthesized via reverse transcription using canine liver RNA. The full-length cDNA of CSA was amplified by PCR using Prime STAR Max DNA polymerase (Takara Bio Inc.) and the following oligonucleotide primer set: forward [5′-TCGAAACGAGGAAT TCATGAAGTGGGTAACTTTTATTT-3′] and reverse [5′-TGTCTAAGGCG AATTCTCAGACTAAGGCAGCTTGA-3′]. The amplified fragment was cloned into the *Eco*RI site of pHIL-D2 plasmid using an In Fusion reaction with Cloning Enhancer (Takara Bio Inc.), thereby generating pHIL-D2-CSA. Insertion of the entire CSA coding region was confirmed using DNA sequencing analysis. Then the obtained pHIL-D2-CSA plasmid vector was linearized by *Sal*I digestion and was used to transform the GS115 strain of *Pichia pastoris* (Thermo Fisher Scientific K.K.) by electroporation using an electroporator (MicroPulser; Bio-Rad Laboratories, Inc.).

Expression of rCSA was conducted by standard protocols provided by Thermo Fisher Scientific K.K. with some modifications. The transformed GS115 clone was first grown upon buffered mineral glycerol-complex (BMGY) medium (total 4 L) in a shaking incubator (Bio-Shaker G·BR-200; Taitec Corp.) with vigorous shaking (200 rpm) at 30 °C. When the culture medium reached an OD_{600} = 30–35, it was centrifuged (3000 × *g*, 10 min) to harvest the cells at 25 °C. To induce expression, the cell pellet was resuspended in a buffered mineral methanol complex (BMMY) medium (total 0.8 L). The expression was continued with vigorous shaking (200 rpm) at 30 °C for 10 days. During the cultivation, 100% methanol (1% volume of medium) was added every 24 h.

The secreted rCSA was isolated as follows. The growth medium was centrifuged (11,000 × *g*, 4 °C) to obtain the supernatant, which was brought to 50% saturation with ammonium sulfate at 25 °C. After leaving for 30 min at 4 °C, the slightly turbid solution was centrifuged (11,000 × *g*, 4 °C) to remove the dark brown precipitate. Subsequently, the supernatant fluid was brought to 95% saturation with ammonium sulfate and was stirred slowly for 30 min at 4 °C. The precipitated protein was collected by centrifugation (11,000 × *g*, 4 °C) and was dissolved in water. The brown solution was dialyzed against deionized water at 4 °C. Then 11% volume of 500 mM sodium phosphate (pH 5.8) was added. The resultant protein solution in 50 mM sodium phosphate (pH 6.0) was filtered using a membrane filter (C020A047A, 0.2 μm pore; Toyo Roshi Kaisha Ltd.).

Next, the obtained sample was loaded onto an affinity chromatography with a Toyopearl AF-Blue HC-650M (Tosoh Corp.) column. After washing with five-bed volumes of 50 mM sodium phosphate (pH 6.0), elution of rCSA was performed with 50 mM sodium phosphate (pH 7.4) containing 3 M NaCl. The eluent was dialyzed against deionized water at 4 °C, followed by addition of 25% volume of 100 mM Tris-HCl (pH 8.0). The resulting 20 mM Tris-HCl solution (pH 8.0) of the protein was filtered using a membrane filter (C020A047A, 0.2 μm pore). The sample was then applied to anion exchange chromatography with a Q Sepharose Fast Flow (GE Healthcare UK Ltd.) column using 20 mM Tris-HCl (pH 8.0) as the running buffer. After washing adequately with five-bed volumes of 20 mM Tris-HCl (pH 8.0) containing 100 mM NaCl, rCSA was eluted with 20 mM Tris-HCl (pH 8.0) containing 200 mM NaCl. The eluent was dialyzed against deionized water at 4 °C. Thereafter, 11% volume of 10× phosphate-buffered saline (PBS, pH 7.4) was added. Finally, the rCSA solution was concentrated to 30 mL using an ultraholder (Advantec UHP-76K; Toyo Roshi Kaisha Ltd.) with an ultrafilter (Advantec Q0100, 10 kDa MWCO; Toyo Roshi Kaisha Ltd.), and was sterilized with a sterile membrane filter (DISMIC-25CS, 0.2 μm pore; Toyo Roshi Kaisha Ltd.). All purification processes were followed by SDS-PAGE or Native-PAGE analysis. The concentration of rCSA was measured using a protein assay kit (Pierce 660 nm; Thermo Fisher Scientific K.K.). The cysteinyl thiol assay of rCSA was performed by reaction with 4,4′-dithiopyridine (4,4′-DTP), which binds the sulfhydryl group of the protein to generate 4-thiopyridinone (λ_{max}: 324 nm)50.

Crystallization, Data Collection, and Structure Determination of rCSA

The obtained rCSA was used directly for crystallization without further purification, such as powder-activated charcoal treatment. Colorless crystals (0.15 × 0.05 × 0.05 mm) grew spontaneously from a 30 mg/mL protein solution in the Morpheus screen condition H1 (Molecular Dimensions Ltd.) at 293 K within two weeks. The crystals were flash-frozen in liquid nitrogen. X-ray diffraction data were collected using an ADSC Quantum CCD detector with a Synchrotron radiation source at the Photon Factory AR-NW12A (Tsukuba, Japan). All data were collected at 100 K and were processed with HKL200051. Data collection statistics are presented in Table S1. The rCSA structure was ascertained using molecular replacement with Auto-MR in PHENIX52. The structure of HSA with palmitic acid (PDB ID: 4BKE) was used as the search model53. In the initial electron density map, several helical features were clearly identified. Sequence assignment of the atomic model was conducted using the amino acid sequence of CSA33. Further iterative model building and refinement were performed using PHENIX and COOT54. For the poor electron density region (562–565), a poly-alanine model was built using COOT. Electron densities corresponding to the residues 1–2 (IA), 79–87 (IA-h4–h5), 114–120 (IA-h6–IB-h7), and 363–365 (IIB-h9–h10) were disordered in the model because of conformational flexibility at *N*-terminal, turns, and stretches of extended loops. Atomic coordinates of the rCSA structure reported in this paper have been deposited in the Protein Data Bank (PDB) under accession code 5GHK. All molecular images were produced using PyMOL (Schrödinger K.K.).

Preparation of Hb-rCSA$_3$

A DMSO solution of heterobifunctional cross-linker, succinimidyl-4-(*N*-maleimidomethyl)-cyclohexane-1-carboxylate (SMCC; Wako Pure Chemical Industries Ltd.) (20 mM, 0.2 mL), was added dropwise into PBS solution (pH 7.4) of carbonyl Hb (0.1 mM, 2 mL) in a round-bottom flask (20 mL volume). The mixture was stirred for 30 min in CO atmosphere at 4 °C. After removing unreacted excess cross-linker using gel filtration chromatography (GFC) with a Sephadex G25 (superfine) column (GE Healthcare UK Ltd.), the obtained SMCC-bound Hb (maleimide-activated Hb) was concentrated to 2 mL ([Hb] = 0.1 mM) using a centrifugal concentrator (Vivaspin 20 ultrafilter, 10 kDa MWCO; GE Healthcare UK Ltd.).

Before coupling of maleimide-activated Hb to rCSA, partially oxidized Cys-34 of rCSA was reduced using dithiothreithol (DTT). Aqueous DTT solution (10 mM, 0.1 mL) was added to the PBS solution (pH 7.4) of rCSA (1 mM, 2 mL) ([DTT]/[rCSA] = 0.5 mol/mol). Then the mixture was incubated for 1 h at 25 °C. The excess DTT was removed using GFC with a Sephadex G25 (superfine) column. The pooled eluent was condensed to 2 mL ([rCSA] = 1 mM) using a Vivaspin 20 ultrafilter (10 kDa MWCO). The increase of the mercapto-ratio of the Cys-34 was confirmed as ca. 80% using the 4,4′-DTP procedure50.

Then the maleimide-activated Hb solution (0.1 mM, 2 mL) was added slowly into rCSA solution (1 mM, 2 mL) in PBS, followed by stirring in CO atmosphere under dark conditions for 48 h at 4 °C. An aliquot of the reaction mixture was applied to size-exclusion chromatography (SEC) on an HPLC system (Prominence LC-20AD/CTO-20A/SPD-20A; Shimadzu Corp.) with an SEC column (YMC-Pack Diol-300 S-5; YMC Co. Ltd.) using 50 mM sodium phosphate (pH 7.4) as the mobile phase. The elution curve exhibited new triple peaks at the high molecular weight region. The reaction mixture was subjected to GFC with a Superdex 200 pg (GE Healthcare UK Ltd.) using PBS as the running buffer. All the fractions corresponding to the clusters were collected together. The entire protein concentration and Hb concentration were measured, respectively, using a protein assay kit (Pierce 660 nm) and Hb assay kit (Nescoat Hemokit-N; Alfresa Pharma Corp.). The average rCSA/Hb ratio of the product was estimated as 3.0 ± 0.2. We designated this cluster as Hb-rCSA$_{3}$, with italicized subscript 3. The Hb-CSA$_{3}$cluster was prepared using the same procedure using native CSA obtained from canine plasma.

Immunogenicity

Quick Chaser Occult Blood Test kit (Mizuho Medy Co., Ltd.), which detect human Hb (HbA) by immuno-chromatographic method, was used to evaluate the reaction between HbA-rCSA$_{3}$ cluster and anti-HbA antibody. The HbA-rCSA$_{3}$ cluster was synthesized by the same procedure as Hb-rCSA$_{3}$ using HbA30. Two drops of the HbA-rCSA$_{3}$solution (0.1–1 μM) in PBS (pH 7.4) was applied to the assay plate and the appearance of red test-line signal was evaluated. Identical experiments with PBS, rCSA, and HbA were also conducted.

Immunological reactivity of HbA-rCSA$_3$ cluster against anti-CSA antibody was measured using latex immuno-agglutination reagents (Canine albumin in urea; Shima Laboratories Co., Ltd.). The Hb-rCSA$_3$cluster solution ([rCSA unit] = 27–300 µg/mL, 30 µL) in PBS (pH 7.4) was first diluted with saline (120 µL), and aliquot (2.25 µL) of the solution was further diluted with special buffer (60 µL). Then the obtained sample solution was mixed with anti-CSA antibody sensitized latex (13 µL) at 37 °C, and the absorbance at 694 nm of the reactant was measured by a clinical chemistry analyzer (BioMajesty JCA-BM2250; Jeol Ltd.). Identical experiments with rCSA were also conducted.

O_2-Binding Property

To prepare the oxy Hb-rCSA$_3$ cluster, the PBS solution (pH 7.4) of carbonyl Hb-rCSA$_3$ ([Hb]: ca. 10 µM, 3 mL) in a round-bottom flask (20 mL volume) was rotated using a rotary evaporator and was exposed to 100% O_2 gas under an ice-water bath. A liquid thin film formed on the inner wall of the flask was illuminated by a 500 W halogen lamp that was fixed above the rotating flask at a distance of 10 cm. After complete oxygenation, the oxy Hb-rCSA$_3$ cluster solution was transferred to an optical quartz cuvette (10 mm path length) sealed with a rubber septum. To prepare deoxy Hb-rCSA$_3$ cluster, N_2 gas was blown to the O_2-complex solution.

The O_2 affinity (P_{50}: O_2-partial pressure where Hb is half-saturated with O_2) and Hill coefficient (n) were obtained using an automatic recording system for O_2-equilibrium curve (Hemox Analyzer; TCS Scientific Corp.) at 37 °C. The Hb-rCSA$_3$ cluster in PBS solution (pH 7.4) was deoxygenated by flushing with N_2 and was oxygenated by increasing the O_2-partial pressure. The visible absorption spectra, P_{50}, and n values of the Hb-CSA$_3$ cluster were obtained using the same procedures.

O_2-Complex Stability

The O_2-complex stability of the Hb-rHSA$_3$ cluster was evaluated using the first-order autoxidation rate constant (k_{ox}) of the core Hb. The oxy Hb-rCSA$_3$ in PBS solution (pH 7.4, [Hb] = 10 µM, 2 mL) was put into an optical quartz cuvette (10 mm path length). The top of the cuvette was sealed with gas-permeable film (AeraSeal Film MAF710; Gel Co.), which facilitates air exchange but prevents water evaporation. The absorbance increase at 630 nm, which is based on metHb [Fe(III)-state] formation, was observed under aerobic conditions at 37 °C. After monitoring, the completely oxidized metHb-rCSA$_3$ cluster was prepared by addition of small excess $K_3[Fe(CN)_3]$. From the absorption intensity change, the k_{ox}value was calculated using nonlinear least-squares curve fitting method. The O_2-complex stability of Hb and Hb-CSA$_3$ cluster was evaluated using the same procedure.

Additional Information

How to cite this article: Yamada, K. *et al.* Artificial Blood for Dogs. *Sci. Rep.* **6**, 36782;

Publisher's note: Springer Nature remains neutral with regard to jurisdictional claims in published maps and institutional affiliations.

ACKNOWLEDGEMENTS

The authors acknowledge Dr. Natsuhiko Sugimura, Dr. Toshimichi Shibue, Materials Characterization Central Laboratory, Waseda University, for skillful experiments related to mass spectroscopy, Dr. Hiromi Asakawa, Dr. Masakatsu Hashimoto, Shima Laboratories for skillful experiments related to immunogenicity measurements, and Mr. Ryo Yoneyama, Chuo University, for experiments related to rCSA expression. This work was supported by a Joint Research Grant from the Institute of Science and Engineering, Chuo University.

REFERENCES

1. Japan Pet Food Association. Research results on dog and cat population in Japan, 2015. http://www.petfood.or.jp/data/chart2015/01.html (2016) (Date of access: 11/07/2016).
2. Statistics Bureau, Ministry of Internal Affairs and Communications. Population Estimates (May 2016). www.stat.go.jp/english/data/jinsui/tsuki/index.htm (2016) (Date of access: 11/07/2016).
3. Peters, Jr. T. All about Albumin, Biochemistry, Genetics, and Medical Applications (Academic Press, 2006).
4. Fanali, G. et al. Human serum albumin: from bench to bedside. Mol. Asp. Med. 33, 209–290 (2012).
5. Pandjaitan, B. et al. Escherichia coli expression and purification of recombinant dog albumin, a cross-reactive animal allergen. J. Allergy Clin. Immunol. 105, 279–285 (2000).

6. Giger, U., Gelens, C. J., Callan, M. B. & Oakley, D. A. An acute hemolytic transfusion reaction caused by dog erythrocyte antigen-1.1 incompatibility in a previously sensitized dog. J. Am. Vet. Med. Assoc. 206, 1358–1362 (1995).
7. Squires, J. E. Artificial blood. Science 295, 1002–1005 (2002).
8. Kim, H. W. & Greenburg, A. G. Artificial oxygen carriers as red blood cell substitutes: a selected review and current status. Art. Organs 28, 813–827 (2004).
9. Jahr, J. H., Sadighi, A., Doherty, L., Li, A. & Kim, H. W. Hemoglobin-based oxygen carriers: history, limits, brief summary of the state of the art, including clinical trials in Chemistry and biochemistry of oxygen therapeutics: from transfusion to artificial blood (eds Bettati, S. & Mozzarelli, A.) p. 301–316 (John Wiley & Sons, 2011).
10. Kluger, R. & Lui, F. E. HBOCs from chemical modification of Hb in Hemoglobin-based oxygen carriers as red cell substitutes and oxygen therapeutics (eds Kim, H. W. & Greenburg, A. G.) p. 159–183 (Springer-Verlag, 2013).
11. Vandegriff, K. D., Malavalli, A., Wooldridge, J., Lohman, W. & Winslow, R. M. MP4, a new nonvasoactive PEG-Hb conjugate. Transfusion 43, 509–516 (2003).
12. Manjula, B. N. et al. Site-specific PEGylation of hemoglobin at Cys-93(β): correlation between the colligative properties of the PEGylated protein and the length of the conjugated PEG chain. Bioconjugate Chem. 14, 464–472 (2003).
13. Li, D., Hu, T., Manjula, B. N. & Acharya, S. A. Extension arms facilitated pegylation of αα-hemoglobin with modifications targeted exclusively to amino groups: functional and structural advantages of free Cys-93(β) in the PEG-Hb adduct. Bioconjugate Chem. 20, 2062–2070 (2009).
14. Buehler, P. W. et al. Structural and functional characterization of glutaraldehyde-polymerized bovine hemoglobin and its isolated fractions. Anal. Chem. 77, 3466–3478 (2005).
15. Chatterjee, R. et al. Isolation and characterization of new hemoglobin derivative cross-linked between the α chains (lysine 99α1 → lysine 99α2). J. Biol. Chem. 261, 9929–9937 (1986).
16. Hu, D. & Kluger, R. A. Functional cross-linked hemoglobin bis-tetramers: geometry and cooperativity. Biochemistry 47, 12551–12561 (2008).
17. Pearce, L. B., Gawryl. M. S., Rentko, V. T., Moon-Massat, P. F. & Rausch, C. W. HBOC-201 (hemoglobin glutamer-250) (bovine), Hemopure®): clinical studies in Blood substitutes (ed Winslow, R. M.) p. 437–450 (Elsevier, 2006).
18. Winslow, R. M. Cell-free oxygen carriers: scientific foundations, clinical development and new directions. Biochim. Biophys. Acta 1784, 1382–1386 (2008).
19. Chen, J. Y., Scerbo, M. & Kramer, G. A review of blood substitutes: examining the history, clinical trial results, and ethics of hemoglobin-based oxygen carriers. Clinics 64, 803–813 (2009).
20. Natanson, C., Kern, S. J., Lurie, P., Banks, S. M. & Wolfe, S. M. Cell-free hemoglobin-based blood substitutes and risk of myocardial infarction and death. J. Am. Med. Assoc. 299, 2304–2312 (2008).
21. Doherty, D. H. et al. Rate of reaction with nitric oxide determines the hypertensive effects of cell-free hemoglobin. Nat. Biotechnol. 16, 672–676 (1998).
22. Yu, B., Bloch, K. D. & Zapol, W. M. Hemoglobin-based red blood cell substitutes and nitric oxide. Trends Cardiovasc. Med. 19, 103–107 (2009).
23. Kluger, R. Red cell substitutes from hemoglobin–Do we start all over again? Curr. Opin. Chem. Biol. 14, 538–543 (2010).
24. Alayash, A. I. Blood substitutes: why haven't we been more successful? Trends Biotechnol. 32, 177–185 (2014).
25. Driessen, B., Jahr, J. H., Lurie, F., Griffery, S. M. & Gunther, R. A. Effects of haemoglobin-based oxygen carrier haemoglobin glutamer-200 (bovine) on intestinal perfusion and oxygenation in a canine hypovolaemia model. Br. J. Anaesth. 86, 683–692 (2001).
26. Buehler, P. W. et al. Structural and functional characterization of glutaraldehyde-polymerized bovine hemoglobin and its isolated fractions. Anal. Chem. 77, 3466–3478 (2007).

27. European Medicines Agency. Oxyglobin. http://www.ema.europa.eu/docs/en_GB/document_library/EPAR_-_Summary_for_the_public/veterinary/000045/WC500068122.pdf#search='oxyglobin+side+effects' (2007) (Date of access: 11/07/2016).
28. Tomita, D. et al. Covalent core–shell architecture of hemoglobin and human serum albumin as an artificial O2 carrier. Biomacromolecules 14, 1816–1825 (2013).
29. Hosaka, H., Haruki, R., Yamada, K., Böttcher, C. & Komatsu, T. Hemoglobin–albumin cluster incorporating a Pt nanoparticle: artificial O2 carrier with antioxidant activities. PLoS One 9, e110541 (2014).
30. Kimura, T., Shinohara, R., Böttcher, C. & Komatsu, T. Core–shell clusters of human haemoglobin A and human serum albumin: artificial O2-carriers having various O2-affinities. J. Mater. Chem. B 3, 6157–6164 (2015).
31. Haruki, R. et al. Safety evaluation of hemoglobin–albumin cluster "HemoAct" as a red blood cell substitute. Sci. Rep. 9, 12778 (2015).
32. Cohn, L. A., Kerl, M. E., Lenox, C. E., Livingston, R. S. & Dodam, J. R. Response of healthy dogs to infusion of human serum albumin. Am. J. Vet. Res. 68, 657–663 (2007).
33. UniProt KB-P49822 (ALBU_CANLF) for CSA, and UniProt KB-P02768 (ALBU_HUMAN) for HSA.
34. He, X. & Carter, D. C. Atomic structure and chemistry of human serum albumin. Nature 358, 209–215 (1992).
35. Curry, S., Mandelkow, H., Brick, P. & Franks, N. Crystal structure of human serum albumin complexed with fatty acid reveals an asymmetric distribution of binding sites. Nat. Struct. Biol. 5, 827–835 (1997).
36. Bhattacharya, A. A., Curry, S. & Franks, N. P. Binding of the general anesthetics propofol and halothane to human serum albumin. J. Biol. Chem. 275, 38731–38738 (2000).
37. Sugio, S., Kashima, A., Mochizuki, S., Noda, M. & Kobayashi, K. Crystal structure of human serum albumin at 2.5 Å resolution. Protein Eng. 12, 439–446 (1999).
38. Ghuman, J. et al. Structural basis of the drug-binding specificity of human serum albumin. J. Mol. Biol. 353, 38–52 (2005).
39. Majorek, K. A. et al. Structural and immunologic characterization of bovine, horse, and rabbit serum albumins. Mol. Immunol. 52, 174–182 (2012).
40. Antonini, E. & Brunori, M. Hemoglobin and myoglobin in their reactions with ligands. North-Holland Research Monographs. Frontiers of Biology, vol. 21 (eds Neuberger, A. & Tatum, E. L.) p. 13–39 (North-Holland Pub. Co., 1971).
41. Mueser, T. C., Rogers, P. H. & Arnone, S. Interface sliding as illustrated by the multiple quaternary structures of liganded hemoglobin. Biochemistry 39, 15353–15364 (2000).
42. Zhang, Y., Bhatt, V. S., Sun, G., Wang, P. G. & Palmer, A. F. Site-selective glycosylation of hemoglobin on Cys 93. Bioconjugate Chem. 19, 2221–2230 (2008).
43. Kluger, R. & Zhang, J. Hemoglobin dendrimers: functional protein clusters. J. Am. Chem. Soc. 125, 6070–6071 (2003).
44. Hu, D. & Kluger, R. A. Functional cross-linked hemoglobin bis-tetramers: geometry and cooperativity. Biochemistry 47, 12551–12561 (2008).
45. Kosa, T., Maruyama, T. & Otagiri, M. Species differences of serum albumins: I. drug binding sites. Pharmaceut. Res. 14, 1607–1612 (1997).
46. Rohlfs, R. J. et al. Arterial blood pressure responses to cell-free hemoglobin solutions and the reaction with nitric oxide. J. Biol. Chem. 273, 12128–12134 (1998).
47. Intaglietta, M. Microvascular transport factors in the design of effective blood substitutes in Microcirculatory effects of hemoglobin solutions (eds Messmer, K., Burhop, K. E. & Hutter, J.) p. 8–15 (Karger AG, 2004).
48. Yamada, K. et al. Influence of molecular structure on O2-binding properties and blood circulation of hemoglobin-albumin clusters. PLoS One 11, e0149526, 1–15 (2016).

49. Rennen, H. J. J. et al. The effect of molecular weight on nonspecific accumulation of 99mT-labeled proteins in inflammatory foci. Nucl. Med. Biol. 28, 401–408 (2001).
50. Grassetti, D. R. & Murry, Jr. J. F. Determination of sulfhydryl group with 2,2′- or 4,4′-dithiopyridine. Arch. Biochem. Biophys. 119, 41–49 (1967).
51. Otwinowski, Z. & Minor, W. Processing of X-ray diffraction data collected in Oscillation mode. Methods Enzymol. 276, 307–326 (1997).
52. Adams, P. D. et al. PHENIX: a comprehensive Python-based system for macromolecular structure solution. Acta Cryst. D 66, 213–221 (2010).
53. Sivertsen, A. et al. Synthetic cationic antimicrobial peptides bind with their hydrophobic parts to drug site II of human serum albumin. BMC Struct. Biol. 14, 4 (2014).
54. Emsley, P., Lohkamp, B., Scott, W. G. & Cowtan, K. Features and development of Coot. Acta Cryst. D 66, 486–501 (2010).

CHAPTER 13

CRISPR/Cas9-mediated Mutation of Tyrosinase (Tyr) 3′ UTR Induce Graying in Rabbit

Yuning Song[1], Yuxin Xu[1], Jichao Deng[1], Mao Chen[1], Yi Lu[1], Yong Wang[1], Haobin Yao[1], Lina Zhou[1], Zhiquan Liu[1], Liangxue Lai[1,2] & Zhanjun Li[1]

[1] *Jilin Provincial Key Laboratory of Animal Embryo Engineering, Jilin University, Changchun, 130062, China.*

[2] *CAS Key Laboratory of Regenerative Biology, South China Institute for Stem Cell Biology and Regenerative Medicine, Guangzhou Institutes of Biomedicine and Health, Chinese Academy of Sciences, Guangzhou, 510530, China.*

ABSTRACT

The 3′ untranslated regions (UTRs), located at the end of mRNA molecules, are believed to play a role in RNA replication and/or protein translation. Mutations in the *tyrosinase* (*Tyr*) gene are known to cause recessive albinism in humans and other species. In this study, to test whether the CRISPR/Cas9 system works on the mutation of the UTRs regulatory region in rabbit, the 3′ UTR of the rabbit *Tyr* gene was deleted by a dual sgRNA directed CRISPR/Cas9 system. As expected, gray coat color and reduced melanin in hair follicles and irises was found in the mutated rabbit, thus increasing confidence in the association of the mutation of the *Tyr* 3′ UTR with graying in rabbit. The graying phenotype was also found in the F1 generation, suggesting that the mutated allele can be stably inherited by the offspring. Thus, we provide the first evidence that reduced melanin and graying can be caused by deletion of the *Tyr* 3′ UTR in rabbits. Additionally,

CRISPR/Cas9-mediated large fragment deletions can facilitate genotype to phenotype studies of UTRs or non-coding RNAs in future.

INTRODUCTION

Tyrosinases (*Tyr*) are essential enzymes in melanin biosynthesis and are responsible for pigmentation of skin and hair in mammals1. Mutations in the *Tyr* gene result in white coat color and lack of pigmentation due to absence of melanin production. Melanin biosynthesis in these mutants is disrupted at the critical first and second reactions: hydroxylation of tyrosine to L-DOPA and the oxidation of L-DOPA to DOPA-quinone1. Previous studies have reported that albinismis associated with a *Tyr* mutation in American mink2, cattle3, mouse4 and rabbit5. In an updated list of *Tyr* mutations, oculocutaneous albinism was also reported in humans1.

The rabbit tyrosinase gene (Chromosome 1: 127,562,607–127,667,237) has alength of 136 kb and is composed of five exons. In albino rabbits, a homozygous mutation (T373K) was identified that resulted in alteration of the last N-glycosylation site of the *Tyr* coding sequence5. Furthermore, the CRISPR/Cas9-mediated *Tyr* knockout rabbit also displayed typical albinism phenotype6, 7.

It is known that mammalian 5′ UTRs, which are non-coding DNA regulatory regions, mediate post- or co-transcriptional autoregulation of gene expression through direct interaction with proteins8. On the other hand, the 3′ UTRs direct an extensive range of alternative post-transcriptional, including regulation of mRNA decay and translation9, contributing significantly to gene regulation. It is well established that 3′ poly(A) tail of eukaryotic mRNAs is critical for the proper regulation of gene expression, and enhances the *in vivo* and *in vitro* translation in mammas10. In addition, several studies have highlighted the polyadenylation and alternative polyadenylation are important in gene expression and associated with a risk of clinical diseases11, 12, 13, 14, 15. However, all the studies were carried out *in vitro* or clinical cases, with very little studies conducted at the animal level. Therefore, there is a need to further characterize the association between the 3′ UTR and mRNA in animals.

A previous study has shown that the 5′ UTR sequence is a potentially important regulatory element for appropriate expression of the human *Tyr* gene16. However, very little is known about the role of 3′ UTR in *Tyr* gene expression. More recently, an insertion/deletion of 13 bp in the 3′ UTR of tyrosinase-related protein 1 (*Tyrp1*) was found in rabbits with brown coats17, implicating an important role of the 3′ UTR in gene expression. This study aimed to further characterize the relationship between the 3′ UTR and gene expression. Therefore, we deleted the 3′ UTR of the rabbit *Tyr* gene by dual sgRNA-directed CRISPR/Cas9 system, and the genotype to phenotype of the *Tyr* 3′ UTR were also validated in this animal model.

RESULTS

Dual sgRNA-directed deletion of *Tyr* 3′ UTR in zygotes

Initially, a search for poly(A), AU-rich elements (AREs) and potential miRNA binding sites in the rabbit *Tyr* 3′ UTR was performed using online tools. As shown in Figure 1A, we found two poly(A) sites located at 290 and 380 bp downstream of the stop codon. We then designed two guide RNAs (gRNAs) to target the 3′ UTR of the *Tyr* gene and determine the efficiency of the CRISPR/Cas9 system in zygotes. A mixture of *in vitro*-transcribed mRNA of Cas9 (100 ng/μL) and sgRNA (25 ng/μL) was injected into 21 pronuclear-stage zygotes by cytoplasm injection. Of the 17 blastocysts obtained, ten were tested by PCR amplification and sequence analysis. As shown in Figure 1B and C, mutations were found in nine of these ten blastocysts. A 330 bp long deletion was found in B1, B3 and B6 samples, indicating that deletion of the *Tyr* 3′ UTR can be achieved in zygotes via the CRISPR/Cas9 system.

Generation of *Tyr* 3′ UTR knockout (KO) rabbits

To generate *Tyr* 3′ UTR KO rabbits, 157 injected zygotes from Lianshan black rabbits were implanted into four pseudo-pregnant recipient females (Table 1). Two of the recipients carried the pregnancies to term and gave birth to four live pups (Figure 2A). T-cloning and PCR-sequencing results revealed that a monoallelic large-fragment deletion of *Tyr* 3′ UTR was presentin pups #3 (−26 and −345 bp) and #4 (−29 and −355 bp), while an insertion (+165 bp) and a deletion (−27 bp) were detected in pup #1 (Figure 2B and F).

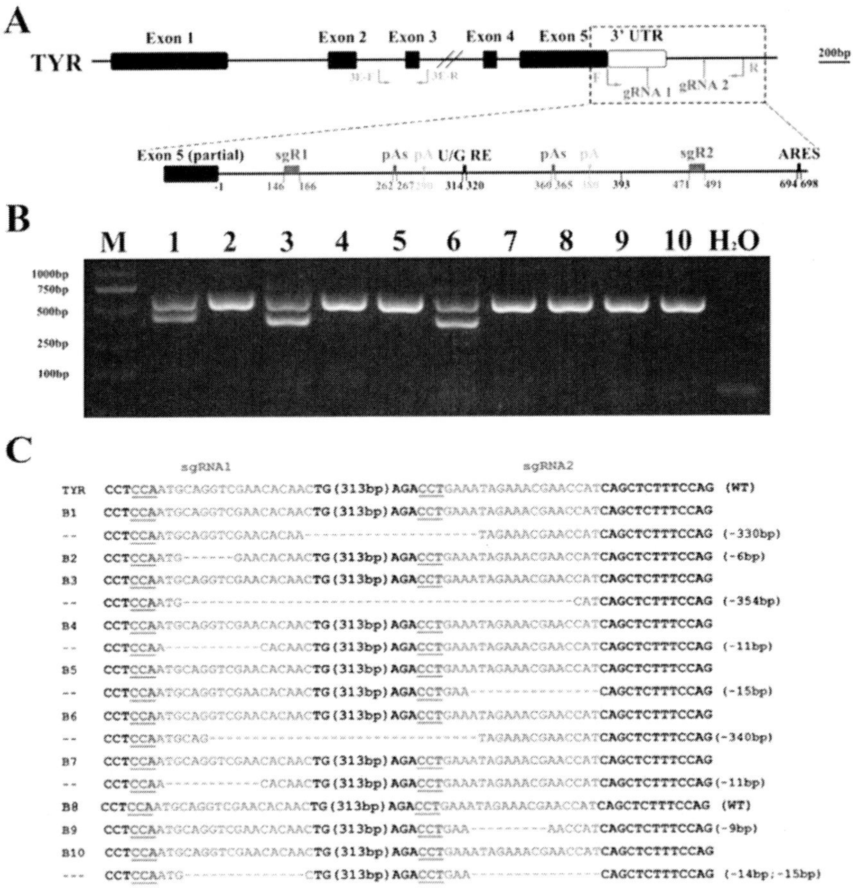

Figure 1. Dual sgRNA-directed deletion of the *Tyr* 3′ UTR in zygotes. (**A**) Schematic diagram of 3′ polyA tail, sgRNAs and primers targeting the rabbit TYR 3′ UTR loci. Two sgRNAs are marked in red and the primers are presented in blue. pAs: poly (**A**) addition signal; pA: poly (**A**) site; U/G RE: U/G rich area. AREs: AU-rich elements. (**B**) Mutation detection in blastocyst by PCR. M, marker; 1–10 represent different blastocysts used in this study. Original images are included "Authors' original file for Figure 1B". (**C**) Mutation detection in blastocyst by T-cloning and Sanger sequencing. The WT sequence is shown at the top of the targeting sequence. sgRNA sequences are marked in green and the protospacer adjacent moti (PAM) sequences are in red with underline. WT: wild type; deletions "–"; insertion "+".

Table 1: Generation of *Tyr* 3′ UTR KO rabbits using CRISPR/Cas9.

	gRNA/Cas9 mRNA (ng/uL)	Embryos injected	Embryos transferred	Pregnancy	Pups obtained (% transferred)	3′ UTR KO pups (% pups)	Pups with Color change
1	25/100	40	35	No			
2	25/100	45	40	Yes	2(5%)	1(50%)	0
3	25/100	40	38	No			
4	25/100	47	44	Yes	2(4.5%)	2(100%)	2

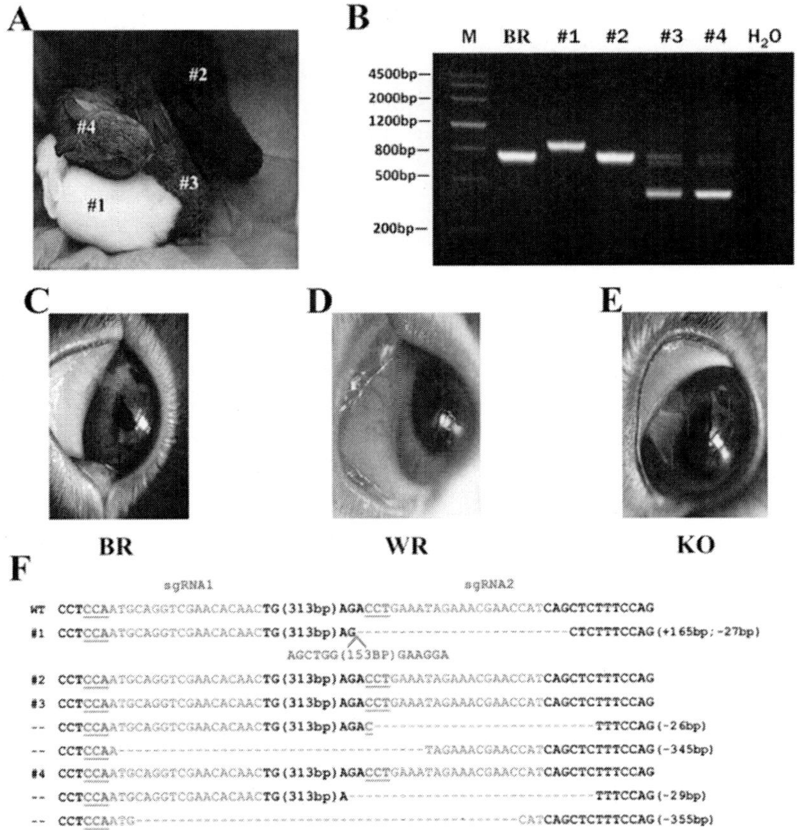

Figure 2. Generation of *Tyr* 3′ UTR deleted rabbit by CRISPR/Cas9. (**A**) Phenotype of *Tyr* mutant rabbits; rabbit #1, albinism; rabbit #2, black; rabbit #3 and #4, graying. (**B**) The mutation determination in founder rabbits by PCR. (**C**) The eye of black rabbit (BR). Original images are included "Authors' original file for Figure 2B". (**D**) The eye of white rabbit (WR). (**E**) The eye of *Tyr* 3′ UTR deleted rabbit (KO). (**F**) Mutation detection in founder by T-cloning and Sanger sequencing. The PAM sites are underlined and highlighted in red; the target sequences are green; deletions (−) and insertions (+) are shown. WT, wild-type.

Next, we examined the coat and eye color of the 3′ UTR KO rabbits. We observed the typical albino phenotype with loss of dark pigment in the skin and eyes in rabbit #1, and graying in rabbits #3 and #4, as compared to black skin and eyes in rabbit #2 (Figure 2C–E). The information about the genotype (T373K and 3′ UTR deletion) and skin color of F0 pups is listed in Table S3.

In addition, we observed that rabbit #1 that carried an insertion (+165 bp) showed the typical albino phenotype in the skin and eyes (Figure 2A). Further analysis revealed that rabbit #1 had homozygous A/A genotype, while heterozygous A/C genotype was identified in rabbit #3, #4 (Table S3). The 165 bp inserted sequence was neither derived from the rabbit genome nor from the cloning vector. We therefore speculated that the non-pigmented phenotype in rabbit #1 is due to SNP (A/C) at 1118 bp of rabbit *Tyr* gene while not the insertion (+165 bp).

To test whether off-target effects occurred in these genetically modified rabbits, we screened the rabbit genome and identified five Potential off-target sites (POTs) for each sgRNA. The primers and mismatch sites are listed in Table S2. POTs were amplified by PCR from the genomic DNA of the mutant rabbits and subjected to T7E1 analysis (Figure S1). The results revealed that none of these POTs were mutated, indicating that the sgRNAs used in this study are locus-specific.

Reduced expression of melanin in *Tyr* 3′ UTR KO rabbits

We further examined whether deletion of the 3′ UTR lead to reduced gene expression and related phenotypes. As shown in Figure 3A, *Tyr* mRNA level was significantly reduced in 3′ UTR KO rabbits, when compared with the New Zealand White rabbits (WR) and Lianshan black rabbits (BR). These results were also confirmed by Western blot and gray-scale analysis at the protein level (Figure 3B,C), indicating that both *Tyr* mRNA as well as protein levels were reduced in the 3′ UTR KO rabbits. Furthermore, histological HE staining showed that melanin amounts were reduced in hair follicles and irises of 3′ UTR KO rabbits and absent in WR, when compared with BR (Figure 3D). These results confirmed that graying and reduced melaninin hair follicles and irises was observed in the *Tyr* 3′ UTR KO rabbits.

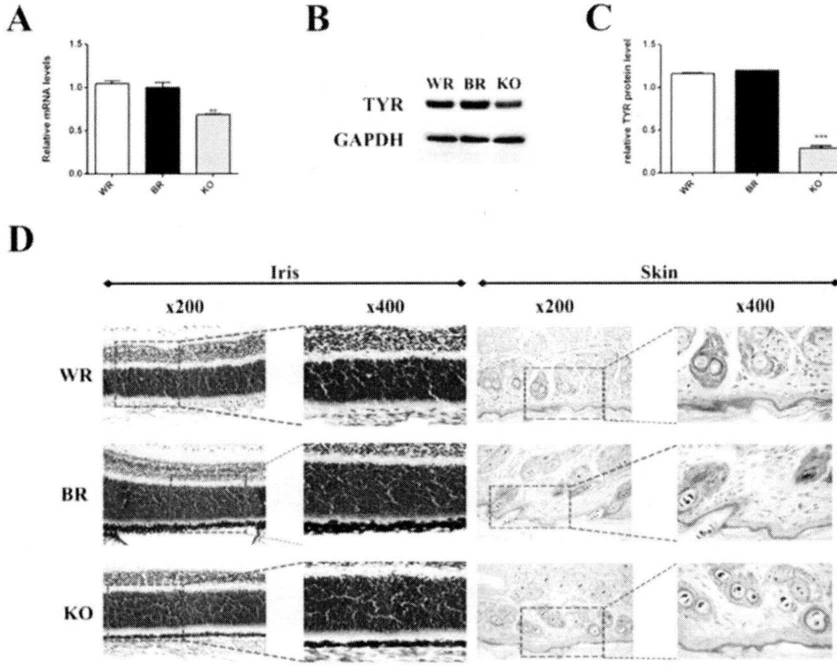

Figure 3. Reduced expression of melanin in *Tyr* 3′ UTR deleted rabbit. (**A**) Expression of *Tyr* gene was determined by qRT-PCR. (**B**) *Tyr* protein was determination by Western Blot. (**C**) Gray-scale analysis of the *Tyr* protein by ImageJ software. (**D**) H&E staining of the iris and skin from the WR, BR and KO rabbits. All the data are expressed as the mean ± SEM. *P < 0.005; **P < 0.001; ***P < 0.005. BR, black rabbit; WR, white rabbit; KO, *Tyr* 3′ UTR deleted rabbits.

Heritability of *Tyr* 3′ UTR KO mutation

Next, we then examined whether the *Tyr* 3′ UTR KO mutation and the graying phenotype could be stably transmitted to progeny. Rabbit #3 (female, 3′ UTR KO gray rabbits) was mated with Rabbit #2 (male, BR). As shown in Figure 4A, one (*04) of five newborn rabbits showed the typical gray phenotype with reduced dark pigment in the skin and eyes. T-cloning and PCR analysis confirmed the presence of the *Tyr* 3′ UTR deletion in pups of rabbits *04 and *05 (Figure 4B–D), demonstrating that 3′ UTR deletions are heritable and that the graying phenotype could be transmitted through the germ-line. In addition, to segregate the T373K and 3′ UTR modifications, more F1 progeny were obtained from Rabbits #3 and #2. As shown in

Table S4, the rabbits carrying at least one dominant mutation (C/C, C/A) of *Tyr* are fully pigmented (*02, *03, *05, *07, *08, *11, *12), whereas the dominant mutation containing haplotype was linked to the 3′ UTR deletion and the other recessive mutant allele (A nucleotide at SNP), conferred gray coat color to the rabbits (*04, *09, *10). Similar observation was also made in the F0 generation (Table S3).

DISCUSSION

Previous studies have demonstrated that *Tyr* plays key roles in melanin biosynthesis and albinism1. We have previously described the feasibility of applying the Cas9/gRNA system to produce a large deletion (105 kb) in the *Tyr* gene, resulting in loss of pigmentation, thus leading to albinism in rabbits6. In this study, we presenta functional validation the *Tyr* 3′ UTR KO and albinismin rabbits. Interestingly, gray coat color and reduced melanin in hair follicles and irises was observed in rabbits with the 3′ UTR KO mutation. Therefore, we believed that the graying phenotype in these rabbits was due to targeted degradation of *Tyr* mRNAs, as has been demonstrated in previous studies9.

Poly(A) is presumed to play a significant role in mRNA synthesis, translation, and/or metabolism, and is potentially involved in any of the four stages of mRNA metabolism: splicing, nucleus-to-cytoplasm transport, translation, and stability18, 19. It has been demonstrated that the poly(A) protects mRNAs from rapid, indiscriminate degradation and also affects the half-lives of mRNA20, 21. In our work, we found that two poly(A) sites were destroyed in the *Tyr* 3′ UTR KO rabbits, that showed gray coat color. Here we hypothesize that destruction of the poly(A) sites affects the polyadenylation and stability of *Tyr* mRNA, resulting in lower expression of tyrosinase.

In addition, progeny from rabbit #3 appear to inherit only one of the two detected deleted alleles. Tissue mosaicism has been widely reported in genetically modified animals generated by co-injection of Cas9 mRNA and single-guide RNA (sgRNA) into zygotes22. In our study, the genotype of the F0 rabbits was determined from ear tissues. Therefore, we speculated that only one deleted allele was detected in the F1 progeny generated from rabbit #3 owing to mosaicism between germ cells and ear tissues.

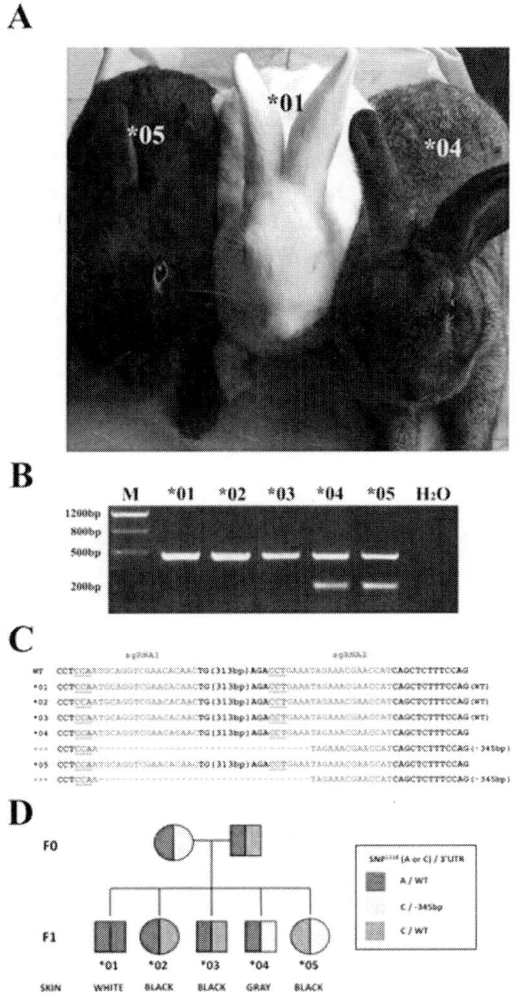

Figure 4. Heritability of *Tyr* 3′ UTR deleted rabbit. (**A**) The F1 pups with different skin color. (**B**) Mutation detection in F1 rabbits by PCR. The results demonstrated that the *Tyr* 3′ UTR deletions were determined in *04 and *05. M, DNA marker III. Original images are included "Authors' original file for Figure 4B". (**C**) T-cloning sequence analysis of the F1 rabbits. The PAM sites are underlined and highlighted in red; the target sequences are green; deletions (−) and insertions (+) are shown. BR, black rabbit; WR, white rabbit; KO, *Tyr* 3′ UTR deleted rabbits. (**D**) A pedigree obtained from analyses in (**C**). The identical alleles are shown by the same colors. The large deletion of *Tyr* 3′ UTR in F0 is heritability transmitted to the F1 pups. SNP[1118], the SNP sites (A/C) located in the *Tyr* gene.

MicroRNAs (miRNAs) are a class of endogenous non-coding small RNAs that contribute to the regulation of gene expression at the post-transcriptional level by binding to the 3′ UTR. Previous work has showed that miR-196a-2 can inhibit the expression of *Tyr* by attenuating melanin biosynthesis in the melanocytes23. Furthermore, six new binding sites for miRNAs have been identified, suggesting that reduced expression of *Tyr* is associated with miRNA-directed mRNA decay9. Unexpectedly, no miRNAs binding site was found in the rabbit *Tyr* 3′ UTR region, suggesting that the graying and reduced tyrosinase production phenotypes in the *Tyr* 3′ UTR KO rabbit is not regulated by miRNAs. The underlying molecular mechanism of this observation remains to be explored.

In summary, we provide the first evidence that a *Tyr* 3′ UTR mutation causes reduced melanin production and graying in rabbits. Moreover, functional analysis of the 3′ UTR in animal models can be achieved by dual sgRNAs directed CRISPR/Cas9 system. Our novel animal model will be beneficial to understand the relationship between the 3′ UTR region and gene expression, as well as the underlying mechanisms of genetic diseases.

MATERIAL AND METHOD

Ethics statement

Rabbits involving in this experiments were New Zealand white rabbit and Lianshan black rabbit. All animal protocols were approved by the Animal Care Center and Use Committee of Jilin University. All experiments were performed according to the guidelines approved by Jilin University. New Zealand white rabbit and Lianshan black rabbit were housed under half-light condition in individual cages and were fed twice a day with commercial rabbit basic diet and water ad libitum in the Laboratory Animal Center of Jilin University.

DNA constructs and *in vitro* transcription

The Cas9 expression construct, 3 × FLAG-NLS-SpCas9-NLS (Addgene ID 48137), was synthesized and cloned into the vector. The vector was linearized with NotI and transcribed *in vitro* according to the manufacturer's using the mMessage mMachine SP6 Kit (Ambion, USA) and the RNeasy Mini Kit (Qiagen) to purify the mRNA.

A pair of complementary oligonucleotides encoding the 20-nt guide sequences were annealed at 95 °C for 5 min and ramped down to 25 °C to generate the dsDNA fragment, which was then cloned into the BbsI-digested pUC57-Simple vector (Addgene ID 51306) under the control of T7 promoter. The sgRNAs were transcribed using the T7 RNA Synthesis Kit (Ambion) and purified using the miRNeasy Mini Kit (Qiagen) according to the manufacturer. The concentration and quality of the synthesized mRNAs were determined by Nanodrop 2000 and agarose gel electrophoresis, respectively.

Microinjection and embryo transfer

The protocol for microinjection of pronuclear-stage embryos has been described in detail in our published protocols24. Briefly, zygotes were collected from sexually matured New Zealand white rabbits, which had been superovulated by six times and a 12-h interval of intravenous injection of follicle-stimulating hormone (FSH). The rabbits were mated after intravenously injection of 100 IU human chorionic gonadotrophin (hCG). The oviducts were then flushed with 5 mL DPBS-BSA for the collection of pronuclear-stage embryos. Mixtures of *in vitro*-transcribed mRNA derived from the gRNAs (25 ng/ul) and Cas9 (100 ng/ul) were injected into the cytoplasm of pronuclear stage embryos. Then the injected embryos were transferred to embryo culture medium for 30–60 min, followed by transfer into the oviduct of the recipient mother (approximately 30–50 embryos).

Mutation detection in embryos and pups by PCR

Each injected zygote was collected at the blastocyst stage and the DNA was extracted with embryo lysis buffer (contining1% NP40, 50 mM Tris-HCl, 150 mM NaCl, 0.5% sodium deoxycholate, 0.1% SDS) at 50 °C for 20 min and 90 °C for 5 min in a BIO-RAD PCR machine. Genomic DNA from BR, WR and KO rabbits was isolated using the TIANamp Genomic DNA Kit (TIANGEN, Beijing, China) according to the manufacturer's instructions. PCR primers used for amplification and mutation detection are listed in Supplementary Table S1. PCR products were gel purified with TIAN gel Midi Purification Kit (TIANGEN, Beijing, China) and cloned into pGM-T (Tiangen, Beijing, China). Ten positive plasmid clones were sequenced and the sequences were analyzed by DNAMAN software.

Off-target assay

Potential off-target sites (POTS) of the sgRNAs were predicted using the online CRISPR design tool (http://crispr.mit.edu/). The top five POTS that were selected. The primers (supplementary Table S2) were used to test whether the sgRNAs have off target mutations and the products were analyzed by both T7E1 assay and sanger sequence cloned into the pGM-T (Tiangen, Beijing, China).

T7 endonuclease I (T7EI) assay

A T7 endonuclease I (T7EI) assay was performed as described previously25. Briefly, the genomic DNA of each Cas9/gRNA-injected blastocyst and its pups was amplified by PCR with primers supplied in Table S2 under following conditions: 95 °C for 5 min, 95 °C for 5 min, 95–85 °C at −2 °C/s, 85–25 °C at −0.1 °C/s, hold at 4 °C. The PCR product were digested with T7E1 (NEB M0302L) and analyzed on an ethidium bromide-stained TAE gel.

Histology analysis

Skin and eye tissues of BR, WR and 3′ UTR KO rabbits were fixed with 4% paraformaldehyde for 48 h, embedded in paraffin wax, and slide sectioned. Skin and eye sections were stained with hematoxylin and eosin (H&E) and analyzed by microscopy (Nikon ts100).

Real-time quantitative PCR (qRT-PCR) and Western Blotting

Total RNA from Con (n = 5) samples was isolated with TRNzol-A+ reagent (TIANGEN, Beijing, China) according to the manufacturer's instructions. cDNA was synthesized with DNAse I (Fermentas) treated total RNA using the BioRT cDNA First Stand Synthesis Kit (Bioer Technology, Hangzhou, China). Primers used for RT-PCR are listed in Table S1. Q-PCR was performed using the BioEasy SYBR Green I Real Time PCR Kit (Bioer Technology, Hangzhou, China) with the BIO-RAD Iq5 Multicolor Real-Time PCR Detection System. The relative gene expression normalized to the

GAPDH was determined by $2^{-\Delta\Delta CT}$formula. All the data of gene expression were performed three times.

For Western blotting, the eyes tissues from BR, WR and 3′ UTR KO rabbits were homogenized in 150 lL of lysis buffer. The protein concentrations were measured by the Braford method (Bio- Rad). In this study anti-TYR polyclonal antibody (1:2000; abcam) and anti-GAPDH monoclonal antibody (1:2000; Beyotime) were used as primary and internal control respectively. Images were quantified using ImageJ software (NIH) and all data are expressed as mean ± SEM.

Statistical analyses

All data are expressed as mean ± SEM, with at least three individual determinations in all experiments. The data were analyzed by t-test using GraphPad Prism software 6.0. A probability of $p < 0.05$ was considered statistically significant.

Additional Information

Publisher's note: Springer Nature remains neutral with regard to jurisdictional claims in published maps and institutional affiliations.

ACKNOWLEDGEMENTS

We thank Peiran Hu at the Embryo Engineering Center for the critical technical assistance. This work was financially supported by the National Natural Science Foundation of China (Grant Nos 31201080 and 31272394).

REFERENCES

1. Simeonov, D. R. et al. DNA Variations in Oculocutaneous Albinism: An Updated Mutation List and Current Outstanding Issues in Molecular Diagnostics. Human Mutation 34, 827–835,
2. Anistoroaei, R., Fredholm, M., Christensen, K. & Leeb, T. Albinism in the American mink (Neovison vison) is associated with a tyrosinase nonsense mutation. Animal Genetics 39, 645–648,

3. Schmutz, S. M. et al. A form of albinism in cattle is caused by a tyrosinase frameshift mutation. Mammalian Genome 15, 62–67,
4. Beermann, F., Orlow, S. J. & Lamoreux, M. L. The Tyr (albino) locus of the laboratory mouse. Mammalian Genome 15, 749–758,
5. Aigner, B., Besenfelder, U., Müller, M. & Brem, G. Tyrosinase gene variants in different rabbit strains. Mammalian Genome 11, 700–702,
6. Song, Y. et al. Efficient dual sgRNA-directed large gene deletion in rabbit with CRISPR/Cas9 system. Cellular and molecular life sciences: CMLS 73, 2959–2968,
7. Honda, A. et al. Single-step generation of rabbits carrying a targeted allele of the tyrosinase gene using CRISPR/Cas9. Experimental animals/Japanese Association for Laboratory Animal Science 64, 31–37,
8. Livny, J. & Waldor, M. K. Mining regulatory 5′ UTRs from cDNA deep sequencing datasets. Nucleic acids research 38, 1504–1514,
9. Geissler, R. & Grimson, A. A position-specific 3′ UTR sequence that accelerates mRNA decay. RNA Biol. 0, 1075–1077,
10. Brown, J. T. & Johnson, A. W. A cis-acting element known to block 3′ mRNA degradation enhances expression of polyA-minus mRNA in wild-type yeast cells and phenocopies a ski mutant. Rna 7, 1566–1577 (2001).
11. Farashi, S. et al. Homozygosity for the AATAAA > AATA–Polyadenylation Site Mutation on the alpha2-Globin Gene Causing Transfusion-Dependent Hb H Disease in an Iranian Patient: A Case Report. Hemoglobin 39, 355–358,
12. Deshpande, P. et al. Characterization of Clinical and Laboratory Profiles of the Deletional alpha2-Globin Gene Polyadenylation Signal Sequence (AATAAA > AATA–) in an Indian Population. Hemoglobin 39, 415–418,
13. Nair, S. B., Nadkarni, A. H., Ghosh, K. & Colah, R. B. Variable presentation of HB H disease due to homozygosity for the rare polyadenylation signal A T(Indian) (AATAAA > AATA–) mutation in four Indian families. Hemoglobin 37, 277–284,
14. Yu, X. et al. Insertion/deletion polymorphism in IL1A 3′-UTR is associated with susceptibility to endometrial cancer in Chinese Han women. The journal of obstetrics and gynaecology research 42, 983–989,
15. Jiang, Z. S. & Hao, Z. H. An insertion/deletion polymorphism within the 3′ untranslated region of COL1A2 confers susceptibility to osteoporosis. Molecular medicine reports 14, 4415–4421,
16. Seruggia, D., Fernandez, A., Cantero, M., Pelczar, P. & Montoliu, L. Functional validation of mouse tyrosinase non-coding regulatory DNA elements by CRISPR-Cas9-mediated mutagenesis. Nucleic acids research 43, 4855–4867,
17. Utzeri, V. J., Ribani, A. & Fontanesi, L. A premature stop codon in the TYRP1 gene is associated with brown coat colour in the European rabbit (Oryctolagus cuniculus). Anim Genet 45, 600–603,
18. Proudfoot, N. New perspectives on connecting messenger RNA 3′ end formation to transcription. Current opinion in cell biology 16, 272–278,
19. Minvielle-Sebastia, L. & Keller, W. mRNA polyadenylation and its coupling to other RNA processing reactions and to transcription. Current opinion in cell biology 11, 352–357,
20. Edmonds, M. A history of poly A sequences: from formation to factors to function. Progress in nucleic acid research and molecular biology 71, 285–389,
21. Martin, G. & Keller, W. Mutational analysis of mammalian poly(A) polymerase identifies a region for primer binding and catalytic domain, homologous to the family X polymerases, and to other nucleotidyltransferases. The EMBO journal 15, 2593–2603 (1996).
22. Zhou, X. et al. Generation of CRISPR/Cas9-mediated gene-targeted pigs via somatic cell nuclear transfer. Cellular and molecular life sciences: CMLS 72, 1175–1184,
23. Cui, T. T. et al. miR-196a-2 rs11614913 polymorphism is associated with vitiligo by affecting heterodimeric molecular complexes of Tyr and Tyrp1. Archives of dermatological research 307, 683–692,

24. Wang, Y. et al. Generation of knockout rabbits using transcription activator-like effector nucleases. Cell regeneration 3, 3,
25. Guschin, D. Y. et al. A rapid and general assay for monitoring endogenous gene modification. Methods in molecular biology 649, 247–256,

CHAPTER 14

The Use of Microbial Biotechnology in Value Addition to Cassava Wastes as Feed for Pigs: Prospects and Challenges

Samuel Olanrewaju Aro[1*], Andrew Bamidele Falowo[2], Olusola Olufisayo Awoneye[3], Valentine Ayobare Aletor[1]

[1]*Department of Animal Production and Health, The Federal University of Technology, Akure, Nigeria*

[2]*Department of Livestock and Pasture Science, University of Fort Hare, Alice, South Africa*

[3]*Department of Animal Production, Rufus Giwa Polytechnic, Owo, Nigeria*

ABSTRACT

Cassava peels were fermented through a combination of two lactic acid bacteria (Lactobacillus coryneformis and Lactobacillus delbrueckii) and a fungus (Aspergillus fumigatus) using solid substrate fermentation (SSF) technique for five consecutive days. The enhanced product designated as microbially fermented cassava peel (MFCP) was used at graded levels in the feed formulation for growing pigs. Four iso-nitrogenous and iso-caloric diets were formulated with MFCP at 0% (T1), 20% (T2), 40% (T3) and 60% (T4) inclusion levels. Twenty-four boarlings of an average initial weight of 37.15 kg were used in a fourteen-week trial to test the effect of these graded levels of MFCP on growth performance, digestive and reproductive physiology of the experimental animals. The result showed significant differences ($P < 0.05$) in mean daily feed consumption. Pigs in T3 had the highest feed intake (2.24 ± 0.05 kg), while pigs in T4 had the lowest (1.78 ± 0.04 kg). The final highest live weight was in T3 (72.00 kg) and the least in T4 (58.17 kg). The feed conversion ratios also showed significant differences ($P < 0.05$) among treatment means. The result obtained from gut morphometry showed that animals in T1 had the highest stomach weight (full and empty), while animals in T3 had the least rate of gastric emptying (81.50%). Animals in T3 and T4 had the longest small intestine (16.21 m and 15.35 m, respectively), while those in T4 had the longest colon (4.24 m). Animals in T1 and T2 had more numbers of gastric glands, while animals in T3 and T4 had more parietal cells. The reproductive indices revealed that animals in T1 were better in terms of the weight and relative weight of vesicular and Cowper's gland, while animals in T4 had the highest value of the weight and relative weight of prostate gland. Proximate analyses of the ileal digesta revealed that the rate of nutrient absorption was highest in T3 and T1 and lowest in T4. Histological examination of the testes revealed normal testicular architecture in the control diet. The testes of animals fed 0%, 20% and 40% MFCP diets exhibited normal and intact lumen, Sertoli cells, and germ cells. However, animal fed 60% MFCP diet revealed massive disruption and degeneration of germinal epithelium, and complete erosion of the seminiferous tubules. It could be concluded that feeding pigs at 40% MFCP would enhance feed consumption, nutrient absorption and improve the growth rate of pigs. Using MFCP beyond 40% level on swine diet could compromise gut health through diarrhoeagenic disposition of the digesta in the distal colon and modification of the parietal and chief cells of the gastric glands. The hypertrophy of the prostate gland and vacuolization of the seminiferous tubules of pigs fed the 60% MFCP diet indicated a probable reduced reproductive performance.

Keywords: Biotechnology, Cassava, Fermentation, Microbes, Physiology

1. INTRODUCTION

Globally, livestock production is growing more dynamically than any other agricultural sector. As countries become more affluent, demand for livestock-derived food has substantially increased, leading to a major transformation of animal food production. Pig and poultry production are the fastest growing and industrialized livestock subsectors with annual production growth rates of 2.6% and 3.7% respectively over the past decade [1] . Sources of animal protein in Nigeria include beef, milk, pork, poultry, sheep, goats, fish and game animals. Of all the sources, pork represents one of the fastest ways of increasing animal protein, since pigs grow at a faster rate and are more prolific than cattle, sheep and goats [2]. Also pigs excel other animals in converting feed to flesh [3] and they display a unique ability to adapt and survive in areas where they have been found. However, the development of the swine industry has been plagued by the increasing unavailability and consequent high cost of the conventional feeds. Feed cost, according to Sastry and Thomas [4], accounts for 67.2% of the total pig production cost. It is therefore pertinent to look for cheaper and readily available sources for compounding swine rations. As a way of confronting this challenge, the use of alternative feed sources such as cassava peel and cassava starch residue has been canvassed [5].

The potential use of cassava peel to adequately fill this nutritional gap is heavily constrained by intrinsic factors like their high crude fibre level, low nitrogen content and the presence of anti-nutrients [6]. There has been a concerted effort by scientists to reduce or totally remove all these constraints militating against the utilization of cassava peels. The most viable option for circumventing these constraints is via the biotechnological route using generally regarded as safe (GRAS) organisms [7]. A host of micro-organisms which can be used for value addition to agro-industrial by-products abound and have been used by researchers with varying degrees of success. Thus, this research was designed to investigate the effect of using graded levels of cassava peel fermented with a combination of two lactic acid bacteria (Lactobacillus coryneformis and Lactobacillus delbrueckii) and a fungus (Aspergillus fumigatus) on the growth performance, gut and gonadal morphometry of growing pigs.

2. MATERIALS AND METHODS

Fresh cassava peels were collected from cassava processing mills within Akure town, Ondo State, Nigeria. The peels were processed by the methods described by Aro [5] prior to their incorporation into the compounded grower pigs' diets. A total of 24 grower pigs (all males) with average initial weight of 37.15 kg were purchased for the trial. Four experimental diets were formulated to contain microbially fermented cassava peel (MFCP) at 0%, 20%, 40% and 60% as diet 1, diet 2, diet 3 and diet 4 respectively. The diets composition is shown in Table 1. The animals were randomly divided into four treatments, comprising of six animals per treatment with each animal serving as a replicate of its treatment group.

Data were collected for weekly weight change, daily feed intake and feed conversion ratio throughout the 14 weeks of the trial. At the end of the experiment, all the animals were mechanically stunned and slaughtered. Immediately after death, the abdominal cavity was opened and the entire

Table 1. Gross composition (g/100g) of the experimental diets.

Ingredients	T1	T2	T3	T4
Maize	40.00	35.00	15.00	0.00
Rice bran	11.00	7.20	10.00	0.00
Palm kernel cake	26.30	16.50	14.50	21.30
Groundnut cake	17.00	18.30	17.50	15.70
MFCP	---------	20.00	40.00	60.00
Vegetable oil	2.50	----------	--------	--------
Bone meal	1.50	1.50	1.50	1.50
Oyster shell	0.50	0.50	0.50	0.50
Premix	0.50	0.50	0.50	0.50
Salt	0.50	0.50	0.50	0.50
Total	100.00	100.00	100.00	100.00
Crude protein (%)	18.07	18.09	18.09	18.04
Crude fibre (%)	6.20	11.35	17.87	24.01
M. E. (Kcal/kg)	3022.82	3012.30	3003.15	3015.74
HCN Conc. (mg/kg)	0.00	5.88	11.16	16.75

MFCP = Microbially fermented cassava peel; HCN = Hydrogen cyanide; M. E. = Metabolizable energy; T1 = Control diet; T2 = Diet with 20% inclusion of MFCP; T3 = Diet with 40% inclusion of MFCP; T4 = Diet with 60% inclusion of MFCP.

gastrointestinal tract was immediately removed. Different sections of the gastro-intestinal tract were ligated and cut into the following segments: stomach, small intestine, large intestine and caecum for measurement. The length of small and large intestine were determined using a linear measuring tape. The weights of caecum and stomach (full and empty) were also determined.

Tissue samples from the stomach and testis were collected for histological examination. Sections were then observed with a light microscope and photographs were taken at 600× magnification. Also, gonadal parameters like the weight of the accessory glands and testicular histology were determined as response criteria to the treatment diets.

3. DATA ANALYSIS

Data generated were subjected to analysis of variance (ANOVA) using SPSS version 16 statistical package. Mean separation was done with Duncan's multiple range test of the same statistical package.

4. RESULTS AND DISCUSSION

The result of the performance of grower pigs fed varying levels of MFCP is shown in Table 2. The average feed intake values differed significantly ($P < 0.05$). The animals on T3 had highest feed intake (2.24 ± 0.05 kg), while animals on T4 had the least (1.78 ± 0.04 kg). Since the experimental diets were both iso-caloric and iso-nitro- genous, any increase in feed intake represents an increase in the intake of energy and protein requirement [8] , which was translated to better protein accretion and hence increased growth rate by the animals on T3. This same trend was reflected on other performance parameters like final body weight feed conversion ratio and daily weight gain. The poor performance recorded for animals on T4 as far as these parameters are concerned could be attributed to high fiber level and the toxigenic effect of residual cyanide at 60% level of inclusion. The better feed conversion ratio and feed efficiency for animals on the control diet could be attributed to better digestibility as a result of lower level of crude fibre and probable absence of hydrogen cyanide (HCN) on the diet [5] . HCN intake per body weight was highest in animal on diet T4 and least on diet T2. The lethal dose of HCN for human has been reported to be between 0.5 - 3.5 mg per kg body weight or 30 and 210 mg for 60 kg adult human [9] . Since pig is similar to human in body physiology, the high oral intake of HCN per kg body weight (0.51 mg/kg)

Table 2. Growth performance of pigs fed microbially fermented cassava peel diet.

Parameters	T1	T2	T3	T4
Average feed intake (kg/d)	1.84 ± 0.56^b	1.94 ± 0.18^b	2.24 ± 0.05^a	1.78 ± 0.04^b
Initial live weight (kg)	37.00 ± 4.49	37.67 ± 3.23	37.50 ± 2.03	37.17 ± 2.17
Final live weight (kg)	70.50 ± 9.27	64.00 ± 7.21	72.00 ± 3.76	58.00 ± 0.89
Total weight gain (kg)	33.50 ± 5.47	26.33 ± 4.26	34.50 ± 5.22	20.83 ± 2.14
Daily weight gain (kg)	0.34 ± 0.06	0.27 ± 0.04	0.35 ± 0.06	0.21 ± 0.02
FCR	5.41 ± 1.40^b	7.19 ± 1.62^b	6.40 ± 1.75^b	8.48 ± 0.80^a
Feed efficiency	0.19 ± 0.03^a	0.14 ± 0.04^a	0.16 ± 0.02^a	0.12 ± 0.01^b
HCN intake/kg body weight (mg/kg)	0.00 ± 0.00^d	0.18 ± 0.24^c	0.35 ± 0.16^b	0.51 ± 0.07^a

Mean ± Standard Deviation. $^{a, b, c, d}$ = Means with different superscripts within the same rows and for the same parameters are significantly different ($P < 0.05$); MFCP = Microbially fermented cassava peel; T1 = 0% MFCP inclusion as control diet; T2 = 20% MFCP inclusion of the diet; T3 = 40% MFCP inclusion of the diet; T4 = 60% MFCP inclusion of the diet; FCR = Feed conversion rate; HCN = Hydrogen cyanide.

recorded for animals on T4 could have resulted in poor growth of the animals. Hence HCN/kg body weight in MFCP for growers pig should not 0.35 mg/kg for better growth performance.

The morphometric data (Table 3) of the pigs' gastrointestinal tract showed that animals on the control diet (T1) had the highest full stomach weight (1.42 ± 0.61 kg) while T4 had the least full stomach weight (0.88 ± 0.17 kg).

The weight of stomach content (ingesta) was highest in T3 than in all other treatments. This probably depicts that animals on T3 retained dietary nutrients in the stomach for digestion/processing longer than animals on T1, T2, and T4. This could also mean that dietary nutrients were retained longer and made more accessible to enzymatic digestion and hence responsible for the presence of more number of parietal cells (Figure 1). The greater accessibility of dietary nutrients to enzymatic digestion may explain why animal on T3 had longest small intestine (16.21 ± 0.86 m) and largest caecum (0.83 ± 0.17 kg) compared to T1, T2, and T4. The caecum was reported to play an important role with respect to the microbial degradation of some carbohydrates [10] , absorption of water, microbial synthesis of vitamins and degradation of nitrogenous compounds. Also, the increase in length of the small intestine of pigs on T3 may be to accommodate more ingested nutrients for processing and absorption since small intestine has been reported as the site for nutrient absorption [11] .

Histological examination of the fundic region of the stomach (Figure 1) revealed gastric glands comprising mainly of parietal (oxyntic) cells and chief (zymogenic) cells, gastric pit and gastric lumen (Figure 1). The result showed that animals fed T1 and T2 had more number of gastric glands than animals fed T3 and T4. The number of gastric glands in animals fed T1 was similar to those fed T2. This shows that the probable toxigenic effect of residual cyanide did not cause major upset at 20% inclusion level. The reduction in the number of gastric glands in animal fed T3 and T4 could probably be due to high toxigenic effect of residual cyanide in the diet at 40% and 60% inclusion reducing the proliferation of gastric glands for acid secretion.

In animals fed T4, there were alterations in shape and structure of gastric glands with parietal cells unevenly distributed in the glands. The chief cells were not fully formed. The toxigenic nature of residual cyanide at 60% inclusion may have caused the malformation of chief cells and gastric glands observed in stomach of animals fed T4. Hence most of the nutrients in the diet (T4) must have escaped the stomach without much action of hydrochloric acid secretion for digestion.

The rate of gastric emptying and proximate composition of the ileal content (Table 4) showed that there were significant ($P < 0.05$) treatment effects. T4 had the highest rate of gastric emptying (88.89%), followed by T2

(83.87%) and the lowest value (81.02%) was observed on T1. The high rate of gastric emptying on T4 was contrary to the work of many researchers which revealed that high dietary fibre slows down the rate of gastric emptying [12] . The reason for this disparity may be due to the fact that the fibre on T4 act as emulsifiers by stabilizing the gastric chyme thus preventing the separation of the solid from the liquid phase thereby impairing selective retention of the largest particles and increasing the rate of passage into small intestine thus making the nutrients inaccessible to gastric enzymes for digestion.

The ileal content of T4 had the lowest dry matter (DM) value (16.36 g/100 g) while T3 had the highest value (19.42 g/100 g). Conversely, T4 had the highest moisture content. The high moisture content in T4 may be as a result of the water binding ability of dietary fibre [13] . It also agrees with the report of Varel and Yen [14] that the amount of digesta flow at the terminal ileum is greater in pigs fed diets with high level of dietary fibre. T4 had the highest CP value (18.33 g/100 g) while the lowest crude protein content of the ileum was recorded in T3 (13.74 g/100 g). The high crude protein content in T4 is an indication that crude protein was not well absorbed and this could be as a result of the high crude fibre content of this diet. This is in consonance with the report of Bindelle [15] that an average of 31% of nitrogen is bound to the fibre and it is not available to the animal, more so, the anti-nutritional factors interfere with the digestive processes. The ileal crude fibre content was generally high in all the treatments, this showed that there was little or no digestion of fibre in the ileum, this could be as a result of the inability of the digestive enzymes of the ileum to digest crude fibre components [16] .

Table 3. Gut Morphometry of grower pigs fed microbially fermented cassava peels diet.

Parameters Weight of Stomach	T1	T2	T3	T4
Full (kg)	1.42 ± 0.61	0.92 ± 0.14	1.20 ± 0.15	0.88 ± 0.17
Empty (kg)	0.87 ± 0.37^a	0.43 ± 0.33^b	0.36 ± 0.18^b	0.48 ± 0.06^b
Weight of stomach content (kg)	0.55 ± 0.16^b	0.49 ± 0.13^b	0.84 ± 0.22^a	0.40 ± 0.13^b
Weight of empty Caecum (kg)	0.67 ± 0.07^{ab}	0.62 ± 0.04^b	0.83 ± 0.17^a	0.51 ± 0.02^c
Length of small intestine (m)	15.28 ± 0.75	14.59 ± 1.11	16.21 ± 0.86	15.35 ± 0.43
Length of large intestine (m)	4.11 ± 0.56	4.53 ± 0.25	4.22 ± 0.15	4.24 ± 0.17

Mean ± Standard Error. $^{a, b, c, ab}$ = Means with different superscripts within the same rows and for the same parameters are significantly different ($P < 0.05$); MFCP = Microbially fermented cassava peel; T1 = Control diet; T2 = Diet with 20% inclusion of MFCP; T3 = Diet with 40% inclusion of MFCP; T4 = Diet with 60% inclusion of MFCP.

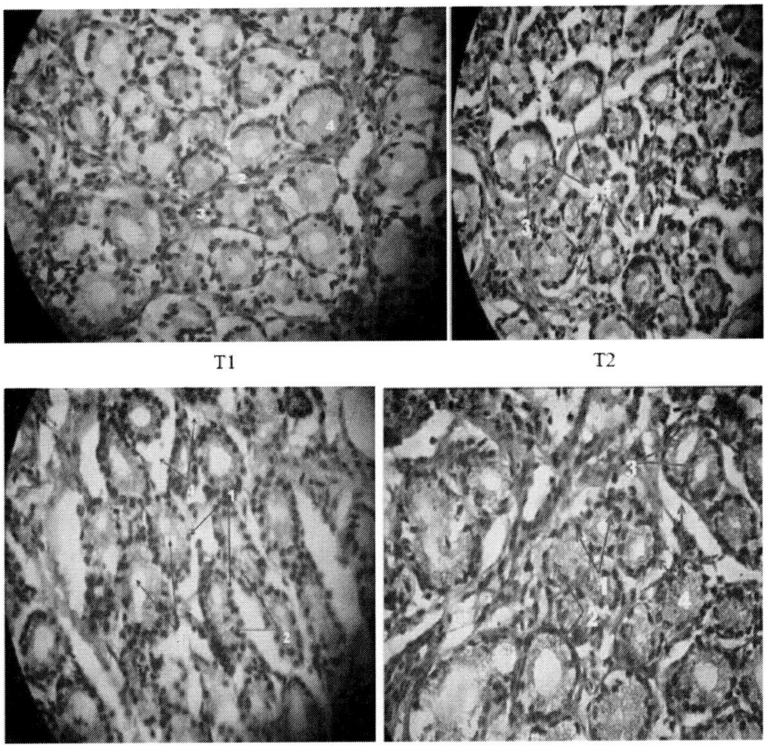

1: Parietal cells; 2: Chief cells; 3: Gastric pit; 4: Gastric lumen. T1: Stomach section of animals fed the control diet; T2: Stomach section of animals fed diet with 20% MFCP; T3: Stomach section of animals fed diet with 40% MFCP; T4: Stomach section of animals fed diet with 60% MFCP. MFCP: Microbially fermented cassava peel. Magnification 600× (H & E stains).

Figure 1. Cross sections of Gastric glands from the fundic region of pigs fed microbially fermented cassava peels.

The low fat content of T1 is an indication that fat was well absorbed in T1, while high fat content in T4 is an indication of poor absorption and this may be as a result of the fact that dietary fibre hinders the absorption of all other nutrients [17] . The lowest NFE value of ileal samples was observed in T4 (43.57 g/100 g), while the highest value was observed in T3 (54.57 g/100 g). The highest NFE recorded in T3 indicates that digestion of soluble carbohydrates was highest as a result of its slow transit or long retention time in the ileum there by enhancing the time of exposure of the diets to the host's digestive enzymes [18] .

Table 4. Proximate composition (g/100 g) of ileal digesta and rate of gastric emptying (%) of pigs fed graded levels of fermented cassava peel-based diets.

Parameters	T1	T2	T3	T4
DM	19.05 ± 0.42^{ab}	17.95 ± 0.30^{b}	19.42 ± 0.42^{a}	16.36 ± 0.30^{c}
Moisture	80.95 ± 0.42^{ab}	82.05 ± 0.30^{b}	80.58 ± 0.42^{c}	83.64 ± 0.30^{a}
Crude fibre	11.20 ± 0.09^{c}	13.18 ± 0.30^{b}	11.37 ± 0.04^{c}	15.64 ± 0.34^{a}
CP	15.60 ± 0.19^{b}	18.10 ± 0.35^{ab}	13.74 ± 0.39^{c}	18.33 ± 0.22^{a}
EE	9.51 ± 0.07^{c}	9.76 ± 0.06^{b}	9.70 ± 0.04^{bc}	10.31 ± 0.07^{a}
Ash	11.07 ± 0.41^{b}	9.25 ± 0.13^{c}	10.62 ± 0.10^{b}	12.14 ± 0.34^{a}
NFE	52.63 ± 0.22^{b}	49.69 ± 0.20^{c}	54.57 ± 0.35^{a}	43.57 ± 0.62^{d}
RGE	81.02 ± 8.51^{b}	83.87 ± 4.35^{b}	81.50 ± 6.99^{b}	88.89 ± 3.67^{a}

$^{a, b, c, d, ab}$ = Means on the same rows but with different superscripts are statistically ($P < 0.05$) significant; DM = Dry matter; CF = Crude fibre; CP = Crude protein; EE = Ether extract; NFE = Nitrogen free extract. T1 = No inclusion of MFCP (Control); T2 = 20% inclusion of MFCP; T3 = 40% inclusion of MFCP; T4 = 60% inclusion of MFCP; RGE = Rate of gastric emptying.

Table 5 shows the proximate composition of the content of gastro-intestinal tract from the distal colon. T4 had the lowest DM value (29.36 g/100 g) and the highest moisture content (70.64 g/100 g while T1 had the highest DM value of 38.75 g/100 g the lowest moisture value (61.25 g/100 g). The high moisture content in T4 is an indication that water was less absorbed and this corroborates the findings of Spiller [19] that dietary fibre has a role in changing the consistency of the stool by increasing the water content and the plasticity thus increasing stool frequency and output, as a result, the high moisture content recorded in T4 is a predisposition to diarrhoeagenic condition [20] .

Table 6 shows the accessory sex glands morphometry of the pigs fed microbially fermented cassava peel- based diets. There were significant differences (P < 0.05) in the vesicular glands due to treatment effects. The highest weight of vesicular glands (0.35 kg) was recorded in T1 while T2 had the lowest value (0.15 kg). The heaviest prostates (0.15 kg) were recorded in T2 and T4 while T1 had the heaviest Cowper's glands.

Pigs that were fed T1 had the heaviest vesicular glands when compared with pigs fed MFCP diets. The toxigenic effect of the residual cyanide of the cassava peel could have been responsible for the low weight of the vesicular glands in the MFCP-fed pigs. The heavier vesicular glands in pigs that were fed T1 is an indication of greater fluid volume of semen to ensure optimum motility and fertility [21] . T2 and T4 had heavier prostate glands while the lowest was recorded in T3. The general observation was that there is the tendency of bigger prostate glands with pigs fed MFCP-based diets. The abnormal enlargement of the prostate gland has been implicated in prostate cancer that is currently ravaging human population [22] .

Table 7 shows that the dietary treatments resulted in a higher relative weight of the vesicular glands and Cowper's glands in the control relative to the MFCP diets. Components of the secretions from the glands are known to ensure nutrition of the sperm cells as well as acting as buffer on the maintenance of the semen's acid- base balance aside from contributing to the electrolyte stability of the seminal plasma. The reduced weight of

Table 5. Proximate composition (g/100g) of the digesta from distal colon (dry matter basis) of pigs fed graded levels of fermented cassava peel-based diets.

Parameters	T1	T2	T3	T4
DM	38.75 ± 0.42^{a}	33.37 ± 0.30^{b}	37.78 ± 0.39^{a}	29.36 ± 0.41^{c}
Moisture	61.25 ± 0.42^{c}	66.63 ± 0.30^{b}	62.22 ± 0.39^{c}	70.64 ± 0.41^{a}
CF	8.18 ± 0.08^{c}	9.47 ± 0.09^{b}	8.45 ± 0.06^{c}	12.92 ± 0.26^{a}
CP	15.18 ± 0.21^{b}	16.85 ± 0.33^{a}	15.07 ± 0.27^{b}	16.97 ± 0.43^{a}
EE	8.76 ± 0.07^{bc}	8.93 ± 0.08^{b}	8.31 ± 0.07^{c}	10.11 ± 0.29^{a}
Ash	11.06 ± 0.37^{bc}	10.28 ± 0.15^{c}	11.46 ± 0.22^{ab}	12.28 ± 0.38^{a}
NFE	56.82 ± 0.55^{a}	54.48 ± 0.26^{b}	56.71 ± 0.58^{a}	47.73 ± 0.40^{c}

$^{a, b, c, ab, bc}$ = Means on the same rows but with different superscripts are statistically ($P < 0.05$) significant; DM = Dry matter; CF = Crude fibre; CP = Crude protein; EE = Ether extract; NFE = Nitrogen free extract; T1 = 0% inclusion of MFCP (Control); T2 = 20% inclusion of MFCP; T3 = 40% inclusion of MFCP; T4 = 60% inclusion of MFCP.

Table 6. Weight of accessory sex gland (kg) of pigs fed graded levels of microbially fermented cassava peel-based diets.

Parameters	T1	T2	T3	T4
Vesicular gland	0.35 ± 0.03^{a}	0.15 ± 0.03^{d}	0.30 ± 0.11^{b}	0.23 ± 0.04^{c}
Prostate gland	0.10 ± 0.00^{b}	0.15 ± 0.03^{a}	0.08 ± 0.01^{c}	0.15 ± 0.06^{a}
Cowper's gland	0.20 ± 0.00^{a}	0.13 ± 0.01^{b}	0.13 ± 0.01^{b}	0.13 ± 0.01^{b}

$^{a, b, c, d}$ = Means on the same rows but with different superscripts are statistically ($P < 0.05$) significant; T1 = 0% inclusion of MFCP (control); T2 = 20% inclusion of MFCP; T3 = 40% inclusion of MFCP; T4 = 60% inclusion of MFCP.

Table 7. Relative weights (g/100 g) of the accessory sex glands of pigs fed graded levels of microbially fermented cassava peel-based diets.

Parameters	T1	T2	T3	T4
Vesicular gland	0.53 ± 0.05^a	0.23 ± 0.02^c	0.44 ± 0.17^b	0.41 ± 0.09^b
Prostate gland	0.15 ± 0.01^b	0.25 ± 0.07^a	0.11 ± 0.02^c	0.28 ± 0.11^a
Cowper's gland	0.30 ± 0.02^a	0.20 ± 0.01^b	0.19 ± 0.03^b	0.23 ± 0.03^b

$^{a, b, c}$ = means on the same rows but with different superscripts are statistically ($P < 0.05$) significant; T1 = 0% inclusion of MFCP (control); T2 = 20% inclusion of MFCP; T3 = 40% inclusion of MFCP; T4 = 60% inclusion of MFCP.

these two glands in pigs fed MFCP in comparison with those in the control would therefore mean less secretion and hence reduced ovum fertilizing capacity of the seminal plasma of pigs fed MFCP-based diets. The relatively lower weights of the vesicular and Cowper's glands would have been caused by the adverse effect of residual cyanide in the peels [5] [23] .

The histopathology of the testes of animals on T1 (control diet) in Figure 2 revealed normal testicular architecture with distinct seminiferous tubule composed of Sertoli cells, germ cells with active stages of spermatogenesis with intact lumen. The interstitial cells of Leydig were also prominently interspersed and occupied most of the space in between the seminiferous tubule. Vacuolization of the peritubular tissue surrounding the seminiferous tubule were rarely observed. In T2 (Figure 3), testes of the animals showed evidence of intact lumen, there was active spermatogenesis marked by the characteristic layered arrangement of cells from the germinal epithelium to the lumen. There was vacuolization of the peritubular tissue surrounding seminiferous tubule. The interstitial cells of Leydig were interspersed between the seminiferous tubule. Few spermatogenic cells were ob served compared with testis of animals on T1. In T3 (Figure 4), the testes of animals had intact lumen. The spermatogenic cells were at different stages of cell differentiation. Vacuolization were also observed within and between the seminiferous tubule. There was much keratinization of the myofibrils while the Leydig cells were much displaced and disrupted because of vacuolization.

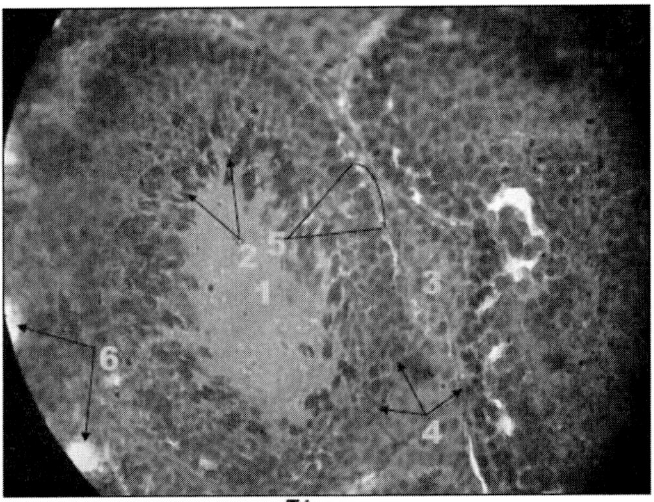

T1

1: Lumen of seminiferous tubules; 2: Spermatids; 3: Leydig cells; 4: Myo-fibroblasts/peritubular tissue; 5: Germinal epithelium; 6: Vacuolization of myofibroblast. Magnification: 600× (H & E stains).

Figure 2. Cross section of the seminiferous tubules of pig fed 0% microbially fermented cassava peel-based diet (T1).

In T4 (Figure 5), the animals' testis showed evidence of massive vacuolization, disruption and degeneration of the seminiferous tubule and primordial germ cells. No distinct lumen and Sertoli cells were visible. There was sloughing of apical seminiferous tubule epithelium into the lumen. There was disruption of peritubular cells while the interstitial cells of Leydig were totally shrunken.

These Histological examinations revealed progressive changes in testicular architecture according to treatment effects (Figures 2-5). This disruption and degeneration of the testicular tissues could be attributed to the toxic effect of residual cyanide (16.75 mg/kg) on the testes at 60% level of inclusion. This was similar to the alteration observed in spermatogenesis due to inclusion of gossypol [24] , Tryotergium walfordii [25] and Achillea millefolium [26] in mice diet which were considered to be anti-spermatogenic agents.

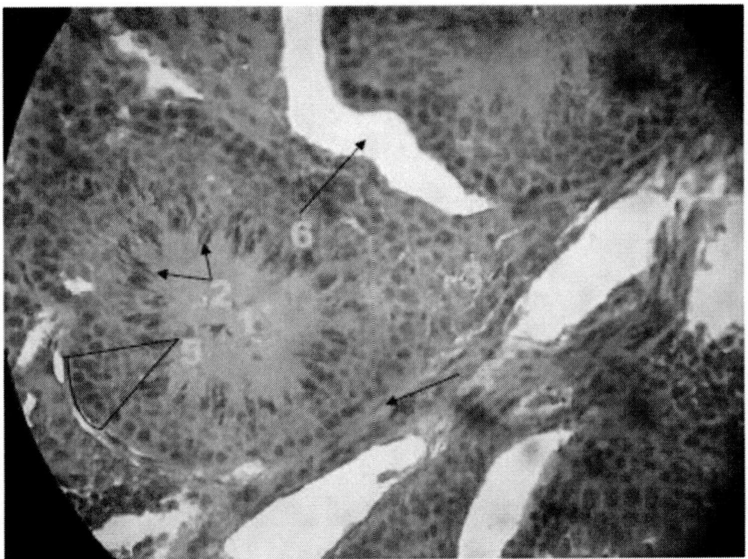

T2

1: Lumen of seminiferous tubules; 2: Spermatids; 3: Leydig cells; 4: Myo-fibroblasts/peritubular tissue; 5: Germinal epithelium; 6: Vacuolization of myofibroblast Magnification: 600× (H & E stains).

Figure 3. Cross section of the seminiferous tubules of pig fed 20% microbially fermented cassava peel-based diet (T2).

The sloughing of apical seminiferous tubule epithelium into the lumen observed in the testis of animals fed T4 (Figure 5) could be have occurred when germ cell-to-Sertoli cell adhesion are lost as reported by Creasy [27] . The loss of germ and Sertoli cells confirmed the inhibitory effect of residual cyanide in MFCP (16.75 mg/kg) at 60% level of inclusion on spermatogenesis since Sertoli and germ cell adhesion are essential for successful spermatogenesis in animals [28] .

Vacuolization and keratinization of seminiferous tubule and peritubular tissue of the seminiferous tubules were increased in the testes of animal fed MFCP diets. The vacuolization of the seminiferous tubule may be as a result of toxigenic nature of residual cyanide in MFCP that could probably engender reduced fertility. This is similar to the vacuolization and degeneration of seminiferous tubule observed in rat due to obstruction or dysfunction of epididymis [29] [30] causing abnormal fluid reabsorption.

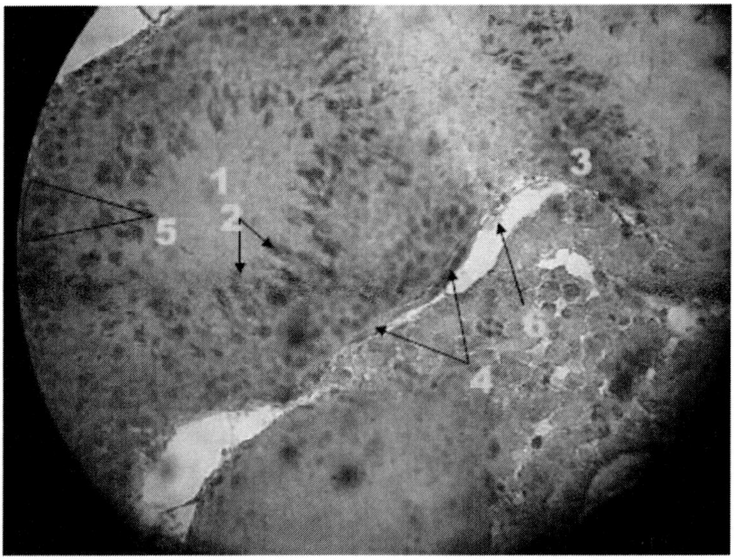

T3

1: Lumen of seminiferous tubules; 2: Spermatids; 3: Leydig cells; 4: Myo-fibroblasts/peritubular tissue; 5: Germinal epithelium; 6: Vacuolization of myofibroblast. Magnification: 600× (H & E stains).

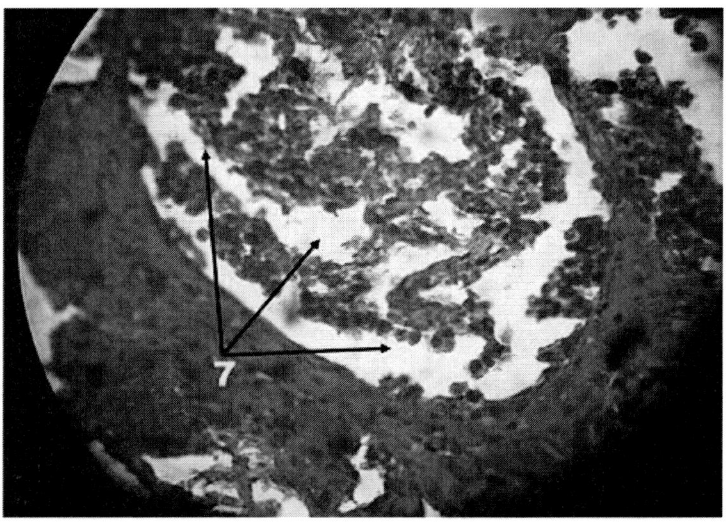

T4

Figure 5. Cross section of the seminiferous tubules of pig fed 60% microbially fermented cassava peel-based diet (T4).

The interstitial cells of Leydig were prominently interspersed in testes of animals fed diet T1 (Figure 2), sparsely interspersed in T2 (Figure 3), massively displaced and disrupted in T3 (Figure 4), and totally lost in T4 (Figure 5). It has been reported that Leydig cells secrete testosterone [31] , and rostenedione and dihydroandrosterone when stimulated by pituitary luteinizing hormone. Kukucka and Misra [32] also reported that Leydig cells synthesize testicular oxytocin de novo and vitamin C endogenously in rat and guinea pig. Thus, the harmful effect of residual cyanide in MFCP could be advanced for decrease in distribution of Leydig cells in testes animals fed diets T2 (Figure 3) and T3 (Figure 4) to loss in T4 (Figure 5). This may result in decrease in testosterone and oxytocin secretion thereby affecting effective libido and total sperm production in the animals.

5. CONCLUSIONS

The performance of the grower pigs was better in terms of total weight gained of the pigs in favour of the 40% level of inclusion of MFCP. This means that the inclusion of MFCP at 40% level in the diets of grower pigs will not adversely affect growth performance and the digestive physiology of the pigs. The result however reveals that nutrient and water absorption in the GIT are affected by dietary inclusion of microbially fermented cassava peel. These performance indicators of the normal gastro-intestinal physiology were decreased at 60% inclusion level of MFCP. Feeding pigs with inclusion of MFCP above 40% could therefore adversely affect the rate of nutrient absorption, gut health and general health status of growing pigs. Therefore, within the limit of this study MFCP should not be fed above 40% to growing pigs. Also, feeding boars with more than 40% inclusion MFCP could depress reproductive performance. The high relative weight of the prostate glands in pigs fed on the MFCP diets calls for more empirical evaluation, as it is suggestive of a pathological condition in the similitude of an abnormal enlargement of the gland in response to the diets. Hence, predisposition to prostate cancer could result from prolonged exposure to diets containing more than 40% MFCP as observed in this study. The testicular architecture observed in this study showed that feeding microbially fermented cassava peel higher than 40% inclusion in diet would lead to reduction in reproductive performance of growing pigs.

The prospects of value addition to cassava peels through the intervention of micro-organisms within the limit of this experiment are in the areas of better growth performance and feed conversion up to 40% inclusion level. The challenges or constraints are lowered reproductive performance and reduced growth rate that are experienced at levels of inclusion higher than 40%, probably due to the crude fibre and residual cyanide levels in the microbially fermented peels. The focus of further research in this field should therefore be directed at those micro-organisms with better crude fibre and cyanide degrading ability than the ones used in this trial. However, from this experiment, tolerant levels of 0.35 mg of HCN intake/kg body weight, 11.16 mg of HCN/kg of feed and crude fibre level of 17.87% were established for growing pigs fed dietary inclusion of cassava peels fermented with a consortium of Lactobacillus coryneformis, Lactobacillus delbrueckii and Aspergillus fumigatus.

ACKNOWLEDGEMENTS

Samuel Olanrewaju Aro, Andrew Bamidele Falowo, Olusola Olufisayo Awoneye, Valentine Ayobare Aletor The authors express their gratitude to the Federal University of Technology, Akure under the Centre of Excellence for Food Security, for providing financial support for this research work.

REFERENCES

1. Otte, J. and Roland-Holst, D. (2007) Industrial Livestock Production and Global Health Risks. A Living from Livestock. Research Report. 1. RR Nr 07-09. Pharmacology, 42, 539-551.
2. Osaro, O.M. (1995) Enhancing Production Performance of Smallholder Pig Farmers. Pig Production Workshop Training Manual, Zaria, 100-130.
3. Ikani, I.E. and Dafwang, I.I. (1995) Pig Production Technology for Piggery Farmers. Extension Bulletin No. 25, Livestock Series No. 1, NAERLS, ABU Zaria, Nigeria.
4. Sastry, N.S.R. and Thomas, C.K. (1976) Farm Animal Management. Vikkas Pub. House PVT. Ltd., 181.
5. Aro, S.O. (2010) Growth and Reproductive Response of Swine Fed Fermented Cassava Tuber Wastes. Ph.D. Thesis, University of Ibadan, Ibadan, 1-176.
6. Makkar, H.P.S. (1993) Anti-nutrient Factors in Food for Livestock. British Society for Animal Production, 16, 69-85.
7. Israelides, C.J., Smith, A., Harhill, J.E., Bambalor, G. and Scallon, B. (1998) Pullulan Content of the Ethanol Precipitated from Agro-Industrial Wastes. Applied Microbiology and Biotechnology, 49, 613-617.
8. Bi, Y. and Peter, W.S.C. (1996) Effect of Crude Fibre Level in the Diet on the Intestinal Morphology of Growing Rabbits. Laboratory Animal, 30, 143-148.
9. Nhassico, D., Muquigue, H., Cliff, J., Cumbana, A. and Bradbury, J.H. (2008) Rising African Cassava Production, Diseases Due to High Cyanide Intake and Control Measures. Journal of the Science of Food and Agriculture, 88, 2043- 2049.
10. Jorgensen, H., Zhao, X.Q., Bach Knudsen, K.E. and Eggum, B.O. (1996) The Influence of Dietary Fibre Source and Level on the Development of Gastrointestinal Tract, Digestibility and Energy Metabolism in Broiler Chickens. British Journal of Nutrition, 75, 379-395.
11. Darcy, B. (1982) Transit et Digestion dans L'intestin Grêle Chez le Porc. In: Laplace, J.P., Corring, T. and Rérat, A., Eds., Proceedings of International Seminar on Digestive Physiology of Pig, Les Colloques de l' INRA No. 12, Paris, 45-59.

12. Wenk, C. (2001) The Role of Dietary Fibre in the Digestive Physiology of the Pig. Animal Feed Science and Tech- nology ,90, 21-33.
13. Asp, N.G. (1996) Dietary Carbohydrates: Classification by Chemistry and Physiology. Food Chemistry, 57, 9-14.
14. Varel, V.H. and Yen, J.T. (1997) Microbial Perspective on Fibre Utilisation by Swine. Journal of Animal Science, 75, 2715-2722.
15. Bindelle, J. (2005) A Rapid Estimation of Nitrogen Bound to Neutral Detergent Fibre in Forages by Near Infrared Reflectance Spectroscopy. Proceedings of the XXth International Grassland Congress, University College Dublin, 259.
16. Cummings, J.H. and Englyst, H.N. (1995) Gastro-Intestinal Effects of Food Carbohydrate. American Journal of Clini- cal Nutrition, 61, S928-S945.
17. Sauer, W.C., Mosenthin, R., Ahrens, F. and Den Hartog, L.A. (1991) The Effect of Source of Fiber on Ileac and Fecal Amino Acid Digestibility and Bacterial Nitrogen Excretion in Growing Pigs. Journal of Animal Science, 69, 4070- 4077.
18. Leterme, P. (2005) Chemical Composition, Nutritive Value and Voluntary Intake of Tropical Tree Foliage and Cocoyam in Pigs. Journal of the Science of Food and Agriculture, 85, 1725-1732.
19. Spiller, G., Margo, N.W., Sherwood, L. and Gorbach, A. (2001) Influence of Fibre on the Ecology of the Intestinal Flora. CRC Press, Boca Raton, 257 p.
20. Drummond, K. and Lisa, M. (2007) Nutrition for Food Service and Culinary Professionals. John Wiley and Sons, Hoboken, 114 p.
21. Eliasson, R. and Lindholmer, C. (1976) Functions of Male Accessory Genital Organs. In: Hafez, E.S.E., Ed., Human Semen and Fertility Regulation in Man, Mosby Co., St. Louis, 44-50.
22. Green, C.M., Cockle, S.M. and Watson, P.F. (1996) Fertilization Promoting Peptide, a Tripeptide Similar to Thyro- trophin-Releasing Hormone, Stimulates the Capacitation and Fertilizing Ability of Human Spermatozoa in Vitro. Human Reproduction, 11, 830-836.
23. Egbunike, G.N., Dawodu, M.O. and Eboreime, A.O.A. (1999) Effect of Gossypol Acetic Acid on Sperm Production and Fertility in the Rat. Tropical Animal Production Investigation, 2, 111-117.
24. Haider, S.G., Passial, D. and Chen, K. (1985) Reversible Changes in Rat Spermatogenesis Induced by an Antifertility Substance (Gossypol): A Histochemical Report. Acta Histochemica, 77, 185-191, IN3?IN4.
25. Qian, S.Z., Zhong, C.Q. and Xu, Y. (1986) Effect of Trypterygium wilfordii on The Fertility of Rats. Contraception, 33, 105-110.
26. Montanari, T., De Carvalho, J.E. and Dolder, H. (1998) Antispermatogenic Effect of Achillea millefolium L. in Mice. Contraception, 58, 309-313.
27. Creasy, D.M. (2002) Histopathology of the Male Reproductive System II-Evaluation. In: Chapin, R.E. and Boekel- heide, K., Eds., Chapter 16, Unit 16.4, John Willy and Sons, Inc., New York.
28. Rodriguez, I., Ody, C., Araki, K., Garcia, I. and Vassalli, P. (1997) An Early and Massive Wave of Germinal Cell Apoptosis Is Required for the Development of Functional Spermatogenesis. The EMBO Journal, 16, 2262-2270.
29. Altay, B., Hekimgil, M., Kefi, A. and Cikili, N. (2001) Histopathological Examination of both Ipsilateral and Contralateral Testes with Different Obstructive Models in Prepubertal and Adult Rats. BJU International, 88, 84-89.
30. Flickinger, C.J., Baran, M.L., Howards, S.S. and Herr, J.C. (1999) Degeneration of the Seminiferous Epithelium Following Epididymal Obstruction in Prepubertal Rats. Anatomical Record, 254, 76-86.
31. Cook, J.C., Kiinefeiter, G.R., Hardisty, J.F., Sharpe, R.M. and Foster, P.M. (1999) Rodent Leydig Cells Tumorigenesis: A Review of the Pathology, Physiology, Mechanism and Relevance to Humans. Critical Reviews in Toxicology, 29, 169-261.
32. Kukucka, M.A. and Misra, H.P. (1992) HPLC Determination of an Oxytocin-Like Peptide Produced by Isolated Guinea Pig Leydig Cells: Stimulation by Ascorbate. Systems Biology in Reproductive Medicine, 29, 185-190.

CHAPTER 15

Current and future prospects for nanotechnology in animal production

Emily K. Hill[1,2] and Julang Li[1,2*]

[1] *School of Life Science and Engineering, Foshan University, Foshan, Guangdong, China*

[2] *Department of Animal Biosciences, University of Guelph, 50 Stone Road East, Building #70, Guelph, ON N1G 2 W1, Canada*

ABSTRACT

Nanoparticles have been used as diagnostic and therapeutic agents in the human medical field for quite some time, though their application in veterinary medicine and animal production is still relatively new. Recently, production demands on the livestock industry have been centered around the use of antibiotics as growth promoters due to growing concern over microbial antibiotic resistance. With many countries reporting increased incidences of antibiotic-resistant bacteria, laws and regulations are being updated to end in-feed antibiotic use in the animal production industry. This sets the need for suitable alternatives to be established for inclusion in feed. Many reports have shown evidence that nanoparticles may be good candidates for animal growth promotion and antimicrobials. The current status and advancements of nanotechnological applications in animal production will be the focus of this review and the emerging roles of nanoparticles for nutrient delivery, biocidal agents, and tools in veterinary medicine and reproduction will be discussed. Additionally, influences on meat, egg, and milk quality will be reviewed.

Keywords: Animal production Antibiotic replacements Artificial insemination Biocides Feed Nanoparticles Veterinary medicine

BACKGROUND

Nanotechnology is the study of materials at the nanoscale. With at least one dimension generally ranging between 1 and 100 nm (10^{-9} – 10^{-7}m), nanomaterials are best referred to as particles [1, 2]. These nanoparticles are particularly appealing as they take up very little space yet have relatively large surface areas, and therefore an increased ratio between surface atoms and interior atoms. As a result, when bulky materials are scaled down to nanosizes, their surface chemistries become more influential and alter the physical properties of the material [2]. For example, copper is known for its malleability, a useful feature for wiring and piping. However, when copper is scaled down into a nanoform, it loses its malleability as its surface atoms resist bending [3]. The interior copper atoms in a bulkier form facilitate bending but are out-numbered by surface atoms in the nanoform. Enlarging the ratio of surface area to volume allows for nanoparticles to be more versatile, whether as a single functional unit, or as a carrier for functional units which can be adhered to their surfaces or encapsulated within (Figure 1). Nanoparticles are becoming more attractive as novel uses, from medical diagnostics to gene therapy vehicles, are discovered.

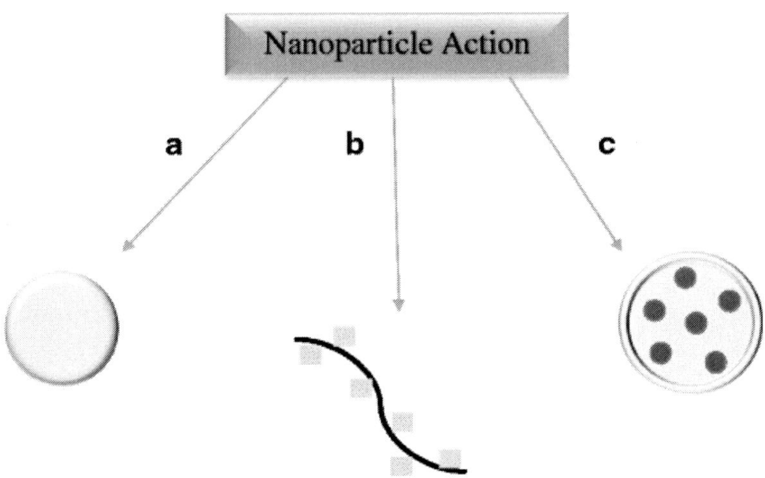

Figure 1. Three basic systems of nanoparticles in their applications. Nanoparticles can serve as the functional unit (a) but can also act as a delivery vehicle for materials conjugated to their surface (b) or encapsulated within (c)

TYPES OF NANOPARTICLES

Nanoparticles, currently available or under development, can be categorized into four groups: metals, polymers, natural compounds, and nanostructured materials. Although different engineering techniques are required depending on the group, nanoparticles can facilitate an array of biotechnical functions through different mechanisms of action (Table 1). Metal nanoparticles are the powdery version of solid metal, after large pieces have been ground down to nanosizes, effectively changing its associated physical properties [4]. These particles have drawn the attention of the medical field for their use in imaging and as antimicrobial therapies that lyse Gram positive and Gram negative bacterial cell walls [5]. External or topical applications may be more suitable for some metal nanoparticles to avoid accumulation in the body, as certain species can elicit harmful dosage-toxicity responses, although this is not always the case [6, 7, 8]. The non-biodegradable nature of metal is a major drawback for these particles. Polymeric nanoparticles, or nano-polymers, are polymers that have been synthesized or fragmented into pieces that are nanometers long. Nano-polymers have the ability to be grafted onto other materials, potentially improving their biocompatibility and degradation while expanding their utility [8]. Biocompatibility is highly advantageous for the medical and food industries as working concentrations of biocompatible nanoparticles will have few to no negative side effects on patients or consumers [9, 10, 11, 12]. Similar to the metal varieties, polymeric nanoparticles with a fluorescent or radiolabeled component may be used for medical imaging, although dosage toxicity would still have to be considered [13]. Nanoparticles made of natural compounds are materials that come from nature with limited manipulation, such as natural polymers or proteins. With few alterations, natural compounds are more likely to be biocompatible, distributable in the body, and biodegradable. Nanostructured materials are synthesized nanoparticles that originate from many sources, including natural compounds such as lipid- and protein-based nanoparticles. Natural and nanostructured nanoparticles share many advantages and can serve as the sole functioning unit or carriers for functional groups, such as drugs and nutrients, via encapsulation or superficial adhesion. While nature-derived nanomaterials may seem a safer choice, these particles could elicit toxic or immunogenic responses if not carefully engineered or appropriately distributed in a biological system. Despite these potential limitations, the advantages of employing nanotechnology are far greater.

Table 1. Mechanisms of action for nanoparticles performing a function useful to animal production

Function	Type of Nanoparticle	Mechanism of Action	References
Nutriceuticals	Metal	- feed supplements at the nanoscale are more bioavailable to animals than at a microscale, allowing more interactions to occur in the gut and better absorbance	[27, 38, 47, 51, 52]
	Natural	- nanoparticle additives to food products for human consumption can increase bioavailability	[55]
Drug and Nutrient Delivery	Polymer	- can be loaded with traditional antibiotics and may act as a shuttle to release them when in close proximity to a pathogen	[41–43]
		- metal nanoparticles may be conjugated to polymers for a combined nutrient/biocide delivery approach	[48, 49]
	Natural	- enclose around nutrients to protect against their degradation in the stomach for maximum intestinal absorption	[21]
	Nanostructured	-designed to carry nutrients or pharmaceuticals through the gastrointestinal tract for targeted release	[26, 66, 67]
Biocides	Metal	- lyse negatively charged Gram + and Gram − bacterial cell walls	[5, 31, 32, 53]
	Polymer	- destabilize bacterial cell walls such that homeostasis is disrupted to a lethal extent	[33, 44, 45]
Diagnostic Tools	Metal	- magnetic metal nanoparticles can disperse throughout the body and be imaged via MRI	[60]
	Nanostructured	- fluorescence can be initiated via light activation or two-photon excitation	[61, 62, 65]
Reproductive Aids	Nanostructured	- purification of sperm through the removal of damaged spermatozoa via surface markers recognized by nanoparticle-bound antibodies or lectins	[75, 76]
Molecular Biology Agents	Nanostructured	- gene transfer mediation through interactions between nucleic acids, nanoparticles, and sperm	[81]
	Polymer	- DNA transfection vehicles (as above)	[82]

APPLICATIONS OF NANOTECHNOLOGY

Current applications of nanotechnology in the medical and food industries are relatively analogous. Medical diagnostics and food safety testing are being improved through lab-on-a-chip technologies, capitalizing on the large surface area and small volume of nanoparticles. These molecular technologies require less samples, shorter run times, and provide a simpler user experience without the need for large bench top machines [14, 15]. A major advantage of lab-on-a-chip technologies is the ability to generate real-time data in the field. Therapeutic nanoparticles and nanoparticles as therapy delivery systems are both growing interests in the medical field and parallel the food industry where nanoparticles are being investigated as nutraceutical delivery systems, and biocides to better preserve food stuffs. Germicidal metal and polymeric nanoparticles attack the integrity of microbial cell walls, while natural and nanostructured materials can entrap and protect nutrients for delivery. The complementary nature of the medical and food industries is best exemplified in animal production where livestock health is directly linked to food safety. Livestock producers require their herds and flocks to reach ideal slaughter weights rapidly to maximize profitability. To achieve this, antibiotics are currently used prophylactically as feed additives to prevent illness and accelerate growth, thus shortening animal production cycles [16]. While helpful from the production standpoint, this global practice has led to the rise of drug-resistant bacteria that can cause illness in livestock and contaminate meat [17]. Trends in antibiotic use from various countries show that the majority is used in agriculture and less in humans [18]. This has prompted several countries to pass legislation restricting the use of antibiotics in animal production [19]. Nanoparticles not only have the potential to fill the gap created by these restrictions, but can also do so without driving antibiotic-resistance in microbes. Further to this, nutrient delivery, biocides, veterinary medicine, and reproduction are areas of the animal production industry that could benefit from nanotechnology, and the advantages and limitations of their inclusion will be discussed in this review (Table 2).

NUTRIENT DELIVERY

Casein micelles are naturally occurring nanoparticles in milk, where casein phosphoproteins make up approximately 80% of the protein profile in cow's milk [20, 21]. Some casein isoforms assemble themselves around calcium, proteins, and other nutrients to allow for transport from mother to young. Manipulation of these micelles has led to the incorporation of choice hydrophobic nutrients [22]. Vitamin D was dispensed to human volunteers within these casein nanoparticles, increasing vitamin D bioavailability in vivo, as casein particles undergo proteolytic cleavage in the stomach, releasing their encapsulated vitamins [21]. A similar strategy could be investigated to help producers assist neonates through weaning, a sensitive time for young as their digestive and immune systems are still maturing. In the absence of milk from the mother, weanlings must adapt to a complex carbohydrate diet and reduced immune support. This is an important stage not only for animal welfare reasons, but also from a production standpoint as weanlings that maintain their growth rates through the weaning process are healthier and heavier by slaughter [23]. Benefits from nutritional supplements may help weaned animals and poultry increase body size as well [24]. Nanoparticles engineered for nutrient delivery could facilitate this supplementation and bolster growth rates of livestock by increasing nutrient cargo bioavailability.

Table 2. Current and future applications of nanotechnology in animal production with their advantages and limitations

	Advantages	Limitations	References
Current applications			
Medical diagnostics	- disease diagnosis and prognosis - small sample volume requirements - faster sample analyses; faster action time - suitable for field use	- greater sample preparation measures - sensitivity concerns from small sample sizes - device pricing	[14, 15, 60–64]
Medical supplies	- antibacterial wound dressings - prevention of catheter-caused infections - inhibits biofilm formation	- patient sensitivities and allergies - potential free radical formation	[85, 86]
Food safety	- contaminant and pathogen identification and indication - contamination prevention	- food and food packaging safety concerns	[53, 55]
Biocides	- alternatives to antibiotics and conventional cleaners - internal and external uses - antimicrobial coatings - multi-purpose (i.e. bactericidal and growth promoting)	- little in vivo evidence to support many internal applications demonstrated in vitro - cytotoxicity	[31–34, 44, 45, 85, 86]
Future applications			
Nutraceuticals	- increased nutrient bioavailability - extra support for weanlings - growth and performance enhancement - solution for nutrient deficiencies	- nanoparticles must be not be degraded in the GI tract before absorption in the intestines	[27, 38, 47–52, 59]
Nutrient delivery systems	- increased nutrient bioavailability - extra support for weanlings - growth and performance enhancement	- carrier system must be designed to withstand GI tract challenges - bioaccumulation - biocompatibility concerns	[21, 26]
Therapeutics	- alternatives to antibiotics - enhanced pathogen/organ targeting	- biocompatibility considerations - efficacy against different pathogens - relative MIC	[41–43, 53]
Drug delivery systems	- platforms to enhance drug specificity and delivery - reduce MIC and kill antibiotic-resistant strains of pathogens	- biocompatibility considerations - potential for bodily accumulation - confidence in specificity	[61, 62, 65–67]
Medical Imaging	- additional modes of tracing organ systems and tracking drug therapies in the body	- depth of tissue - biocompatibility considerations	[61, 62, 65]
Nanopurification of sperm	- isolate desired sperm based on biomarkers (i.e. healthy from unhealthy) - increase fertilization efficiency, more females fertilized from a single collection	- biomarker library to be developed - purebred restrictions on artificial insemination	[75, 76, 83, 84]
Cryopreservation of sperm	- sperm preserved for longer - protecting against freeze-thaw cycles - to replace antibiotics from extenders; lower risk of reduced sperm motility	- spermatotoxicity - oocyte toxicity	[76–80, 83, 84]
Genetic manipulation	- can carry DNA of interest from cytoplasm into cell nucleus - benefits to sperm-mediated gene transfer approach - transduction replacement strategy (no viruses involved in DNA administration)	- cytotoxicity - must not disturb cellular functioning	[81–84]

Orally administering additional nutrients to livestock via feed brings inherent challenges that must be considered when designing a carrier nanoparticle. Each compartment of the gastrointestinal tract has a unique environment that includes its own complement of enzymes and specific pH level. Nanoparticles must be able to overcome these obstacles to deliver their nutritional cargo at the appropriate location, i.e. the small intestine [25]. Akbari and Wu [26] investigated a nanoparticle made from the canola protein cruciferin, and demonstrated that these nanoparticles could encapsulate both hydrophobic and hydrophilic bioactive compounds, protect them from a simulated stomach environment, and release them in a simulated intestinal environment. Once nutrients are released from nanoparticles, they must reach the intestinal epithelia and carry out their function as if they were to have originated from the feed. The nanoparticle cannot remain in the intestine, as accumulation may prevent the absorption of other nutrients in the lumen. Cellular uptake, degradation, or removal from the body must occur. Depending on the type of nanoparticle, cellular uptake may not be preferred in the animal production industry where food safety must be considered. Cruciferin nanoparticles are digestible in the small intestine by the pancreatin enzyme, however, indigestible nanoparticle vehicles may be excreted naturally. Before entering the market, a thorough study of nanoparticle action must be done to test for potency and any unwanted biological consequences, like cytotoxicity.

While nutrient delivery at the nanoscale can occur naturally or synthetically, nanoparticles can stabilize bioactive compounds and assist in cellular uptake. To add a bioactive component directly to feed entails a risk of degradation and inaccessibility that can be countered through nanotechnological means. The small size of nanoparticles garners a heightened level of bioavailability compared to microparticles, particularly in the digestive tract, since nanoparticles can more readily pass through the intestinal mucosa. Huang et al. [27] used calcium carbonate and calcium citrate at the nano- and microscales to test bioavailability differences by measuring bone mineral densities of mice. Mice that were administered the calcium compounds at the nanoscale had denser bones compared to mice given micro-calcium and controls. It would be of interest to investigate whether delivering calcium in nanoform may help to strengthen bones of production animals such as turkeys, as body mass is heavily selected for over leg strength, making it difficult for turkeys to support themselves [28].

BIOCIDES

Nanoparticles may present a feasible alternative to antibiotics and may help bar pathogens from entering animal production sites. The unregulated use of antibiotics, a common practice in many countries, provides the impetus for bacteria to become drug-resistant. New legislation for the restriction of prophylactic antibiotic use in agriculture is gaining ground as a method to combat this growing problem [29]. Limiting antibiotic use necessitates the development of alternatives due to the high-density nature of modern animal production facilities which invites and expedites disease transmission [30]. Metal nanoparticles with net positive charges are drawn to negatively charged bacterial membranes, resulting in leakage and bacterial lysis [31]. Kim et al. [32] found that silver nanoparticles could inhibit the growth of hemorrhagic enteritis-inciting *E. coli* O157:H7 and yeast isolated from a case of bovine mastitis with an estimated MIC of 3.3-6.6 nmol/L and 6.6-13.2 nmol/L, respectively. At present, silver is used in medicine and dentistry to prevent wound infections and the formation of biofilms on catheters and dental appliances. Similarly, positively charged and/or quaternized polymeric nanoparticles can limit microbe growth [33]. Qi et al. [34] tested the antimicrobial activity of chitosan, a biopolymer with a positive surface charge (Zeta potential of approximately +51 mV), against a variety of bacterial pathogens and concluded that the MIC for all species tested was smaller than 0.25 µg/mL. Current biosecurity measures can impede pathogen entry but feed, water, air, and personnel may still serve as accession points [35]. The use of antimicrobial nanoparticles at these points may create additional hurdles for potential pathogens seeking entry.

Copper is regularly added to feeds for its ability to promote animal growth and performance in addition to its antimicrobial properties [36, 37]. As copper is delivered to animals orally in feed, the same challenges seen in nutrient delivery can be applied here. It has been reported that copper nanoparticles pass intestinal mucosa more easily than microforms, aiding in absorption. Gonzales-Eguia et al. [38] demonstrated that nanoform copper could better improve piglet energy and crude fat digestion through the augmentation of lipase and phospholipase A activity in the small intestine compared to a basal diet supplemented with $CuSO_4$. These piglets experienced an increase in daily weight gain, metabolic rates, and immune capacity. Analysis of nanocopper influence on the immune system showed significantly greater total globulin and superoxide dismutase concentrations in blood serum, although hematology was unchanged among experimental groups. These results suggest that the nutrient value of feed can be enhanced through the addition of antimicrobial metal supplements in nanoform. However, further investigation is required before determining whether antibiotics in feed can be completely replaced by nano-antimicrobials.

Despite the expansion of antibiotic-resistance in bacteria, antibiotics have not yet been rendered totally ineffective against them. Their delivery and efficacy can be enhanced by nanoparticle carriers, potentially decreasing the dosage of antibiotics required for treatment. Susceptible bacteria lack at least one mechanism of resistance where resistant bacteria have, at minimum, one of the many known resistance mechanisms [39]. β-lactamases are resistance enzymes that cleave the β-lactam ring in β-lactam drugs such as penicillin and methicillin, destroying drug activity [40]. Loading penicillin onto polymeric polyacrylate nanoparticles has been identified as an effective method of making the β-lactam ring insusceptible to β-lactamase binding and cleavage, restoring penicillin function against methicillin-resistant *Staphylococcus aureus* (MRSA), a penicillin- and methicillin-resistant bacterial strain [41]. In a study by Turos et al., different penicillin-polyacrylate nanoparticle preparations were examined and it was found that two of them could inhibit MRSA just as well as the positive control, Penicillin G, and that one of the two had an MIC 8× lower than that of Penicillin G (2 μg/mL versus 16 μg/mL) [41]. No evidence of toxicity was seen during either the systemic or topical application of these nanoparticles in a murine model, and when applied topically to a dermal abrasion, the healing time was shortened by 3–5 d [42]. A similar study was performed using tetracycline-bound polymeric chitosan nanoparticles against a tetracycline-resistant *Escherichia coli* strain, requiring a minimum concentration of 700 μg/mL for inhibition of bacterial growth [43].

Nanoparticles made from natural materials that target Gram negative bacteria are most desired for feed, since bacterial pathogens are predominately Gram negative. This is due to the Gram-negative endotoxin, lipopolysaccharide (LPS), that is absent in Gram positive bacteria. It is important to monitor the materials provided to food animals as the bodily accumulation of toxins could pose a threat to consumers and potentially dampen public opinion towards the inclusion of nanotechnology in animal production, highlighting the importance of only selecting biocompatible nanoparticles for use in feed. However, the addition of nanoparticles in feed is not the only solution to limit livestock exposure to pathogens; the external application of hydrogels could assist in preventing pathogens from entering production facilities. Quaternization, which gives amine groups a permanent positive charge, of polymer nanoparticles such as poly(2-(dimethylamino)ethyl methacrylate) (qPDMAEMA) permits biocidal activity through electrostatic attraction between polymer and microbe, distorting the cell wall [33]. When qPDMAEMA is grafted onto agarose or cellulose nanocrystals, they can be made into hydrogels and used to cover surfaces to prevent bacterial adhesion, colonization, and biofilm formation. Their ability to bind viruses and virus-like particles has also been reported [44]. Moreau et al. [45] examined the biocompatibility of a few cationic polymers in vitro and in vivo and concluded that toxicity was dependent upon the molar mass of the polymer and the type of polycation. The unquaternized PDMAEMA (193.5 Da) was found to induce hemolysis and cause immediate mortality when administered intravenously in mice, indicating that its use as a biocide should be restricted to external purposes only, and should not be included in feed. Spreading biocidal hydrogels and nano-solutions across thresholds, pens, and feed containers presents yet another use that would benefit from further study [46].

MEAT AND EGG QUALITY

The possibility of using nanoparticles to enhance meat and egg quality has also been investigated. For example, Wang and Xu [47] demonstrated that when finishing pigs destined for market were given chromium nanoparticles (200 μg/kg) in feed, they were 14.06% leaner at slaughter than control pigs fed a basic diet of corn-soybean meal. An increase in skeletal muscle mass and improved pork quality were achieved, with similar effects found when finishing pigs were fed chitosan nanoparticle supplements loaded with chromium [48, 49]. These chromium-loaded chitosan nanoparticles elevated the activity of hormone-sensitive lipase in adipose tissue while decreasing fatty acid synthase activity and boosting blood serum immune components [48, 49]. These data provide a compelling insight into the mechanism of action these nanoparticles have in pigs, and how they affect meat quality. Of further interest is the heightened chromium content in selected tissues, such as 184.11% in the longissimus muscle compared to the control, as some nanoparticles, i.e. $Ag^{+}$and Cu^{2+}, have been found to cross the blood–brain barrier [47, 50].

The inclusion of nanomaterials in livestock feed or water can benefit the quality of product obtained, as well as the production cycle. Chromium nanoparticles added to poultry feed not only positively affected breast and thigh muscle protein content while simultaneously lowering cholesterol, but raised the average daily gain and feed efficiency of the broilers in the experimental group fed 500 μg/kg Cr^{3+} [51]. The implications of these results are shorter production cycles for better quality meat with less feed required to have broilers reach market weight. Conversely, when chromium nanoparticles were supplied to layers, there was no significant effect on body weight or egg production [52]. However, Sirirat et al. [52] did find that egg quality improved from higher chromium and calcium levels in the yolks and shell, respectively. Bioaccumulation of nanoparticles in the liver was noted for the experimental group, an observation shared with Chauke and Siebrits [53] in a study that replaced an antibiotic against coccidiosis with silver nanoparticles in water (0.083 mg/kg of silver compared to 0.001 mg/kg in the control). More information on the interplay between nanoparticle concentration and meat quality would be useful to ensure that quality is not sacrificed after long-term exposure [54]. The inclusion of nutrient supplements in livestock feed, regardless of particle size, will benefit the producer if there is still consumer demand for the final product. If meat and eggs obtained from an animal fed nanoparticle supplements are enhanced, or indiscernible from the original product, they are then likely to still be favourable to consumers. However, it is important to understand the role of the nanoparticle additive in a given biological system and by-products from that system to ensure it is safe for consumption before its application in animal production.

MILK

Mastitis is an example of a common ailment among dairy cows with a variety of inciting factors, often bacterial, that can require the use of antibiotics to clear. Tilmicosin is an example of a drug used in mastitis cases that has negative side-effects if given at too high a concentration. In consideration of this, Han et al. [55] sought to control the release of tilmicosin by using hydrogenated castor oil-solid lipid nanoparticle carriers. Of concern was what the extended half-life of the therapeutic would mean for milk discard times, as tilmicosin was present in mouse blood serum for 5 h without nanocarrier delivery, and 8 d with nanocarrier delivery [55]. However, a lower dosage was required for resolution in a *S. aureus*-induced murine mastitis model (10 mg/kg versus 20 mg/kg) [56]. Careful manipulation of therapeutic nanocarriers to find a balance between dosage and half-life could serve to benefit producers by minimizing milk discard times and the amount of milk wasted.

Nanotechnology can also help to ensure that the quality of milk is safe for human consumption through novel foodborne pathogen detection techniques. Sung et al. [57] developed nanocomposites containing anti-*S. aureus* antibodies, gold nanoparticles, and magnetic nanoparticles to provide a 40 min colorimetric test for the presence of *S. aureus* in milk. An interesting feature of these nanocomposites is the antibody, whose specificity and selectivity could be modified to capture a variety of pathogens [57]. Wang et al. [58] demonstrated a similar technique, employing polyclonal antibodies and gold nanoparticle immunochromatographic strips to detect toxins present in milk within 10 min, using the carcinogenic aflatoxin M1 as an example. While a large focus has been to remove potentially harmful contaminants from milk, there has also been some interest in mixing nanoparticle supplements directly into cow's milk for human consumption. Lee et al. [59] combined nanopowdered oyster shell into milk with the intention of increasing the calcium content from 100 to 120 mg/mL to a level more suitable for growing children and post-menopausal women. Supplementing milk with calcium from nanopowdered oyster shell did not negatively alter its sensory or physicochemical qualities after 16 d of storage at 4 °C [59].

VETERINARY MEDICINE

Nanomedicine is an intriguing discipline in nanotechnology that is showing progress in both diagnostics and therapeutics. Metallic and nanostructured particles are useful diagnostic tools in biomedical research that can be used to visualize the status of a cell or drug distribution in the body. Magnetic nanoform metals, i.e. iron oxide, can be taken up by cells and imaged in vivo at high concentrations using magnetic resonance imaging (MRI) [60]. Nanostructured particles can be made to fluoresce through light activation or two-photon excitation [61, 62]. Further to these diagnostic nanoparticles, exciting developments have been made in molecular-based lab-on-a-chip technologies for qualitative and quantitative biological analyses. Requiring small volumes of analyte and reagents, producing little waste, and shortening wait times make these lab-on-a-chip technologies an attractive option [63]. There are currently a number of products functioning at the microscale available on the market with nanoscale products just emerging; see Tian et al. [64] for more information.

Drug delivery can be monitored through fluorescent nano-carriers. For example, light activated, fluorescent nanostructured glucose- and sucrose-derived nanoparticles can be used to monitor the localization of bound chemotherapies [61]. The biocompatibility of carbohydrate-derived nanoparticles has been demonstrated in a human lung carcinoma cell line by Ajmal et al. [61]. Their findings showed that upon binding methotrexate, a chemotherapeutic drug, the conjugate nanoparticles were reported to have a cytotoxicity close to that of cells treated only with methotrexate. However, the advantage of delivering this drug with a nanoparticle that fluoresces after light activation is the ability to trace the drug. For even better tracking, using a carrier nanoparticle capable of being activated via two-photon excitation can provide a 3D spatial image over a greater tissue depth than a particle activated with visible UV light [62, 65]. As chemotherapies are typically delivered in a high dose regimen, the ability to observe their distribution in the body through fluorescence could help to reduce off-target side effects by better targeting them to desired areas.

Fluorescence is not a shared trait among all nanoparticle drug carriers, and their mechanisms of drug binding and release can be quite varied, especially amongst nanostructured particles. Cylindrical nanotubes can trap pharmaceutical agents within an internal matrix surrounded by an outer layer of poly (L-lactide) or poly (D-lactide) [66]. When these enantiomers come together in solution, they interact with each other to reconfigure their structures from cylinders to spheres, releasing trapped materials in the process. There is no requirement for external stimuli for drug release as there is in some light activated nanoparticles, only that the two nanotubes interact. Other nanostructured particles can be self-loading like albumin-dextran nanoparticles with hydrophobic drugs [67]. Albumin from bovine serum can be stabilized with dextran in aqueous solutions and can bind medicine through hydrophobic and electrostatic interactions. When tested with ibuprofen, albumin-dextran nanoparticles could take up 0.7-unit weight of ibuprofen per 1-unit weight of the conjugated particle [67]. These nanostructured materials present new mechanisms for pharmaceutical uptake and release in nanomedicine, potentially serving as methods to increase release specificity and reduce lag times between drug delivery and effect in the future.

REPRODUCTION

Animal production revolves around animals meant for slaughter. Finishing livestock are the offspring of individuals intended for breeding who have highly ranked genotypes and phenotypes. The traits and reproductive abilities of breeders gives them high value. Some nanoparticles have been demonstrated to enhance fertility and protect spermatozoa through the functional groups they carry. Artificial insemination is widely preferred in animal production as an alternative to live cover strategies due to the lower risk for animals and producers. Commonly done to diversify genetic backgrounds and boost selection of livestock traits, artificial insemination has the potential to be enhanced through the integration of nano-techniques such as the non-invasive bioimaging of gametes, nanopurification, and protectants in cryopreservation.

In order to optimize the efficiency of artificial insemination, livestock gamete biology and reproductive obstacles to fertilization must first be elucidated. Recently, quantum dots have been explored as a research method to improve understanding of mammalian spermatozoon and oocyte movement and their interactions in a physiological setting. These self-illuminating, inorganic nanoparticles are of interest to the field of theriogenology as they are biocompatible, photo-stable, and have a greater signal intensity than organic fluorescent molecules previously used to image gametes and other cell types in vivo [68, 69, 70]. Feugang et al. [68, 71] have demonstrated the real-time tracking ability of bioluminescent resonance energy transfer-conjugated quantum dot (BRET-QD) nanoparticles in vitro, in situ, and ex vivo using pig male gametes (*Sus scrofa domesticus*). Quantum dots can provide targeted or non-targeted imaging as a function of their size, emitted wavelengths, and conjugation possibilities [68, 71]. This engineered nano particle provides a new mean to visualize the molecular and cellular events during fertilization, in a similar way to fluorescent proteins, but at greater tissue depths [71, 72]. Signal strength of quantum dots are dose-dependent and a higher concentration may be required for in vivo imaging on larger animals. Thus, the composition of quantum dots should be further optimized for biocompatibility as many of the current ones include heavy metals, such as cadmium and lead, which may be cytotoxic at high levels [73]. However, if quantum dot concentrations and surface chemistries are carefully selected, cytotoxicity may be decreased or eliminated [74].

Nanopurification of semen can be used to separate damaged sperm from undamaged, healthy sperm. One method is to coat magnetic nanoparticles with antibodies against ubiquitin, a surface marker of defective sperm, for a protein-based removal strategy [75]. A lectin-based strategy features magnetic nanoparticles coated with lectins that bind glycan exposed at the surface of the sperm through acrosomal damage [76]. Nanopurified bull spermatozoa (*Bos taurus*) achieved conception rates equal to those of unpurified semen at half the concentration with no negative impacts reported for inseminated cows or calves [75]. Thus, more females can be inseminated from one sample of nanopurified, diluted ejaculate. Further identification of spermatozoa biomarkers will allow for increased selection ability and fertility improvement, as targeted through the antibody or lectin strategies.

Cryopreservation of sperm can be enhanced by turning to nano-protectant additives in extenders. Used to dilute sperm, extenders are buffering agents and provide sperm with nutrients required for prolonged storage. They serve to protect and contain antibiotics, to prevent bacterial growth from affecting sperm quality and infecting inseminated females [77]. Antimicrobial nanoparticles may serve to replace extender antibiotics in the future as some antibiotics have been shown to inhibit sperm motility and viability in a dose-dependent manner [78]. Nanoparticles may also facilitate the addition of natural products in extenders to increase sperm motility. Research groups have reported that the addition of honey, sugarcane juice, tomato juice, and pineapple juice can increase the survivability of sperm stored at room temperature [76, 79]. While nanoparticles were not involved in those studies, it would be interesting to know how sperm quality would be effected if the functional groups of each product were to be delivered via nanoparticle. As sperm can be shipped internationally over multiple days, extenders with a higher capacity for preserving samples undergoing freeze-thaw cycles would be beneficial [80].

Further advancements in reproductive biotechnology may be possible with the greater inclusion of nanoparticles in molecular biology techniques. Sperm-mediated gene transfer is onesuch approach where mesoporous silica nanoparticles can be loaded with nucleic acids and proteins [81]. These nanoparticles can form strong associations with spermatozoa in vitro, and do not have any diminutive effects on sperm function or quality. Transfections with polymeric nanoparticles, such as PDMAEMA, chitosan, and polyethylenimine, have been reported to be advantageous over traditional viral approaches provided low concentrations of polymers are used [82]. The molecular weight of the nanopolymer has great influence over transfection efficacy and toxicity, i.e. the optimal molecular weight for transfection with PDMAEMA has been determined to be 60 kDa [82].

With continued exploration and refinement, nanoparticles could play a significant role in animal reproduction. However, it should be noted that some nanoparticles are spermatotoxic which may have serious consequences if breeder reproduction is affected. Zinc oxide and titanium oxide nanoparticles are two examples that reduce in vitro sperm viability in a dose- and time-dependent manner by membrane weakening and DNA fragmentation [83, 84]. Barkhordari et al. [83] incubated human sperm with zinc oxide nanoparticles and found that a concentration of 500 μg/mL would significantly increase cell death after 45 min, while a concentration of 100 μg/mL would significantly increase cell death after 180 min. Pawar and Kaul [84] found that buffalo sperm (*Bubalus bubalis*) incubated with 100 μg/mL of titanium oxide nanoparticles would have reduced viability. At 10 μg/mL titanium oxide was found to prematurely increase sperm capacitation, which is the final required step in sperm maturation for oocyte penetration and fertilization. While nanoparticles may be points of advancement for the animal production industry, precautions should also be taken when considering the employment of nanoparticles for assisting reproduction.

FUTURE PROSPECTS

As nanotechnology continues to develop and garner more attention, its applications in the animal production industry will become more expansive. The regular inclusion of nano-supplements to fortify livestock feed is likely possible in the near future; however, it will take longer for nanoparticles to fully replace antibiotics in feed as many biocidal candidates must still be tested in vivo before undergoing clinical trials and food safety tests in accordance with government regulations. External uses for nanoparticles have already been integrated into some aspects of animal production, i.e. antiseptic wound dressings, and more are to follow [85, 86]. For studies interested in nanoparticles with anti-cancer properties, it is important to investigate nanoparticle cytotoxicity in both cancer cell lines and normal, healthy cell lines. Only using cancer cells and claiming the nanoparticle under investigation has anti-cancer properties may be misleading, as the nanoparticle may be cytotoxic to all cell types. In vivo studies are needed for verification of nanoparticle functions seen in in vitro research. Table 3 outlines nanoparticle experiments relevant to the animal production industry and identifies gaps in knowledge where future research will be required.

Table 3. Summary of nanoparticle studies relevant to animal production

Nanoparticle	Type	Experiment	In vitro/ In vivo	Cell line	Animal production application	Reference
Gold and Copper	Metal	Biocides for water treatment	In vitro	N/A	Biocide	[5]
Casein micelles	Natural	Determining storage capacity and stabilization of encapsulated bioactive compounds and their bioavailability	In vivo Humans	N/A	Nutrient delivery	[21]
Lipid nanoparticles	Nanostructured	Simulated digestion assay to test bioavailability of loaded compounds	In vitro	N/A	Nutrient delivery	[25]
Cruciferin	Nanostructured	Encapsulation abilities and nutrient release studies	In vitro	Caco-2 (Human cancer cells)	Nutrient delivery	[26]
Calcium carbonate and calcium citrate	Metal	Bioavailability differences between microparticles and nanoparticles	In vivo Ovariectomized mice	N/A	Nutrient delivery	[27]
Silver	Metal	Eco-friendly biocide synthesis	In vitro *Vibrio cholerae*	N/A	Biocide	[31]
qPDMAEMA-agarose	Polymer	Microbial growth inhibitory properties of qPDMAEMA in solution and hydrogels	In vitro *E. coli* *S. aureus* Biofilms	N/A	Biocide	[33]
Copper	Metal	Enhancing growth promoting effects of copper by nanoscaling	In vivo Piglets	N/A	Nutrient delivery and Biocide	[38]
Gold	Metal	Functionalize with amoxicillin to overcome bacterial resistance.	In vitro Gram + Gram – In vivo Mice	L929 (Mouse fibroblasts)	Biocide	[40]
Polyacrylate	Polymer	Testing protective abilities towards loaded penicillin and aiding its antibacterial activity	In vitro *S. aureus* (methicillin-susceptible and -resistant strains)	N/A	Biocide	[41]
Chitosan	Polymer	Evaluating efficiency of drug loading and release	In vitro *E. coli* (tetracycline-resistant strain)	N/A	Biocide	[43]
qPDMAEMA-CNC	Polymer	Analyzing viral-binding ability for the concentration and extraction of viruses and virus-like particles	In vitro Cowpea chlorotic mottle virus Norovirus-like particles	*Spodoptera frugiperda* Sf9 (Insect cells)	Biocide	[44]
Triclosan	Polymer	Increasing antimicrobial activity of organic agents through aqueous nanodisperive techniques	In vitro *Corynebacterium*	N/A	Biocide	[46]
Iron oxide	Metal	Imaging applications in functional studies in vivo	In vitro	Neural progenitor cells Pheochromocytoma cells (Rat lineage)	Veterinary Medicine	[60]
Carbon (glucose- and sucrose-derived)	Nanostructured	Demonstrating anticancer bioactivity of loaded drugs	In vitro	H157 (Human cancer cells)	Veterinary Medicine	[61]
Mesoporous silica	Nanostructured	Spatial imaging of drug release in the body	In vitro	MCF-7 (Human cancer cells)	Veterinary Medicine	[62]
Poly(L-lactide)- and Poly(D-lactide)-b-poly(acrylic acid)	Nanostructured	Investigation into new controlled delivery of therapeutics	In vitro	N/A	Veterinary Medicine	[66]
Albumin-dextran	Nanostructured	Bind hydrophobic drugs to create aqueous solutions	In vitro	N/A	Veterinary Medicine	[67]
Zinc oxide	Metal	Toxic effects on livestock sperm	In vitro	Sperm (Human)	Reproduction	[83]
Titanium oxide	Metal	Toxic effects on livestock sperm	In vitro	Sperm (*Bubalus bubalis*)	Reproduction	[84]
Antibody-coated or lectin-coated F_2O_3	Metal	Nanopurification of semen	In vitro	Sperm (*Bos Taurus*)	Reproduction	[75]
Mesoporous silica	Nanostructured	Transfer mediator for nucleic acid/protein cargo to sperm	In vitro	Sperm (Boar)	Reproduction	[81]
Silver	Metal	Antimicrobial testing of silver nanoparticles bound to cellulose fibers with alkali lignin	In vitro *E. coli*	N/A	Biocide and Veterinary Medicine	[85]
Biocellulose	Natural	Designing an antiseptic, collagen-stimulating wound dressing	In vitro Infectious bacteria	N/A	Biocide and Veterinary Medicine	[86]

CONCLUSIONS

There are many applications for nanoparticles in animal production and this review serves to highlight these uses and to identify potential opportunities for future applications. Nanoparticles are already available on the market and, with continued development, their properties will be more finely optimized for a wider selection of applications. The use of nanotechnology in animal production is still in its infancy but encouraging results from nutrition, biocidal, remedial, and reproductive studies are driving further investigation.

ACKNOWLEDGEMENTS

Not applicable.

Funding

We would like to thank for the funding support from Foshan University, China, and the Natural Sciences and Engineering Research Council of Canada (NSERC).

Availability of data and materials

Not applicable.

Authors' contributions

EKH wrote the manuscript with the guidance and editing of JL. Both authors read and approved the final manuscript.

Competing interests

The authors declare that they have no competing interests.

Consent for publication

Not applicable.

Ethics approval and consent to participate

Not applicable.

REFERENCES

1. Jiang W, Kim BYS, Rutka JT, Chan WCW. Nanoparticle-mediated cellular response is size-dependent. Nat Nanotechnol. 2008;3:145–50.
2. Albanese A, Tang PS, Chan WCW. The effect of nanoparticle size, shape, and surface chemistry on biological systems. Annu Rev Biomed Eng. 2012;14:1–16.
3. Sharma HS, Sharma A. Neurotoxicity of engineered nanoparticles from metals. CNS Neurol Disord Drug Targets. 2012;11:65–80.
4. Halperin WP. Quantum size effects in metal particles. Rev Mod Phys. 1986;58:533–606.
5. Ramasamy M, Kim S, Lee SS, Kee D. Recyclable photo-thermal nano-aggregates of magnetic nanoparticle conjugated gold nanorods for effective pathogenic bacteria lysis. J Nanosci Nanotechnol. 2016;16:555–61.
6. Asharani PV, Mun GLK, Hande MP, Valiyaveettil S. Cytotoxicity and genotoxicity of silver nanoparticles in human cells. ACS Nano. 2009;3:279–90.
7. Kawata K, Osawa M, Okabe S. In vitro toxicity of silver nanoparticles at noncytotoxic doses to HepG2 human hepatoma cells. Environ Sci Technol. 2009;43:6046–51.
8. Travan A, Pelillo C, Donati I, Marsich E, Benincasa M, Scarpa T, et al. Non-cytotoxic silver nanoparticle-polysaccharide nanocomposites with antimicrobial activity. Biomacromolecules. 2009;10:1429–35.
9. Choi SJ, Oh JM, Choy JH. Biocompatible nanoparticles intercalated with anticancer drug for target delivery: Pharmacokinetic and biodistribution study. J Nanosci Nanotechnol. 2010;10:2913–6.
10. Shim M, Kam NWS, Chen RJ, Li Y, Dai H. Functionalization of carbon nanotubes for biocompatibility and biomolecular recognition. Nano Lett. 2002;2:285–8.
11. Ichikawa S, Iwamoto S, Watanabe J. Formation of biocompatible nanoparticles by self-assembly of enzymatic hydrolysates of chitosan and carboxymethyl cellulose. Biosci Biotechnol Biochem. 2005;69:1637–42.
12. Taylor TM, Davidson PM. Liposomal nanocapsules in food science and agriculture. Crit Rev Food Sci Nutr. 2005;45:587–605.
13. You C, Miranda OR, Gider B, Ghosh PS, Kim I, Erdogan B, et al. Detection and identification of proteins using nanoparticle-fluorescent polymer 'chemical nose' sensors. Nat Nanotechnol. 2007;2:318–23.
14. Jain KK. Nanotechnology in clinical laboratory diagnostics. Clin Chim Acta. 2005;385:37–54.
15. Valdes MG, Gonzalez ACV, Calzon JAG, Diaz-Garcia ME. Analytical nanotechnology for food analysis. Microchim Acta. 2009;166:1–19.
16. Wang JB, Qi LL, Han F. Antibiotic with different antibacterial spectrum changed intestinal microflora structure and reduced immune response of Lingnan yellow broiler. Res J Biotechnol. 2016;11:121–9.
17. Schwarz S, Kehrenberg C, Walsh TR. Use of antimicrobial agents in veterinary medicine and food animal production. Int J Antimicrob Agents. 2001;17:431–7.
18. Silbergeld EK, Graham J, Price LB. Industrial food animal production, antimicrobial resistance, and human health. Annu Rev Public Health. 2008;29:151–69.

19. Casewell M, Friis C, Marco E, McMullin P, Phillips I. The European ban on growth-promoting antibiotics and emerging consequences for human and animal health. J Antimicrob Chemother. 2003;52:159–61.
20. Day L, Williams RPW, Otter D, Augustin MA. Casein polymorphism heterogeneity influences casein micelle size in milk of individual cows. J Dairy Sci. 2015;98:3633–44.PubMed
21. Haham M, Ish-Shalom S, Nodelman M, Duek I, Segal E, Kustanovich M, et al. Stability and bioavailability of vitamin D nanoencapsulated in casein micelles. Food Funct. 2012;3:737–44.
22. Semo E, Kesselman E, Danino D, Livney YD. Casein micelle as a natural nano-capsular vehicle for nutraceuticals. Food Hydrocoll. 2006;21:936–42.
23. Pluske JR, Williams IH, Aherne FX. Maintenance of villous height and crypt depth in piglets by providing continuous nutrition after weaning. Anim Sci. 1996;62:131–44.
24. Nieto CAR, Meza-Herrera CA, Cedillo FDJM, Najera MDJF, Vazquez HGG, Perez FDJV, et al. Vitamin E supplementation of undernourished ewes pre- and post-lambing reduces weight loss of ewes and increases weight of lambs. Trop Anim Health Prod. 2016;48:613–8.
25. Ban C, Park SJ, Lim S, Choi SJ, Choi YJ. Improving flavonoid bioaccessibility using an edible oil-based lipid nanoparticle for oral delivery. J Agric Food Chem. 2015;63:5266–72.
26. Akbari A, Wu J. Cruciferin nanoparticles: preparation, characterization and their potential application in delivery of bioactive compounds. Food Hydrocoll. 2016;54:107–18.
27. Huang S, Chen JC, Hsu CW, Chang WH. Effects of nano calcium carbonate and nano calcium citrate on toxicity in ICR mice and on bone mineral density in an ovariectomized mice model. Nanotechnology. 2009;20:375102.
28. Emmerson DA, Anthony NB, Nestor KE, Saif YM. Genetic association of selection for increased leg muscle and increased shank diameter with body composition and walking ability. Poult Sci. 1991;70:739–45.
29. Yang Z, Guo Z, Qiu C, Li Y, Feng X, Liu Y, et al. Preliminary analysis showed country-specific gut resistome based on 1267 feces samples. Gene. 2016;581:178–82.
30. Huijskens EGW, Smit LAM, Rossen JWA, Heederik D, Koopmans M. Evaluation of patients with community-acquired pneumonia caused by zoonotic pathogens in an area with a high density of animal farms. Zoonoses Public Health. 2016;63:160–6.
31. Gahlawat G, Shikha S, Chaddha BS, Chaudhuri SR, Mayilraj S, Choudhury AR. Microbial glycolipoprotein-capped silver nanoparticles as emerging antibacterial agents. Microb Cell Fact. 2016;
32. Kim JS, Kuk E, Yu KN, Kim J, Park SJ, Lee HJ, et al. Antimicrobial effects of silver nanoparticles. Nanomedicine. 2007;3:95–101.PubMed
33. Xu LQ, Li NN, Chen JC, Fu GD, Kang E. Quaternized poly(2-(dimethylamino)ethyl methacrylate)-grafted agarose copolymers for multipurpose antibacterial applications. RSC Adv. 2015;5:61742–51.
34. Qi L, Xu Z, Jiang X, Hu C, Zou X. Preparation and antibacterial activity of chitosan nanoparticles. Carbohydr Res. 2004;339:2693–700.
35. Colles FM, Cain RJ, Nickson T, Smith AL, Roberts SJ, Maiden MCJ, et al. Monitoring chicken flock behaviour provides early warning of infection by human pathogen Campylobacter. Proc R Soc B. 2015;283:20152323.
36. Zhao J, Harper AF, Estienne MJ, Webb KE, McElroy AP, Denbow DM. Growth performance and intestinal morphology responses in early weaned pigs to supplementation of antibiotic-free diets with an organic copper complex and spray-dried plasma protein in sanitary and nonsanitary environments. J Anim Sci. 2007;85:1302–10.
37. Solaiman SG, Maloney MA, Qureshi MA, Davis G, Andrea GD. Effects of high copper supplements on performance, health, plasma copper and enzymes in goats. Small Ruminant Res. 2001;41:127–39.

38. Gonzales-Eguia A, Fu C, Lu F, Lien T. Effects of nanocopper on copper availability and nutrients digestibility, growth performance and serum traits of piglets. Livest Sci. 2009;126:122–9.
39. Mazel D, Davies J. Antibiotic resistance in microbes. Cell Mol Life Sci. 1999;56:742–54.
40. Kalita S, Kandimalla R, Sharma KK, Kataki AC, Deka M, Kotoky J. Amoxicillin functionalized gold nanoparticles reverts MRSA resistance. Mater Sci Eng C. 2016;61:720–7.
41. Turos E, Reddy GSK, Greenhalgh K, Ramaraju P, Abeylath SC, Jang S, et al. Penicillin-bound polyacrylate nanoparticles: restoring the activity of B-lactam antibiotics against MRSA. Bioorg Med Chem Lett. 2007;17:3468–72.View ArticlePubMedPubMed Central
42. Greenhalgh K, Turos E. In vivo studies of polyacrylate nanoparticle emulsions for topical and systemic applications. Nanomedicine. 2008;5:46–54.PubMed
43. Ghosh D, Pramanik A, Sikdar N, Ghosh SK, Pramanik P. Amelioration studies on optimization of low molecular weight chitosan nanoparticle preparation, characterization with potassium per sulfate and silver nitrate combined action with aid of drug delivery to tetracycline resistant bacteria. Int J Pharm Sci Drug Res. 2010;2:247–53.
44. Rosilo H, McKee JR, Kontturi E, Koho T, Hytonen VP, Ikkala O, et al. Cationic polymer brush-modified cellulose nanocrystals for high-affinity virus binding. Nanoscale. 2014;6:11871–81.
45. Moreau E, Domurado M, Chapon P, Vert M, Domurado D. Biocompatibility of polycations: In vitro agglutination and lysis of red blood cells and in vivo toxicity. J Drug Target. 2002;10:161–73.
46. Zhang H, Wang D, Butler R, Campbell NL, Long J, Tan B, et al. Formation and enhanced biocidal activity of water-dispersable organic nanoparticles. Nat Nanotechnol. 2008;3:506–11.
47. Wang MQ, Xu ZR. Effect of chromium nanoparticle on growth performance, carcass characteristics, pork quality and tissue chromium in finishing pigs. J Anim Sci. 2004;17:1118–22.
48. Wang MQ, Wang C, Li H, Du YJ, Tao WJ, Ye SS, et al. Effects of chromium-loaded chitosan nanoparticles on growth, blood metabolites, immune traits and tissue chromium in finishing pigs. Biol Trace Elem Res. 2012;149:197–203.
49. Wang MQ, Wang C, Du YJ, Li H, Tao WJ, Ye SS, et al. Effects of chromium-loaded chitosan nanoparticles on growth, carcass characteristics, pork quality, and lipid metabolism in finishing pigs. Livest Sci. 2014;161:123–9.
50. Yang Z, Liu ZW, Allaker RP, Reip P, Oxford J, Ahmad Z, et al. A review of nanoparticle functionality and toxicity on the central nervous system. J R Soc Interface. 2010;7:S411–22.View ArticlePubMedPubMed Central
51. Zha LY, Zeng JW, Chu XW, Mao LM, Luo HJ. Efficacy of trivalent chromium on growth performance, carcass characteristics and tissue chromium in heat-stressed broiler chicks. J Sci Food Agric. 2009;89:1782–6.
52. Sirirat N, Lu JJ, Hung ATY, Lien TF. Effect of different levels of nanoparticles chromium picolinate supplementation on performance, egg quality, mineral retention, and tissues minerals accumulation in layer chickens. J Agric Sci. 2013;5:150–9.
53. Chauke N, Siebrits FK. Evaluation of silver nanoparticles as a possible coccidiostat in broiler production. S Afr J Anim Sci. 2012;42:493–7.
54. Panea B, Ripoll G, Gonzalez J, Fernandez-Cuello A, Alberti P. Effect of nanocomposite packaging containing different proportions of ZnO and Ag on chicken breast meat quality. J Food Eng. 2014;123:104–12.
55. Han C, Qi CM, Zhao BK, Cao J, Xie SY, Wang SL, et al. Hydrogenated castor oil nanoparticles as carriers for the subcutaneous administration of tilmicosin: in vitro and in vivo studies. J Wet Pharmacol Therap. 2009;32:116–23.

56. Wang XF, Zhang SL, Zhu LY, Xie SY, Dong Z, Wang Y, et al. Enhancement of antibacterial activity of tilmicosin against Staphylococcus aureus by solid lipid nanoparticles in vitro and in vivo. Vet J. 2012;191:115–20.
57. Sung YJ, Suk HJ, Sung HY, Li T, Poo H, Kim MG. Novel antibody/gold nanoparticle/magnetic nanoparticle nanocomposites for immunomagnetic separation and rapid colorimetric detection of Staphylococcus aureus in milk. Biosens Bioelectron. 2013;43:432–9.
58. Wang JJ, Liu BH, Hsu YT, Yu FY. Sensitive competitive direct enzyme-linked immunosorbent assay and gold nanoparticle immunochromatographic strip for detecting aflatoxin M1 in milk. Food Control. 2011;22:964–9.
59. Lee YK, Ahn SI, Chang YH, Kwak HS. Physiochemical and sensory properties of milk supplemented with dispersible nanopowdered oyster shell during storage. J Dairy Sci. 2015;98:5841–9.
60. Soenen SJH, Himmelreich U, Nuytten N, Pisanic TR, Ferrari A, De Cuyper M. Intracellular nanoparticle coating stability determines nanoparticle diagnostics efficacy and cell functionality. Small. 2010;6:2136–45.
61. Ajmal M, Yunus U, Matin A, Haq NU. Synthesis, characterization and in vitro evaluation of methotrexate conjugated fluorescent carbon nanoparticles as drug delivery system for human lung cancer targeting. J Photochem Photobiol B. 2015;153:111–20.
62. Croissant J, Chaix A, Mongin O, Wang M, Clement S, Raehm L, et al. Two-photon-triggered drug delivery via fluorescent nanovalves. Small. 2014;10:1752–5.
63. Craighead H. Future lab-on-a-chip technologies for interrogating individual molecules. Nature. 2006;442:387–93.
64. Tian T, Li J, Song Y, Zhou L, Zhu Z, Yang CJ. Distance-based microfluidic quantitative detection methods for point-of-care testing. RSC Adv. 2016.
65. Feng Y, Li D, Wang Q, Wang S, Meng X, Shao Z, et al. A carbazole-based mitochondria-targeted two-photon fluorescent probe for gold ions and its application in living cell imaging. Sensor Actuat B-Chem. 2016;225:572–8.
66. Sun L, Pitto-Barry A, Kirby N, Schiller TL, Sanchez AM, Dyson MA, et al. Structural reorganization of cylindrical nanoparticles triggered by polylactide stereocomplexation. Nat Commun. 2014;5:5746–54.
67. Li J, Yao P. Self-assembly of ibuprofen and bovine serum albumin-dextran conjugates leading to effective loading of the drug. Langmuir. 2009;25:6385–91.
68. Feugang JM, Youngblood RC, Greene JM, Fahad AS, Monroe WA, Willard ST, et al. Application of quantum dot nanoparticles for potential non-invasive bio-imaging of mammalian spermatozoa. J Nanobiotechnol. 2012;10:45–52.
69. Long JZ, Lackan CS, Hadjantonakis AK. Genetic and spectrally distinct in vivo imaging: embryonic stem cells and mice with widespread expression of a monomeric red fluorescent protein. BMC Biotechnol. 2005;5:1–11.
70. Druart X, Cognie J, Baril G, Clement F, Dacheux JL, Gatti JL. In vivo imaging of in situ motility of fresh and liquid stored ram spermatozoa in the ewe gential tract. Reproduction. 2009;138:45–53.
71. Feugang JM, Youngblood RC, Greene JM, Willard ST, Ryan PL. Self-illuminating quantum dots for non-invasive bioluminescence imaging of mammalian gametes. J Nanobiotechnol. 2015;13:1–16.
72. Hasuwa H, Muro Y, Ikawa M, Kato N, Tsujimoto Y, Okabe M. Transgenic mouse sperm that have green acrosome and red mitochondria allow visualization of sperm and their acrosome reaction in vivo. Exp Anim. 2010;59:105–7.
73. Vasquez ES, Feugang JM, Willard ST, Ryan PL, Walters KB. Bioluminescent magnetic nanoparticles as potential imaging agents for mammalian spermatozoa. J Nanobiotechnol. 2016;14:1–9.
74. Lin G, Ding Z, Hu R, Wang X, Chen Q, Zhu X, et al. Cytotoxicity and immune response of CdSe/ZnS quantum dots towards a murine macrophage cell line. RSC Adv. 2014;4:5792–7.

75. Odhiambo JF, DeJarnette JM, Geary TW, Kennedy CE, Suarez SS, Sutovsky M, et al. Increased conception rates in beef cattle inseminated with nanopurified bull semen. Biol Reprod. 2014;91:1–10.
76. Petruska P, Capcarova M, Sutovsky P. Antioxidant supplementation and purification of semen for improved artificial insemination in livestock species. Turk J Vet Anim Sci. 2014;38:643–52.
77. Bryla M, Trzcinska M. Quality and fertilizing capacity of boar spermatozoa during liquid storage in extender supplemented with different antibiotics. Anim Reprod Sci. 2015;163:157–63.
78. Hargreaves CA, Rogers S, Hills F, Rahman F, Howell RJS, Homa ST. Effects of co-trimoxazole, erythromycin, amoxycillin, tetracycline and chloroquine on sperm function in vitro. Hum Reprod. 1998;13:1878–86.
79. Akandi A, Ugwu SO, Machebe NS. Survivability of boar sperm stored under room temperature in extenders containing some natural products. Open Access Anim Physiol. 2015;7:57–64.
80. Sariozkan S, Bucak MN, Tuncer PB, Buyukleblebici S, Eken A, Akay C. Influence of fetuin and hyaluronan on the post-thaw quality and fertilizing ability of Holstein bull semen. Cryobiology. 2015;71:119–24.
81. Barkalina N, Jones C, Kashir J, Coote S, Huang X, Morrison R, et al. Effects of mesoporous silica nanoparticles upon the function of mammalian sperm in vitro. Nanomedicine. 2014;10:859–70.PubMed
82. Agarwal S, Zhang Y, Maji S, Greiner A. PDMAEMA based gene delivery materials. Pharmaceut Res. 2007;24:1590–8.
83. Barkhordari A, Hekmatimoghaddan S, Jebali A, Khalili MA, Talebi A, Noorani M. Effect of zinc oxide nanoparticles on viability of human spermatozoa. Iran J Reprod Med. 2013;11:767–71.
84. Pawar K, Kaul G. Toxicity of titanium oxide nanoparticles causes functionality and DNA damage in buffalo (Bubalus bubalis) sperm in vitro. Toxicol Ind Health. 2012;
85. Hu S, Hsieh Y. Synthesis of surface bound silver nanoparticles on cellulose fibers using lignin as multi-functional agent. Carbohydr Polym. 2015;131:134–41.
86. Napavichayanun S, Amornsudthiwat P, Pienpinijtham P, Aramwit P. Interaction and effectiveness of antimicrobials along with healing-promoting agents in a novel biocellulose wound dressing. Mater Sci Eng C. 2015;55:95–104.

CHAPTER 16

Net Protein and Metabolizable Protein Requirements for Maintenance and Growth of Early-weaned Dorper Crossbred Male Lambs

Tao Ma[1] , Kaidong Deng[2], Yan Tu [1], Naifeng Zhang [1], Bingwen Si [1], Guishan Xu1,[3] and Qiyu Diao[1]*

[1] Feed Research Institute/Key laboratory of Feed Biotechnology of the Ministry of Agriculture, Chinese Academy of Agricultural Sciences, Beijing, China.

[2] College of Animal Science, Jinling Institute of Technology, Nanjing, Jiangsu, China.

[3] College of Animal Science, Tarim University, Alar, Xinjiang, China.

ABSTRACT

Background

Dorper is an important breed for meat purpose and widely used in the livestock industry of the world. However, the protein requirement of Dorper crossbred has not been investigated. The current paper reports the net protein (NP) and metabolizable protein (MP) requirements of Dorper crossbred ram lambs from 20 to 35 kg BW.

Methods

Thirty-five Dorper × thin-tailed Han crossbred lambs weaned at approximately 50 d of age (20.3 ± 2.15 kg of BW) were used. Seven lambs of 25 kg BW were slaughtered as the baseline animals at the start of the trial. An intermediate group of seven randomly selected lambs fed ad libitum was slaughtered at 28.6 kg BW. The remaining 21 lambs were randomly divided into three levels of dry matter intake: ad libitum or 70% or 40% of ad libitum intake. Those lambs were slaughtered when the lambs fed ad libitum reached 35 kg BW. Total body N and N retention were measured.

Results

The daily NP and MP requirements for maintenance were 1.89 and 4.52 g/kg metabolic shrunk BW ($SBW^{0.75}$). The partial efficiency of MP utilization for maintenance was 0.42. The NP requirement for growth ranged from 12.1 to 43.5 g/d, for the lambs gaining 100 to 350 g/d, and the partial efficiency of MP utilization for growth was 0.86.

Conclusions

The NP and MP requirements for the maintenance and growth of Dorper crossbred male lambs were lower than the recommendations of American and British nutritional systems.

Keywords: Growth Lamb Maintenance Metabolizable protein Net protein

BACKGROUND

In the intensive livestock industry, protein is commonly the most expensive feed component and therefore, it is necessary to have a precise understanding of protein requirement of livestock not only to ensure farm profitability, but also to help reduce nitrogen (N) emission to the environments [1]. Modern feeding systems, such as Agricultural and Food Research Council (AFRC) [2], Commonwealth Scientific and Industrial Research Organisation (CSIRO) [3], and National Research Council (NRC)

[4], have reported protein and other nutrient requirements of sheep, which are widely adopted for diet formulation in the world.

The Dorper is a popular sire breed for meat production in both South Africa and United States [5]. In recent years, Dorper sheep have been imported to China to improve the growth performance and carcass traits of indigenous breed, among which the thin-tailed Han sheep is one of the most famous native breeds with high prolificacy, as it carries mutations in both the *BMPR-1B* and *BMP15* genes [6]. Thus, the Dorper × thin-tailed Han crossbreed has become one of the dominant breeds for lamb meat production.

Our research team conducted a series of studies on the nutrient requirements (energy, protein, and minerals) of Dorper × thin-tailed Han crossbred sheep [7, 8, 9]. In this paper, we reported the protein requirements of male lambs after weaning ranging from 20 to 35 kg BW by the comparative slaughter technique.

METHODS

The current research was conducted from June to September 2011 at the Experimental Station of the Chinese Academy of Agricultural Sciences (CAAS), Nankou (40°22′N, 116°1′E), Beijing, China. The animals were kept in an enclosed animal house and the mean minimum and maximum room temperatures observed during the experimental period were 15.5 and 26.5 °C (average 21.0 °C), respectively.

Animals, treatments, and experimental procedure

Thirty-five Dorper × thin-tailed Han crossbred male lambs weaned at approximately 50 d of age with 20.4 ± 2.15 kg BW were used in a completely randomized design to measure protein requirements for maintenance and growth. The experimental diet was formulated according to NRC [4] with a concentrate-to-forage ratio of 44:56 on a dry matter (DM) basis (Table 1). The lambs with ad libitum intake were fed once daily at 0800 h and allowed 10% of refusals. The amount of feed provided to the restricted feed intake groups was adjusted daily based on the average DM intake (DMI) of the ad libitum group from the previous day. Feed and refusals were sampled daily and frozen at −20 °C until analysis.

Table 1. Ingredient and chemical compositions of the pelleted mixture diet

Items	Value, %DM
Ingredients, DM basis	
Milled Chinese wildrye hay	55.0
Cracked corn grain	29.4
Soybean meal	14.0
Dicalcium phosphate	0.86
Salt	0.50
Mineral/vitamin premix[a]	0.24
Chemical composition	
ME, MJ/kg DM	8.89
DM, % as fed	95.5
CP, % of DM	11.9
EE, % of DM	2.71
Ash, % of DM	6.32
NDF, % of DM	40.9
ADF, % of DM	15.2
Calcium, % of DM	0.68
Phosphorus, % of DM	0.33

[a]Manufactured by Precision Animal Nutrition Research Centre, Beijing, China. The premix contained (per kg): 22.1 g Fe, 13.0 g Cu, 30.2 g Mn, 77.2 g Zn, 19.2 g Se, 53.5 g I, 9.10 g Co, 56.0 g vitamin A, 18.0 g vitamin D_3, and 170 g vitamin E

A comparative slaughter trial was conducted as described previously [9]. Briefly, the initial body composition was measured on seven lambs slaughtered at 20 kg BW (baseline group). An intermediate slaughter group with seven randomly selected lambs fed ad libitum were slaughtered at 28.6 kg BW. The remaining 21 lambs were randomly assigned to three levels of DMI: ad libitum or 70% or 40% of ad libitum intake. Thus, the lambs were pair-fed in seven slaughter groups, with each group consisting of one lamb from each level of intake. When the lambs fed ad libitum of each slaughter group reached 35 kg BW, all lambs within a slaughter group were fasted and slaughtered. Seven lambs in the baseline and intermediate group were slaughtered in one day, respectively, while the remaining 21 lambs were slaughtered in three consecutive days. All lambs were slaughtered by exsanguination after stunning by CO_2 inhalation. Blood, carcass, head, feet, hide, wool, viscera, and adipose tissue removed from the internal organs were weighed. The empty body weight (EBW) was calculated by subtracting the weight of the digestive tract contents from the shrunk BW (SBW), which was measured as BW after a 16-h fast of feed and water. Carcasses and heads were split longitudinally into two identical halves and the muscle, bone, and fat were dissected from the right-half carcass, head, and feet, while the whole hide and whole viscera were ground and homogenized separately and frozen at −20 °C until analysis. Wool was clipped with electrical clippers after slaughter, and subsamples were collected and stored at 4 °C.

Chemical analyses

Feed and orts were oven-dried at 55 °C for 72 h, ground to pass through a 1-mm screen for analyzing DM (method 930.15) [10], ash (method 924.05) [10], ether extract (EE) (method 920.85) [10], and nitrogen (N) [11]. The gross energy (GE) of feed was measured with a bomb calorimeter (C200, IKA Works Inc., Staufen, Germany). The neutral detergent fiber (NDF) and acid detergent fiber (ADF) of feed were measured according to Van Soest et al. [12] and Goering and Van Soest [13], respectively. Calcium of feed was analyzed using an atomic absorption spectrophotometer (M9W-700, Perkin-Elmer Corp., Norwalk, CT, USA) (method 968.08) [10]. The total P of feed was analyzed by the molybdovanadate colorimetric method (method 965.17) [10] using a spectrophotometer (UV-6100, Mapada Instruments Co., Ltd., Shanghai, China).

The DM of all body components except wool was determined following lyophilization for 72 h to constant weight. The samples were then analyzed for N as described above. Wool samples were cut into 2-mm pieces with scissors and analyzed for N as described above.

Data calculations

Metabolizable protein supply: The ratio of metabolizable protein (MP) to OM intake reported in a previous in vivo study (i.e., 69.4, 88.9, and 102.4 g MP/kg OM intake for ad libitum, 70% and 50% of the ad libitum intake, respectively) [14] with Dorper × thin-tailed Han crossbred sheep subjected to the same feeding regime as the present study was used to calculate the individual MP intake (MPI) by the ram lambs.

Prediction of the initial body N content and N retention (NR): The NR in the body of the lambs in the comparative slaughter trial was calculated as the difference between the final and initial body N content. The initial body N content of each animal was calculated from its initial EBW using a regression equation developed from the relationship between the body N content and EBW of the baseline animals ($R^2 = 0.92$, root mean square error [RMSE] = 0.011, n = 7, $P < 0.001$): $\log_{10}$ empty body N, kg = 1.536 (±0.174) + [0.957 (±0.144) × $\log_{10}$ EBW, kg]. The initial SBW of each animal was computed from its initial BW ($R^2 = 0.97$, RMSE = 0.280, n = 7): SBW, kg = 0.571 (±1.457) + [0.915 (±0.069) × BW, kg], and the initial EBW of each animal was computed from its initial BW ($R^2 = 0.86$, RMSE = 0.429, n = 7): EBW, kg = 2.336 (±2.227) + [0.695 (±0.105) × BW, kg].

Protein requirements for maintenance: A linear regression of daily NR on daily N intake (NI) was used to calculate the net protein (NP) requirement for maintenance. The intercept of the regression was assumed the endogenous and metabolic losses of N multiplied by the factor 6.25, which is assumed the maintenance requirement for NP (NP_m, g/kg$^{0.75}$ SBW). The MP required for maintenance (MP_m, g/kg$^{0.75}$ SBW) was then estimated by regressing the NR on MPI and extrapolating the linear regression to zero NR. The efficiency of MP use for maintenance (k_{pm}) was computed as NP_m/MP_m.

Protein requirements for growth: The NP requirements for body weight gain (NP_g) were calculated as the difference between body protein content at different intervals. For example, the NP_g of a lamb with 20 kg SBW and 250 g of average daily gain (ADG) was computed as the difference between body protein contents at 20.25 and 20 kg SBW. Body protein contents were predicted from EBW using an allometric equation according to ARC (1980): $\log_{10}$ protein, kg = a + [b × $\log_{10}$ EBW, kg]. To estimate the partial efficiencies of MP use for body weight gain (k_{pg}), a regression model through the origin was used to partition the utilization of MPI above maintenance for body protein retention as follows: $MPI_g = b \times RP$, where MPI_g (g/kg$^{0.75}$ of SBW) is the MPI above maintenance calculated as the difference between the total MPI and MP_m, RP (g/kg$^{0.75}$ of SBW) is the daily retention of body protein, and the estimated parameter b is the amount of MP (g) required to retain 1 g of protein, and its inverse was assumed to be k_{pg}.

Statistical analyses

The data were analyzed as a completely randomized design using the SAS statistical software package (version 9.1; SAS Institute, Inc., Cary, NC). Intake, body composition, and growth rate were analyzed using a one-way ANOVA. Pairwise comparisons of means were performed by Tukey's multiple range tests once the significance of the treatment effect was declared at $P < 0.05$. The statistical model is: $y_{ij} = \mu + \alpha_i + \varepsilon_{ij}$, where y_{ij} = dependent variable, μ = overall mean of *Yij*, α_i = effect of the diet (i = 1 to 3), and ε_{ij} = error contribution. Linear regressions were conducted with a GLM, and observations with a studentized residual > 2.5 or < −2.5 were considered outliers. The assumptions of the models, in terms of homoscedasticity, independency, and normality of errors, were examined by plotting residuals against the predicted values.

RESULTS

The intake of OM and N increased with feeding level ($P < 0.001$; Table 2). The lambs fed with ad libitum intake retained greater N than those with either level of restricted feed intake ($P < 0.001$). The lambs fed at 40% of the ad libitum intake had a lower MPI than those in the other two groups ($P < 0.001$), and the MPI did not differ between lambs fed at ad libitum and 70% of the ad libitum intake ($P = 0.286$).

Table 2. Daily protein intake of Dorper × thin-tailed Han crossbred ram lambs at ad libitum (AL) or restricted to 70% or 40% of AL intake[1]

	Level of feed intake[2]			SEM	P[4]
Items	AL	70%	40%		
ADG, g	324.2^a	189.1^b	37.6^c	19.3	<0.001
Initial SBW, kg	19.3	19.5	18.5	0.22	0.271
Final SBW, kg	29.9^a	27.8^a	20.4^b	0.91	<0.001
OM intake, g/d	107.6^a	84.2^b	53.3^c	4.37	<0.001
N intake, g/(kg SBW$^{0.75}$ · d)	1.83^a	1.43^b	1.01^c	0.08	<0.001
N retention, g/(kg SBW$^{0.75}$ · d)	0.47^a	0.29^b	0.11^c	0.03	<0.001
MP intake, g/(kg SBW$^{0.75}$ · d)[3]	7.47^a	7.49^a	5.46^c	0.18	<0.001

[1]ADG: average daily gain; SBW: shrunk body weight; MP: metabolizable protein
[2]Ad libitum or restricted to 70% or 40% of ad libitum of an identical diet
[3]Calculated from the ratio of MP supply to OM intake reported by Ma et al. [14]
[4] a,b,c Means bearing different superscripts differ ($P < 0.05$)

Figure 1 shows the linear relationship between NR and NI: NR, g/(kg $SBW^{0.75}\cdot d) = -0.303$ $(\pm 0.046) + [0.422$ $(\pm 0.040) \times$ N intake, g/(kg $SBW^{0.75}\cdot d)]$ ($R^2 = 0.89$, RMSE = 0.054, n = 28, $P < 0.001$). The endogenous and metabolic loss of N, estimated as the intercept of the linear regression, was 303 ± 46 mg/$kg^{0.75}$ SBW, which corresponds to an NP_m of 1.89 ± 0.29 g/$kg^{0.75}$ SBW.

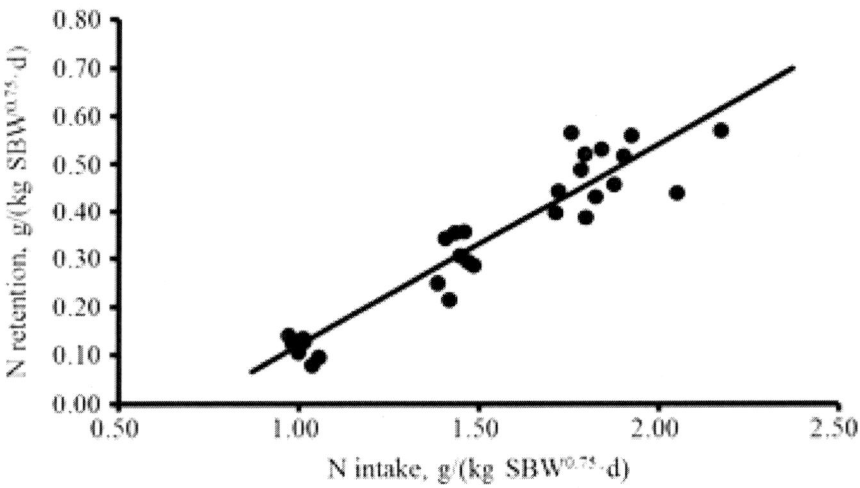

Figure 1. Relationship between N retention (NR) and N intake (NI) of Dorper × thin-tailed Han crossbred male lambs from 20 to 35 kg of BW. NR, g/kg $SBW^{0.75} = -0.303$ $(\pm 0.046) + [0.422$ $(\pm 0.040) \times$ NI, g/kg of $SBW^{0.75}]$, $R^2 = 0.89$, RMSE = 0.054, n = 28

Figure 2 shows the linear relationship between NR and MPI: NR, g/(kg $SBW^{0.75}\cdot d) = -$ 0.628 $(\pm 0.138) + [0.139$ $(\pm 0.020) \times$ MPI, g/(kg $SBW^{0.75}\cdot d)]$, $R^2 = 0.70$, RMSE = 0.096, n = 28, $P < 0.001$). The MP required for maintenance by extrapolating the linear regression to zero N retention was 4.52 g/$kg^{0.75}$ SBW. Consequently, the k_{pm} was 0.42 for Dorper × thin-tailed Han crossbred male lambs from 20 to 35 kg of BW.

The partial efficiencies of MP use for body weight gain were estimated using a regression model and assuming that the MPI above maintenance (MPI_g) is partially recovered as body protein for growth (PR_g, g/$kg^{0.75}$ of SBW). The regression equation was: PR_g, g/($kg^{0.75}$ $SBW\cdot d) = 0.002(\pm 0.052) + [0.864 \pm 0.123) \times MPI_g$, g/($kg^{0.75}$ of $SBW\cdot d)]$ ($R^2 = 0.70$, n = 28, RMSE = 0.081). The slope of the regression equation (0.86) represents k_{pg}.

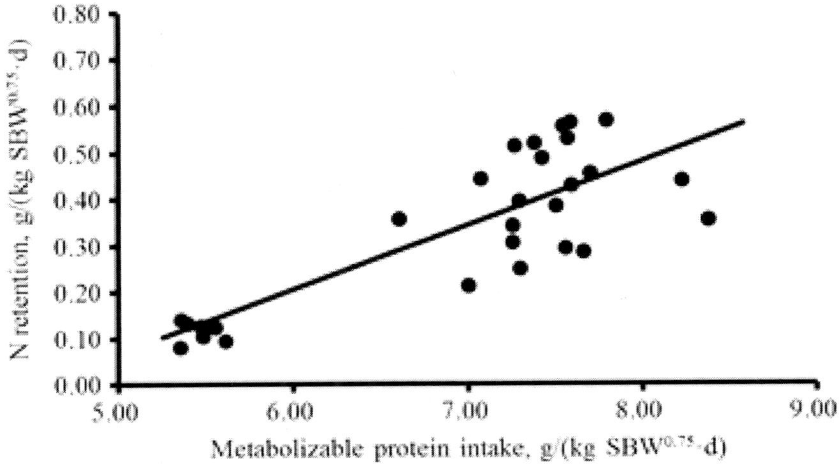

Figure 2. Relationship between N retention (NR) and metabolizable protein intake (MPI) of Dorper × thin-tailed Han crossbred male lambs from 20 to 35 kg of BW. NR, g/kg $SBW^{0.75}$ = − 0.628 (±0.138) + [0.139 (±0.020) × MPI, g/kg $SBW^{0.75}$], R^2 = 0.70, RMSE = 0.096, n = 28

The allometric equation between body protein and EBW (R^2 = 0.96; RMSE = 0.012; n = 20) of the ram lambs with free consumption of feed was proposed as $\log_{10}$ protein, g = 2.385 (±0.072) + [0.922 (±0.054) × $\log_{10}$ EBW, kg]. The NP_g and MP for growth (MP_g) were therefore calculated accordingly (Tables 3 and 4).

Table 3. Requirements of net protein (NP) for growth (g/d) of Dorper × thin-tailed Han crossbred male lambs from 20 to 35 kg of body weight (BW)

ADG, g/d	BW, kg			
	20	25	30	35
100	12.4	12.3	12.2	12.1
200	24.9	24.6	24.4	24.2
300	37.3	36.9	36.6	36.3
350	43.5	43.0	42.7	42.3

ADG: average daily gain

Table 4. Requirements of metabolizable protein (MP) for growth (g/d) of Dorper × thin-tailed Han crossbred male lambs from 20 to 35 kg of body weight (BW)

ADG, g/d	BW, kg			
	20	25	30	35
100	14.4	14.3	14.2	14.1
200	29.0	28.6	28.4	28.1
300	43.4	42.9	42.6	42.2
350	50.6	50.0	49.7	49.2

ADG: average daily gain

DISCUSSION

The NP_m is the quantity of protein to sustain tissue proteins by counterbalancing the inevitable losses of urinary, fecal, and dermal N [3]. AFRC [15] suggested daily endogenous and metabolic N losses of 350 mg/kg $BW^{0.75}$ for lambs nourished by intra-gastric infusions. However, this method might overestimate the endogenous N requirement due to the lack of conservation of protein by the microbial capture of N. CSIRO [3] suggested the NP_m of the sheep is the sum of the endogenous urinary loss (0.147 × BW + 3.375) and fecal loss (15.2 g/kg DM intake). Thus, the NP_m for a lamb of 28 kg consuming 1.02 kg of DM daily (this is the average DM intake of the 28 lambs used in the current study) is about 23.0 g, which is 10% higher than the current value (1.72 g/kg $BW^{0.75}$). In the current study, the NP_m calculated by partial regression NR on NI (1.72 g/kg $BW^{0.75}$ or 1.89 g/kg $SBW^{0.75}$) is slightly higher than that of the Ile de France (1.56 g/kg $SBW^{0.75}$) [16] and Texel crossbreeds (1.52 g/kg $SBW^{0.75}$) [17], but lower than that of Morada Nova (1.83 g/kg $BW^{0.75}$) [18] male lambs measured using the same methods. The lambs fed at the maintenance level had similar NR (around 0.1 to 0.2 g/kg of $SBW^{0.75}$) among all these studies. However, when fed higher than the maintenance level, Dorper crossbred had a relatively higher NR (0.25 to 0.65 g/kg of $SBW^{0.75}$) than the Ile de France [16] and Texel crossbreeds [17], but a lower NR than Morada Nova lambs (0.25 to 1.0 g/kg of $BW^{0.75}$) [18]. Therefore, the variations in NP_m can be explained by the differential utilization efficiency of protein or amino acids by the tissues during the growth of lambs. On the other hand, the above studies were all conducted in a tropical or sub-tropical zone with high humidity, while the current study was conducted in a warm temperate zone where the weather is always dry during summer. As reviewed by Marai et al. [19], both temperature and humidity could have influence on nutrient digestibility and degradation in the rumen; thus, the experimental condition could be another important factor that contributes to the differences in NP_m apart from methods or animal breeds. We concluded that 1.69 g/kg $BW^{0.75}$ or 1.81 g/kg $SBW^{0.75}$ is suitable for the NP_m of Dorper crossbred male lambs from 20 to 35 kg of BW.

The MP_m obtained in the current study was 4.52 g/kg $SBW^{0.75}$. When scaled to BW, our value (4.37 g/kg $BW^{0.75}$) was close to Liu et al. [20], who found an MP_m of 4.41 g/kg $BW^{0.75}$ from a linear multiple regression of MP requirements against the live weight, live weight gain, and wool growth of sheep (n = 213) in a feeding study. Nevertheless, our result is greater than that suggested by AFRC (2.19 g/kg $BW^{0.75}$) [2], INRA (2.50 g/kg $BW^{0.75}$) [21], or NRC (3.72 g/kg $BW^{0.75}$) [22]. In a more recent study, where a comparative slaughter trial was also used, a MP_m of 2.31 g/kg $SBW^{0.75}$ was observed in Texel crossbred lambs [17]. The methods adopted by UK [2], Australia [3], USA [4], and France [21] were all based on a common overall model, although requirements were expressed in different terms. In the current study, MP was calculated based on the method reported by Ma et al. [14], who conducted an in vivo study and measured MP using sheep with ruminal and duodenal cannula. Therefore, the discrepancy in the calculation of MP is inevitably associated with the methodologies adopted. As there is still a lack of simple and robust methods for calculating MP, this area requires further investigation and examination.

The NP_g values (12.4, 24.9, and 37.3 g/d) determined in the current study are extensively lower than those of early maturing growing lambs (23.5, 30.5, and 50.0 g/d) of 20 kg BW gaining 100, 200, and 300 g/d recommended by NRC [4], assuming a k_{pg} of 0.50. AFRC [2] used two equations proposed by ARC [23], $NP_f(g/d) = ADG \times (160.4 - 1.22 \times BW + 0.0105 \times BW^2)$ and NP_w (g/d) = 3 + 0.1 × NP_f, where NP_f is the NP requirement for fleece-free body growth and NP_w is the NP requirement for wool growth, to predict the protein requirement for the growth of body and fleece in lambs, respectively. Using those equations, the NP_g ($NP_f + NP_w$) was approximately 26% to 49% higher (range from 17.4 to 57.0 g/d) than values determined in the current study. Therefore, caution should be taken before applying certain evaluation systems to avoid the overestimation of NP_g of Dorper crossbred lambs. By using the same method, our NP_g results were 26% higher than those of the Texel crossbreed [17] growing from 20 to 35 kg of SBW gaining 100 and 200 g/d, but 20% lower than that of Morada Nova lambs [18] growing from 20 to 30 kg of BW gaining 100, 200, and 300 g/d, respectively. Although the ME (0.76 vs. 0.73 MJ/kg $SBW^{0.75}$), NI (2.12 vs. 2.20 g/kg $SBW^{0.75}$), and N retention/N intake (26.1% vs. 25.9%)

were close, the ADG of ad libitum groups were much higher for the Dorper crossbred (324 g) than Texel crossbred (245 g) lambs reported by Galvani et al. [17]. Many factors could be associated with such discrepancies, including breed, physiological stage, and experimental conditions.

The partial efficiency of use of MP_m for NP_m (k_{pm}) was calculated to be 0.42 in the present study. This value is lower than previously adopted 1.0 by AFRC [15] or 0.67 by CSIRO [3], as well as lower than that of Texel crossbred lambs (0.66) [17]. The partial efficiency of use of MP_g for NP_g (k_{pg}) obtained in the current study (0.86) was greater than that adopted by AFRC (0.59) [2], CSIRO (0.70) [3] and by Galvani et al. [17] in Texel crossbred lambs (0.71). Those discrepancies could be attributed to animal factors, including breed, maturity, and physiological status. The method for calculating or determining MP may be another factor contributing to the variability of the efficiency of MP use. NRC [4] suggested a simple equation, $MP = CP \times (0.64 + 0.16 \times \% \ UDP)/100$, where UDP is undegraded dietary protein, to convert CP to MP. By using this equation, MP/CP was 0.72 in all treatments, and k_{pm} and k_{pg} were 0.59 (1.89/3.22) and 0.58, respectively. Our previous study showed that a decreased feed intake could increase total-tract N digestibility without affecting ruminal N degradability [24], and the increased duodenal N digestibility could be due to the prolonged gastric empty time. Thus, it could not be expected that MP/CP were identical under a different feeding level. In the current study, MP was calculated from OM intake based on the results of our previous study using 6-month-old Dorper × thin-tailed Han male lambs (41.3 ± 2.8 kg BW) with both ruminal and duodenal cannula fed at three different levels (ad libitum, 70%, and 50% of ad libitum) in which MP supply/OM intake (g/kg) was 69.4, 88.9, and 102.4, respectively. As there is still lack of a simple method for the calculation of MP, this could be a reasonable way to calculate MP in the current study. Nevertheless, considering the difference in animal physiology status (BW, age, and cannulation) between our previous and current studies, further study is still needed to examine the utilization efficiency of MP for both the maintenance and growth for early-weaning lambs.

CONCLUSIONS

In conclusion, the current study suggested that the protein requirements for the maintenance and growth of Dorper × thin-tailed Han early-weaned crossbred male lambs were lower than the recommendations of AFRC (1993) and NRC (2007).

ACKNOWLEDGEMENTS

We thank the staff (Y.L. Li, Y.F. Zhang, J. Liu, Y.G. Zhao, S.K. Ji, L.T. Zhang, and C. Lou) of Feed Research Institute of Chinese Academy of Agricultural Sciences for their technical assistance.

Funding

This research was supported by the earmarked fund for the China Agriculture Research System (CARS-39). This manuscript was a part of the 1st International Symposium on Young Ruminant (ISYR)-Physiology and Rearing Strategies jointly organised by Laboratory of Ruminant Physiology and Nutrition, Feed Research Institute, Chinese Academy of Agricultural Sciences, China, The Pennsylvania State University, US, Agri-Food and Biosciences Instute, UK, Beijing Dairy Industry Innovation Consortium of Agriculture Research System, Beijing Key Laboratory for Dairy Nutrition.

Availability of data and materials

Data sharing not applicable to this article as no datasets were generated or analyzed during the current study.

Authors' contributions

GX carried out the whole animal experiment, including sample collection and determination. TM participated in the statistical analysis and wrote the draft. KD helped to revise the manuscript. YT, NZ, and BS participated in the design and coordination of the study. QD conceived of the study. All authors read and approved the final manuscript.

Competing interest

The authors declare that they have no competing interest.

Consent for publication

Not applicable.

Ethics approval

The experimental protocol was approved by the CAAS Animal Ethical Committee, and humane animal care and handling procedures were followed throughout the experiment.

REFERENCES

1. Ma T, Xu GS, Deng KD, Ji SK, Tu Y, Zhang NF, et al. Energy requirements of early-weaned Dorper cross-bred female lambs. J Anim Physiol An N. 2016.
2. AFRC. Energy and protein requirements of ruminants. An advisory manual prepared by the Agricultural and Food Research Council Technical Committee on Responses to Nutrients. Wallingford: CAB International; 1993.
3. CSIRO. Nutrient requirements of domesticated ruminants. Collingwood: Commonwealth Scientific and Industrial Research Organisation Publishing; 2007.
4. NRC. Nutrient Requirements of Small Ruminants: Sheep, Goats, Cervids, and New World Camelids. Washington: National Academy Press; 2007.
5. Snowder GD, Duckett SK. Evaluation of the South African Dorper as a terminal sire breed for growth, carcass, and palatability characteristics. J Anim Sci. 2003;81(2):368–75.
6. Chu MX, Liu ZH, Jiao CL, He YQ, Fang L, Ye SC, et al. Mutations in *BMPR-IB* and *BMP-15* genes are associated with litter size in Small Tailed Han sheep (*Ovis aries*). J Anim Sci. 2007;85:598–603.
7. Deng KD, Diao QY, Jiang CG, Tu Y, Zhang NF, Liu J, et al. Energy requirements for maintenance and growth of Dorper crossbred ram lambs. Livest Sci. 2012;150:102–10.
8. Deng KD, Jiang CG, Tu Y, Zhang NF, Liu J, Ma T. Energy requirements of Dorper crossbred ewe lambs. J Anim Sci. 2014;92:2161–9.
9. Xu GS, Ma T, Ji SK, Deng KD, Tu Y, Jiang CG, et al. Energy requirements for maintenance and growth of early-weane d Dorper crossbred male lambs. Livest Sci. 2015;177:71–8.
10. AOAC. Official methods of analysis. 15th ed. Arlington: AOAC International; 1990.
11. Marshall CM, Walker AF. Comparison of a short method for Kjeldahl digestion using a trace of selenium as catalyst, with other methods. J Sci Food Agri. 1978;29:940–2.
12. Van Soest PJ, Robertson JB, Lewis BA. Methods for dietary fiber, neutral detergent fiber and non-starch polysaccharides in relation to animal nutrition. J Dairy Sci. 1991;74:3583–97.
13. Goering HG, Van Soest JP. Forage fiber analysis. Agricultural Handbook, vol. 379. Washington: UPSDA; 1970.

14. Ma T, Deng KD, Tu Y, Zhang NF, Jiang CG, Liu J, et al. Effect of feed intake on metabolizable protein supply in Dorper × thin-tailed Han crossbred lambs. Small Rumin Res. 2015;132:133–6.
15. AFRC. Technical committee on responses to nutrients, Report 9. Nutritive Requirements of Ruminant Animals: Protein. Nutrition Abstracts and Reviews. Series B. 1992;62:787–835.
16. Silva AMA, Silva Sobrinho AG, Trindade IACM, Resende KT, Bakke OA. Net requirements of protein and energy for maintenance of wool and hair lambs in a tropical region. Small Rumin Res. 2003;49:165–71.
17. Galvani DB, Pires CC, Kozloski GV, Sanchez LMB. Protein requirements of Texel crossbred lambs. Small Rumin Res. 2009;81:55–62.
18. Costa MRGF, Pereira ES, Silva AMA, Paulino PVR, Mizubuti IY, Pimentel PG, et al. Body composition and net energy and protein requirements of Morada Nova lambs. Small Rumin Res. 2013;114:206–13.
19. Marai IFM, El-Darawany AA, Fadiel A, Abdel-Hafez MAM. Physiological traits as affected by heat stress in sheep-a review. Small Rumin Res. 2007;71(1):1–12.
20. Liu SM, Smith TL, Karlsson LJE, Palmer DG, Besier RB. The costs for protein and energy requirements by nematode infection and resistance in Merino sheep. Livest Prod Sci. 2005;97(2):131–9.
21. INRA. Ruminant nutrition: Recommended allowances and feed tables. Paris: John Libbey & Co. Ltd; 1989.
22. NRC. Predicting feed intake of food-producing animals. Washington: National Academy Press; 1987.
23. ARC. The Nutrient Requirements of Ruminant Livestock. Slough: Commonwealth Agricultural Bureaux; 1980.
24. Ma T, Deng KD, Jiang CG, Tu Y, Zhang NF, Liu J, et al. The relationship between microbial N synthesis and urinary excretion of purine derivatives in Dorper × thin-tailed Han crossbred sheep. Small Rumin Res. 2013;112:49–55.

CHAPTER 17

Reproductive performance and gestational effort in relation to dietary fatty acids in guinea pigs

Matthias Nemeth[1*], Eva Millesi[1], Carina Siutz[1], Karl-Heinz Wagner[2], Ruth Quint[2] and Bernard Wallner[1,3]

[1] *Department of Behavioural Biology, University of Vienna, Althanstrasse 14, 1090 Vienna, Austria.*

[2] *Department of Nutritional Sciences, University of Vienna, Althanstrasse 14, 1090 Vienna, Austria.*

[3] *Department of Anthropology, University of Vienna, Althanstrasse 14, 1090 Vienna, Austria*

ABSTRACT

Background

Dietary saturated (SFAs) and polyunsaturated (PUFAs) fatty acids can highly affect reproductive functions by providing additional energy, modulating the biochemical properties of tissues, and hormone secretions. In precocial mammals such as domestic guinea pigs the offspring is born highly developed. Gestation might be the most critical reproductive period in this species and dietary fatty acids may profoundly influence the gestational effort. We therefore determined the hormonal status at conception, the reproductive success, and body mass changes during gestation in guinea pigs maintained on diets high in PUFAs or SFAs, or a control diet.

Results

The diets significantly affected the females' plasma fatty acid status at conception, while cortisol and estrogen levels did not differ among groups. SFA females exhibited a significantly lower body mass and litter size, while the individual birth mass of pups did not differ among groups and a general higher pup mortality rate in larger litters was diminished by PUFAs and SFAs. The gestational effort, determined by a mother's body mass gain during gestation, increased with total litter mass, whereas this increase was lowest in SFA and highest in PUFA individuals. The mother's body mass after parturition did not differ among groups and was positively affected by the total litter mass in PUFA females.

Conclusions

While SFAs reduce the litter size, but also the gestational effort as a consequence, PUFA supplementation may contribute to an adjustment of energy accumulations to the total litter mass, which may both favor a mother's body condition at parturition and perhaps increase the offspring survival at birth.

Keywords: Body mass Female reproduction Gestation Litter size Polyunsaturated fatty acid Saturated fatty acid Total litter mass

BACKGROUND

Reproduction represents the energetically most demanding life history stage in mammalian females. Adequate and balanced dietary intakes of specific macronutrients are of major importance to ensure an appropriate energy supply for maintaining reproductive performances [1]. In this context, dietary fatty acids have been suggested to play a major role in providing energy for an organism as well as by affecting hormone secretions and cell membrane functions in the central nervous system and the reproductive tract. The dietary fatty acid content and composition, particularly the amounts and ratios of specific types of fatty acids in the diet, can therefore directly modulate the physiology of reproduction and ultimately an individual's reproductive success [2, 3].

Among dietary fat types, omega-3 (n-3) and omega-6 (n-6) polyunsaturated fatty acids (PUFAs), including the essential α-linolenic acid (ALA, 18:3 n-3) and linoleic acid (LA, 18:2 n-6) and their long-chain metabolites eicosapentaenoic acid (EPA, 20:5 n-3), docosahexaenoic acid (DHA, 22:6 n-3), and arachidonic acid (AA, 20:4 n-6), can positively influence reproductive processes in females. Diets high in ALA or in the n-3 long-chain metabolites EPA and DHA have been shown to promote ovulation and increase the number of released ova in rats [4], and may further increase conception rates and reduce pregnancy losses in cows [5]. PUFAs in general, including ALA and LA, can also promote the prenatal development of rabbits and mice, resulting in increased body mass and improved physical condition at birth [6, 7]. Although less well studied, dietary intakes of non-essential saturated fatty acids (SFAs) can also improve reproductive performances by increasing the birth mass in rats [8] or the sex ratio at birth towards more male offspring in mice [9]. However, dietary SFAs may not only promote reproductive functions as PUFAs obviously do, since various effects on metabolic processes may be detrimental for an individual [10, 11]. The effects of dietary fatty acids may be simply explained by their high energy content and therefore increased energy allocation for reproductive functions and prenatal development, or by improving a mother's body condition and hormone secretion rates already at the time of conception [3].

By modulating litter size, sex ratio, and the offspring's birth mass, dietary fatty acids not only affect the reproductive output, but also a female's body mass change during pregnancy, reflecting the gestational effort. Especially in precocial mammalian species such as the domestic guinea pig (*Cavia aperea f. porcellus*) pups are born highly developed and gestation is basically characterized by a very high investment of mothers [12]. Larger litter sizes obviously cause higher gestational efforts in guinea pigs and are usually characterized by a lower birth mass of single pups and a higher rate of stillbirths compared to smaller litters [13, 14]. Although lactation in guinea pigs definitely represents an important and energetically demanding period, this may be less pronounced compared to altricial mammalian species. The prolonged gestation period can be assumed to be the energetically more demanding reproductive period in guinea pigs, especially as newborn pups are of relatively high body mass and the lactation period occurs to be rather short [15, 16]. Since mortality in term-born guinea pig pups is relatively low [17], probably due to the high developmental stage at birth, reproduction in guinea pigs can be considered as highly efficient and could probably be further promoted by supplementations with dietary fatty acids.

The aim of this study was therefore to determine and compare the effects of diets high in PUFAs or SFAs on reproductive output, offspring survival and condition, and body mass changes in female guinea pigs during gestation. Due to the relatively long gestation period and the precociality, guinea pigs may represent an adequate model species to study the effects of dietary fatty acids during gestation and prenatal development. Knowledge on such influences may further increase the understanding for the relevance of dietary fatty acids regarding an individual's reproductive success in general, but also how these nutrients may influence the process of gestation in mothers in relation to the offspring's prenatal development.

METHODS

Ethical statement

Experiments were conducted in accordance with EU Directive 2010/63/EU for animal experiments and the Austrian laws for animal experiments and animal keeping. The study has been checked and approved by the internal board on animal ethics and experimentation of the Faculty of Life Sciences of the University of Vienna (# 2014–005) and the Austrian Federal Ministry of Science and Research (BMWF-66.006/0024-II/3b/2013).

Animals and housing conditions

All domestic guinea pigs (30 males and 30 females) used for this study were bred at the Department of Behavioural Biology at the University of Vienna. All animals were adult, sexually intact, and accustomed to the daily contact with humans. Differences in natural fur colorations allowed an individual identification. Animals were housed in single-sexed groups of ten individuals, resulting in three male and three female groups. Each group's enclosure (2 m × 1.6 m) was environmentally enriched with shelters and platforms and the floor was covered with standard bedding material. Animals were housed at a temperature of 20 ± 2 °C, 50 ± 5% humidity, and a light-dark cycle of 12 h with lights on at 0700 h.

The daily provided food consisted of guinea pig pellets (ssniff V2233, ssniff Spezialdiäten GmbH, Soest, Germany) and 50 g of hay per group; for precise nutrient composition of guinea pig pellets see [18]. According to the manufacturer's information, guinea pig pellets contain 3% (w/w) crude fat, which accounts for 8% of the convertible energy provided with the food (gross energy content of guinea pig pellets: 16.6 MJ/kg; see Table 1 for fatty acid composition of the guinea pig pellets). Water was provided in several drinking bottles. In addition to the daily provided standard food, each male and female group received one of three different dietary supplements: walnut oil (high in PUFAs; gross energy content: 38.7 MJ/kg, fat content: 99.6%), coconut fat (high in SFAs; gross energy content: 38.9 MJ/kg, fat content: 100%), or pure water in case of a control group (see Table 1 for fatty acid compositions of walnut oil and coconut fat). Each animal was administered orally with 3 mL of the specific supplement per 1 kg body mass on every day of the experiment using 1 mL syringes. This procedure, including the supplemented amounts, has already been applied previously in studies on the effects of dietary fatty acids in rats and guinea pigs and proved successful in affecting the fatty acid status of the individuals [19, 20] and result approximately in 25% energy intake from dietary fats in these animals. An imbalance in the total energy intake among single individuals and the three dietary groups, which would have been caused by the additional fatty acid supplementations, was counteracted by ad libitum feeding of guinea pig pellets.

Table 1. Most prominent fatty acids (% of total fatty acids based on gas chromatography analyses) of guinea pig pellets, walnut oil, and coconut fat

Fatty acid	Guinea pig pellets[a]	Walnut oil	Coconut fat
C 12:0	n.d.	n.d.	49.26
C 14:0	0.61	n.d.	22.53
C 16:0	16.16	6.55	11.82
C 16:1	0.61	n.d.	n.d.
C 18:0	3.35	2.76	4.60
C 18:1 n-9	18.90	14.16	6.71
C 18:1 n-7	n.d.	1.28	0.53
C 18:2 n-6	50.00	63.14	3.73
C 18:3 n-6	n.d.	0.40	0.26
C 18:3 n-3	10.06	11.03	n.d.
total MUFAs	19.51	15.44	7.39
total PUFAs	60.06	75.25	4.22
total SFAs	20.43	9.31	88.39
P:S ratio	2.94	8.08	0.05

n.d. Not detectable

[a]Ssniff V2233, ssniff Spezialdiäten GmbH, Soest, Germany

Guinea pig pellets were provided ad libitum in each dietary group; walnut oil (PUFA group) and coconut fat (SFA group) were additionally supplemented (3 mL/kg body mass) to animals of the corresponding group

Experimental procedure

Animals of the different groups were compared in their age and body mass in advance to the experiment to exclude possible differences in these variables at the onset of the study. Males were included in these pre-experimental analyses in order to exclude any possible influences of male body condition on female reproductive performance. Using the statistical package R 3.2.2 [21] and two-way analyses of variance, no differences among the dietary and/or sex groups were found in age ($F_{5,54} = 0.566, P = 0.726$; mean age: 21.1 ± 1.2 month) or body mass ($F_{5,54} = 0.746, P = 0.593$; mean body mass: 805 ± 20 g) at the beginning of the experiment.

The experimental procedure started with an initial 100-d feeding phase. All animals were weighed daily at 0900 h and the body mass-based supplementations were carried out, which lasted no longer than 1 min in total per animal. The dietary supplementations for 100 d should ensure that the individuals had highest possible levels of specific fatty acids and a maximum incorporation into neuronal membranes and tissues of the reproductive tracts. After d 100, blood samples were collected from all animals to measure plasma fatty acid levels as indicators of the general fatty acid status [22, 23] and females were mated with heterogeneous males maintained on the same dietary supplements as mentioned above. For this purpose, male and female groups, supplemented with the same fatty acids, were randomly mixed in two mating groups, each consisting of five males and five females. This yielded a total of six mating groups, two for each dietary regime. Mating in small groups with a sex ratio of 1:1 should reduce male competition for females and increase the number of social partners in order to reduce social stress [24] and, therefore, reliably result in frequent pregnancy rates. During the mating period, saliva and plasma samples were collected from females to analyze saliva cortisol and plasma estrogen levels at conception. Both hormones served as indicators of female homeostasis at conception. Once pregnancy was detected, the respective female was removed from the mating group and introduced to its single-sexed group again. Weighing and dietary supplementation procedures continued throughout the experiment and were still carried out daily at 0900 h. until the first day after parturition.

Measurements of gestational and reproductive performance

During mating, the vaginal membrane of each female was inspected visually to monitor receptivity during the estrous cycle, because the vagina appears to be opened at this stage of proestrus/estrus for 1–6 d [25]. To define the day of conception, a time span of approximately 66–69 d was counted back from the day of parturition. This constant period of gestation is well documented in guinea pigs (see for example [26]). The first day with an opened vagina within this time frame represented the day of conception. Due to a relatively long estrous cycle in guinea pigs of about 16 d [27], a misinterpretation was very unlikely, as this would have resulted in unnaturally short (~50 d) or long (~85 d) gestation periods. The number of days from conception to parturition was defined as gestation duration.

Most females gave birth during the night and therefore the offspring of each female was counted, sexed, and weighed 12 h after parturition at the latest. Also pups that were found dead, but fully developed, were weighed and sexed. The body mass of mothers at conception, during gestation, on the last day of gestation (1 d before parturition), and after parturition was recorded using a standard electronic scale (accuracy ± 1 g) to monitor changes in body mass throughout gestation and parturition.

Saliva and blood sampling procedures

Saliva samples were collected by inserting standard cotton buds into the animal's mouth for approximately 1 min. Cotton buds were then centrifuged (14,000 rpm, 10 min) and pure saliva was stored at −20 °C until further analysis. Blood was collected with heparinized micropipettes after punctuation of prominent ear veins. Plasma was separated by centrifugation (14,000 rpm, 10 min) and stored at −20 °C until further analysis (for further information regarding sample collection procedures see [28]).

Hormone analyses

Hormone concentrations in saliva and plasma were analyzed by biotin-strepdavidin enzyme-linked immunoassays [29, 30]. Saliva samples were diluted 1:50 and cortisol concentrations measured in 10 μL aliquots using a cortisol-specific antibody. Extraction of plasma hormones was done by adding 2 mL diethylether to 100 μL plasma, shaking the samples four times for 15 min, and freezing them overnight. After evaporation of the diethylether (30 °C, 10 min), samples were diluted 1:4 and concentrations of total estrogens in plasma were determined in 25 μL aliquots using an antibody against total estrogens. For further information regarding the used antibodies, including cross- reactions with relevant steroids, see Palme & Möstl [29, 30]. All analyses were run in duplicates. The confidence criterion was set at ≤ 15% for the coefficient of variance of the sample duplicates. Intraassay coefficients of variance for saliva cortisol and plasma estrogen concentrations were 11.15 and 8.64%, respectively.

Plasma fatty acid analyses

Proportions of fatty acids in plasma prior to mating and the gestational period were analyzed by gas chromatography. Following the protocols by Wagner et al. [31] and Nemeth et al. [22], fatty acids in 35 µL plasma were transesterificated by adding 1 mL methanolic NaOH, containing butylated hydroxytoluene, and 1 mL boron-trifluoride to obtain fatty acid methyl esters (FAMES). Samples were boiled for 5 min at 100 °C and cooled on ice for 10 min after each step. FAMES were then extracted by adding 500 µL hexane four times, including 5 min of shaking the samples (700 rpm) in between each addition. Samples were evaporated at 40 °C under nitrogen and redissolved in hexane. FAMES were separated by an Auto-System-Gaschromatograph (Perkin Elmer, USA) with flame ionization detector, equipped with an Rtx-2330 30 m × 0.25 mm × 0.20 µm silica column. 1 µL of prepared samples was injected under a 1:25 split at 250 °C and detected at 275 °C; helium was used as carrier gas. Fatty acids were identified using a 37 component FAME Mix Standard (Supelco, Bellafonte, USA) and TotalChrome Workstation 6.3.0 (PE Nelson, Perkin Elmer, USA) was used for peak integration. Fatty acids are expressed as percentage of total fatty acids.

Statistical analyses

Statistical analyses were carried out using R version 3.2.2 [21] and the implemented package 'nlme' [32] for performing linear mixed effect models (LMEs). For post-hoc analyses, the packages 'phia' [33] and 'PMCMR' [34] were used. Package 'effects' [35] was used to extract effect plots.

Based on the distribution of the data, conditions at conception and reproductive parameters were analyzed and compared among groups (control, PUFA, SFA) using one-way analyses of variance (ANOVAs), Kruskal-Wallis, or Pearson's Chi-squared tests. The sex ratio and survival rate at birth were analyzed by generalized linear models (GLMs) with binomial link. Single pup's birth mass was analyzed using LMEs, including 'group', 'sex', and their interaction as fixed effects, and 'mother' as random effect to correct for the relatedness. To control for possible litter size effects, 'litter size' was included in the models as covariate.

The body mass during gestation was analyzed by an LME, including 'group' (control, PUFA, SFA), 'day' as second-order polynomial term, and 'total litter mass', as well as their interactions as fixed effects, and 'individual ID' as random effect to correct for repeated measurements. The second-order polynomial of 'day' was also included as random slope to correct for individual changes in body mass during gestation. In the beginning, models including either 'total litter mass' or 'litter size' as covariate were compared based on the Akaike information criterion (AIC). This analysis suggested to include total litter mass rather than litter size as fixed effect, although litter size is usually used as covariate to analyze the body mass during gestation (see for example [26]). Total litter mass was also preferred due to the relatively low variation of litter size in the SFA group and should definitely represent the litter size effect, as litter size and total litter mass were highly related ($R^2 = 0.87$, $P = 3 \times 10^{-13}$).

To analyze the gestational effort, body mass in females prior to parturition and on the first day after parturition were first compared among groups by ANOVAs. Afterwards, linear models were calculated, including 'group' (control, PUFA, SFA), 'body mass at conception', and 'total litter mass', as well as their interactions as predictor variables, therefore reflecting the gestational effort as the body mass increase during gestation corrected for the body mass at conception.

Model assumptions (linearity, normality, homoscedasticity of residuals) were checked by performing Shapiro-Wilk normality tests and Levene's test for homogeneity of variance as well as by model diagnostic plots of residuals and fitted values. Models were fitted (removal of non-relevant interaction and main effects) based on the AIC. Only the highest significant interaction and/or main effects are considered in the result section. Model statistics are based on type 3 sum of squares. All post-hoc analyses were Bonferroni corrected. The level of significance was set at $P \leq 0.05$.

RESULTS

Plasma fatty acids

The dietary treatments resulted in different plasma fatty acid proportions among the female groups (Table 2). According to the diets, SFA females showed highest percentages in total SFAs, while PUFA females showed lowest percentages in all types of SFAs. PUFA females, in contrast, showed highest percentages in n-3 and n-6 PUFAs, while n-9 MUFAs were highest in control females. The plasma n-6:n-3 ratio was significantly higher in SFA females, while PUFA females showed the highest P:S ratio.

Table 2. Most prominent plasma fatty acids (% of total plasma fatty acids) in female guinea pigs maintained on a control, high-PUFA, or high-SFA diet

Fatty acid	Control	PUFA	SFA	*F*-value	*P*-value
C 12:0	0.06 ± 0.02^{a}	0.07 ± 0.02^{a}	1.27 ± 0.36^{b}	30.059	<0.001
C 14:0	0.93 ± 0.08^{a}	0.59 ± 0.04^{a}	4.64 ± 0.61^{b}	88.240	<0.001
C 16:0	17.12 ± 0.23^{a}	12.70 ± 0.39^{b}	16.80 ± 0.49^{a}	40.270	<0.001
C 18:0	10.09 ± 0.52^{a}	6.84 ± 0.41^{b}	9.05 ± 0.40^{a}	113.780	<0.001
C 18:1 n9	15.53 ± 0.46^{a}	12.85 ± 0.70^{b}	12.49 ± 0.59^{b}	7.901	0.002
C 18:2 n6	39.77 ± 1.00^{a}	49.13 ± 0.86^{b}	41.73 ± 0.60^{a}	34.695	<0.001
C 18:3 n3	6.02 ± 0.22^{a}	7.71 ± 0.14^{b}	4.47 ± 0.36^{c}	19.890	<0.001
total n-9	16.21 ± 0.47^{a}	15.42 ± 0.46^{a}	13.13 ± 0.58^{b}	9.991	<0.001
total n-6	43.09 ± 0.93^{a}	51.41 ± 0.90^{b}	44.62 ± 0.67^{a}	27.577	<0.001
total n-3	6.64 ± 0.24^{a}	8.12 ± 0.45^{b}	4.95 ± 0.36^{c}	19.400	<0.001
total SFA	31.67 ± 0.74^{a}	23.74 ± 0.91^{b}	35.31 ± 0.81^{c}	51.931	<0.001
total MUFA	18.51 ± 0.50^{a}	16.66 ± 0.49^{b}	15.03 ± 0.63^{b}	10.111	<0.001
total PUFA	49.82 ± 1.08^{a}	59.60 ± 0.89^{b}	49.66 ± 0.55^{a}	43.030	<0.001
n-6 : n-3 ratio	6.55 ± 0.22^{a}	6.57 ± 0.51^{a}	9.59 ± 0.88^{b}	9.254	<0.001
P : S ratio	1.59 ± 0.07^{a}	2.56 ± 0.13^{b}	1.42 ± 0.04^{a}	50.46	<0.001

Different superscript letters indicate significant differences between single groups ($P \leq 0.05$)

Body conditions at conception

The body mass at conception, approximately 100 d after the onset of the feeding procedure, differed among the groups ($F_{2,25} = 6.954$, $P = 0.004$) and was significantly lower in SFA females compared to control females (control: 942 ± 22 g; PUFA: 938 ± 50 g; SFA: 841 ± 15 g; control-PUFA: $\chi^2 = 0.006$, $P = 1.000$; control-SFA: $\chi^2 = 13.998$, $P = 0.001$; PUFA-SFA: $\chi^2 = 3.447$, $P = 0.190$). No differences were detected in saliva cortisol concentrations ($F_{2,25} = 1.891$, $P = 0.172$; control: 18.77 ± 4.45 ng/mL; PUFA: 58.97 ± 23.81 ng/mL; SFA: 40.24 ± 12.68 ng/mL) or plasma estrogen levels ($F_{2,17} = 2.472$, $P = 0.114$; control: 0.48 ± 0.09 ng/mL; PUFA: 0.39 ± 0.08 ng/mL; SFA: 0.24 ± 0.07 ng/mL) at conception.

Overall reproductive output

A total of 28 female guinea pigs became pregnant and showed successful gestations, resulting in normal parturitions at term. Most females became pregnant in their first estrus during the mating phase, with no differences among groups in the number of estrous cycles until conception ($\chi^2 = 2.505$, $P = 0.286$). A total of 85 pups were born fully developed (52 ♂ : 33 ♀). Except for a total litter loss of five pups at birth in one control female, the number of fully-developed, but dead-born pups was very low (71 alive : 14 dead). All dead-born pups were found in their amniotic sacs. The overall reproductive output, including pregnancy rates, total number of pups, total sex ratio, and ratio of alive and dead born pups, did not differ among the groups (Table 3).

Table 3. Statistical analysis of the overall reproductive output for groups of female guinea pigs ($n = 10$ per group) maintained on a control, high-PUFA, or high-SFA diet

Reproductive parameter	Control	PUFA	SFA	χ^2	*P*-value
Number of pregnancies	10	8	10	0.286	0.867
Total offspring number	33	28	24	1.435	0.488
Total sex ratio	19 ♂ : 14 ♀	18 ♂ : 10 ♀	15 ♂ : 9 ♀	0.312	0.856
Alive : Dead ratio	24 A : 9 D	25 A : 3 D	22 A : 2 D	4.628	0.099
Alive : Dead ratio ♂	14 A : 5 D	16 A : 2 D	13 A : 2 D	1.726	0.422
Alive : Dead ratio ♀	10 A : 4 D	9 A : 1 D	9 A : 0 D	3.775	0.152

Reproductive parameters

A significant difference among groups was detected in the litter size ($\chi^2 = 6.855, P = 0.032$) and the total litter mass ($F_{2,25} = 4.032, P = 0.030$), while the sex ratio per litter ($\chi^2 = 0.311, P = 0.856$), the gestation duration ($\chi^2 = 0.573, P = 0.751$), the individual birth mass (corrected for the pups' relatedness: $F_{2,25} = 1.568, P = 0.228$), and the survival rate at birth ($\chi^2 = 4.548, P = 0.103$) did not differ (Table 4). A higher litter size was found in PUFA females compared to SFA females and, correspondingly, a lower total litter mass in the SFA group compared to the control group. Although single pup's birth mass did not differ among groups, males were barely heavier at birth than females ($F_{1,56} = 4.059, P = 0.049$; males: 110.5 ± 2.6 g; females: 105.9 ± 2.8 g).

Including litter size as covariate in the respective analyses revealed that neither the sex ratio ($\chi^2 = 0.080, P = 0.777$) nor the gestation duration ($F_{1,25} = 0.871, P = 0.360$) was affected by litter size at all. However, the litter size positively affected the total litter mass ($F_{1,24} = 145.113, P < 0.001$) and had a negative effect on the survival rate at birth ($\chi^2 = 7.706, P = 0.006$), while only a negative tendency was detected regarding the individual birth mass ($F_{1,26} = 3.701, P = 0.065$) (Table 5). Therefore, pups born in larger litters were of lower birth mass and faced a higher mortality rate at birth. Although all these influences did not differ among groups (for all reproductive parameters: $P > 0.240$), the negative effect of litter size on the individual birth mass was only significant in the SFA group, while a significant negative effect of litter size on the survival rate was only detected in the control group (Table 5).

Table 4. Reproductive performance of female guinea pigs maintained on a control (n =10), high-PUFA (n = 8), or high-SFA (n = 10) diet

Reproductive parameter	Control	PUFA	SFA
Litter size[1]	3 (2–5)[ab]	3.5 (2–5)[a]	3 (1–3)[b]
Sex ratio[2]	0.56 (0.36, 0.76)	0.66 (0.44, 0.89)	0.68 (0.49, 0.88)
Gestation duration[1]	69 (67–70)	68 (64–69)	68 (67–73)
Total litter mass[3]	368.8 ± 29.1[a]	355.6 ± 32.5[ab]	260.6 ± 29.1[b]
Individual birth mass[3]	112.1 ± 3.8	102.3 ± 4.2	110.5 ± 4.0
Survival rate[2]	0.73 (0.47, 0.89)	0.89 (0.60, 0.98)	0.92 (0.58, 0.99)

[1]Data given as median and range; analysis via Kruskal-Wallis test
[2]Data given as (mean) effect size and lower and upper confidence limits; analysis via GLM with binomial link
[3]Data given as mean ± s.e.m; analysis via ANOVA or LME
Different superscript letters indicate significant differences between single groups ($P \leq 0.05$)

Table 5. Effect of litter size on the reproductive performance of female guinea pigs maintained on a control (n =10), high-PUFA (n = 8), or high-SFA (n = 10) diet and the general effect of litter size

Reproductive parameter	Control	PUFA	SFA	Litter size[a]
Sex ratio[b]	0.081 (−0.192, 0.356)	−0.195 (−0.638, − 0.055)	−0.042 (−0.398, 0.170)	−0.017 (−0.016, −0.020)
Gestation duration[c]	0.278 ± 0.675	−0.500 ± 0.954	−0.281 ± 0.939	−0.294 ± 0.315
Total litter mass[c]	110.590 ± 14.616***	90.917 ± 20.756***	95.406 ± 20.426***	98.957 ± 8.215***
Individual birth mass[c]	−2.124 ± 4.669	−4.144 ± 6.664	−10.351 ± 6.950*	−4.929 ± 2.562
Survival rate[b]	−0.174 (−0.382, −0.034)*	−0.035 (−0.460, 0.036)	−0.056 (−0.803, −0.028)	−0.120 (−0.020, −0.273)*

*$p \leq 0.05$, **$p \leq 0.01$, ***$p \leq 0.001$ for significant effects of litter size
[a]Effect of litter size determined after removal of non-significant effects based on the AIC. No significant differences between single groups (for all pairwise comparisons $P > 0.05$)
[b]Data given as (mean) effect size and lower and upper confidence limits; analysis via GLM with binomial link
[c]Data given as slope ± standard error of the slope; analysis via linear model or LME

Body mass changes in mothers

From conception until d 65 of gestation (prior to parturition) the body mass increase in mothers differed significantly among groups (group: $F_{2,22} = 0.992, P = 0.387$; day: $F_{2,379} = 0.611, P = 0.544$; total litter mass: $F_{1,22} = 2.774, P = 0.110$; group:day: $F_{4,379} = 3.296, P = 0.011$; group:total litter mass: $F_{2,22} = 2.169, P = 0.138$; day:total litter mass: $F_{2,379} = 28.845$, $P < 0.001$; group:day:total litter mass: $F_{4,379} = 3.337$, $P = 0.011$). The body mass increase in 5-d intervals was generally less pronounced in SFA females compared to control and PUFA females (control-PUFA day[1]: $t = 1.715$, $P = 0.087$; day[2]: $t = 0.567$, $P = 0.571$; control-SFA day[1]: $t = 1.975$, $P = 0.049$; day[2]: $t = 0.349$, $P = 0.728$; PUFA-SFA day[1]: $t = 3.433$, $P = 0.001$; day[2]: $t = 0.862$, $P = 0.389$) (Figure 1). Statistical modeling and effect plotting revealed that PUFA females showed the strongest increase in body mass as total litter mass increased, whereas this effect was least pronounced in SFA females (Figure 2).

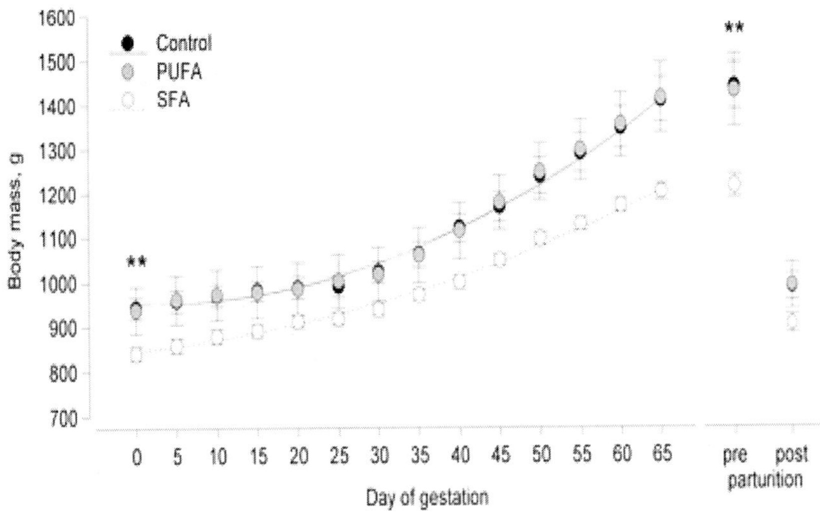

Figure 1. Mean body mass in female guinea pigs maintained on a control, high-PUFA, or high-SFA diet during gestation and after parturition. Circles represent the body mass for each group and day (mean ± s.e.m.); lines for gestation (day 0 to 65) represent the mean fitted values of a linear mixed effect model (corrected for repeated measurements) on the body mass change, including the second-order polynomial term for day and the total litter mass as covariate. Sample sizes: control $n = 10$, PUFA $n = 8$, SFA $n = 10$. ** $P \leq 0.01$ comparing SFA and the remaining groups. Group:day effect during gestation: $P \leq 0.05$

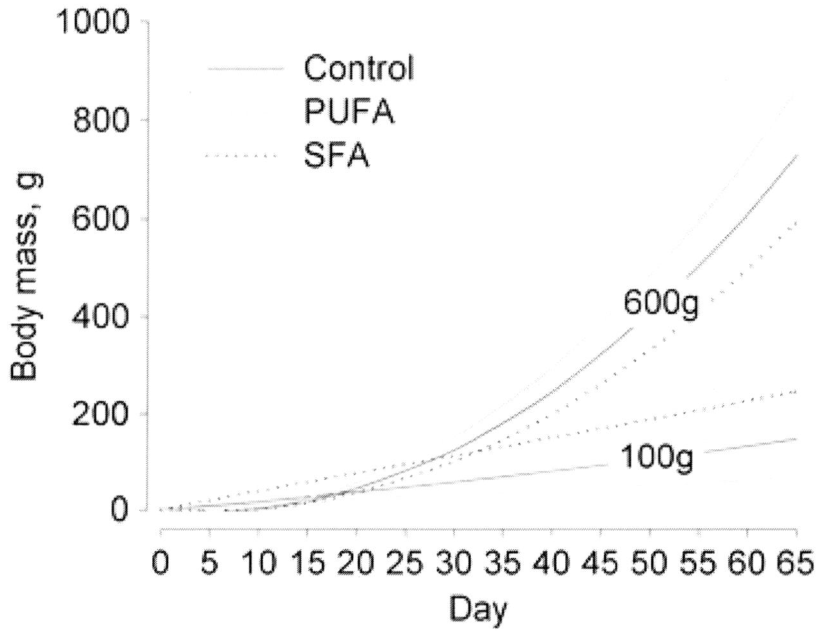

Figure 2. Effect of total litter mass on the body mass gain during gestation in guinea pig females maintained on a control, high-PUFA, or high-SFA diet. Effects were extracted from a linear mixed effect model and are shown for a total litter mass of 100 g and 600 g, respectively. Sample sizes: control $n = 10$, PUFA $n = 8$, SFA $n = 10$. Group:day:total litter mass: $p \leq 0.05$

On the last day of gestation, before parturition, body mass still differed among groups ($F_{2,25} = 5.759, P = 0.009$), with SFA females exhibiting a significantly lower body mass compared to the other groups (control: 1441 ± 52 g; PUFA: 1432 ± 82 g; SFA: 1216 ± 26 g; control-PUFA: $\chi^2 = 0.013, P = 1.000$; control-SFA: $\chi^2 = 9.276, P = 0.016$; PUFA-SFA: $\chi^2 = 7.599, P = 0.032$) (Figure 1). Corrected for the body mass at conception, the effect of total litter mass on the mother's body mass prior to parturition differed among groups (body mass at conception: $F_{1,21} = 38.006, P < 0.001$; group: $F_{2,21} = 3.579, P = 0.046$; total litter mass: $F_{1,21} = 39.070, P < 0.001$; group:total litter mass: $F_{2,21} = 3.990, P = 0.034$) and was more pronounced in PUFA females compared to SFA females (control-PUFA: $F = 2.932, P = 0.305$; control-SFA: $F = 1.864, P = 0.560$; PUFA-SFA: $F = 7.956, P = 0.030$) (Figure 3a). The effect of mother's body mass at conception did not differ among groups and was removed from the model beforehand (group:body mass at conception: $F_{2,19} = 0.258, P = 0.776$).

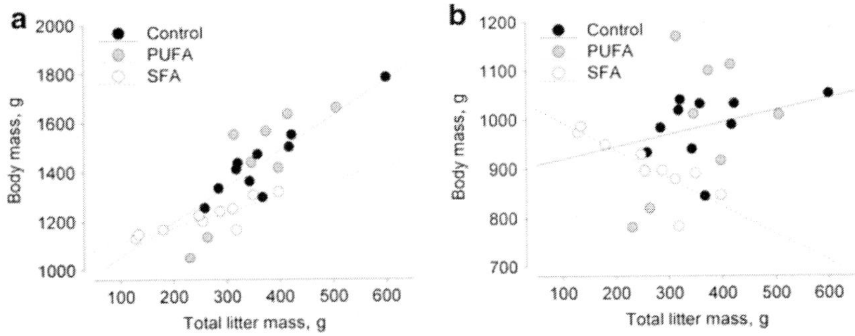

Figure 3. Effects of total litter mass on body mass in guinea pig females maintained on a control, high-PUFA, or high-SFA diet. Effects are corrected for the body mass at conception. **A** Effects on body mass prior to parturition. Control: $p \leq 0.001$, PUFA: $p \leq 0.001$, SFA: $p \leq 0.001$; control vs. PUFA: n.s., control vs. SFA: n.s., PUFA vs. SFA: $p \leq 0.05$. **B** Effects on body mass after parturition. Control: n.s., PUFA: $p \leq 0.05$, SFA: n.s.; all group comparisons: n.s. (n.s.: not significant)

On the first day after parturition, no differences in body mass were detected among the three groups ($F_{2,25} = 2.134, P = 0.140$) (Figure 1). However, compared to the body mass at conception, all groups were significantly heavier after parturition (control: +44.7 g, $t = 2.405, p = 0.024$; PUFA: +52.38 g, $t = 2.521, P = 0.018$; SFA: +62.8 g; $t = 3.379, P = 0.002$). This increase was similar for all groups ($F_{2,25} = 0.239, P = 0.789$). Further analyses revealed that the body mass after parturition, corrected for the body mass at conception, tended to be differently affected by the total litter mass (body mass at conception: $F_{1,21} = 48.454, P < 0.001$; group: $F_{2,21} = 3.118$, $P = 0.065$; total litter mass: $F_{1,21} = 0.445$; $P = 0.512$; group:total litter mass: $F_{2,21} = 3.167, P = 0.063$). While a positive effect of total litter mass on the body mass after parturition was detected for PUFA females ($F = 4.708, P = 0.042$), no effects were detected for control and SFA females (control: $F = 0.445, P = 0.512$; SFA: $F = 1.915, P = 0.181$) (Figure 3b).

The total body mass loss after parturition was not identical to the total litter mass and therefore the difference probably represented placental tissues and amniotic fluid. This part of the females' body mass loss differed significantly among the groups ($F_{2,25} = 3.444, P = 0.048$) and tended to be lower in SFA females (control: 85 ± 10 g; PUFA: 86 ± 14 g; SFA: 52 ± 9 g; post-hoc analyses with Bonferroni corrections remained non-significant). This, however, was positively affected by the total litter mass, which also diminished the group difference (group: $F_{2,22} = 0.914, P = 0.416$; total litter mass: $F_{1,26} = 33.442, P < 0.001$; group:total litter mass: $F_{2,22} = 1.258, P = 0.304$).

DISCUSSION

A mother's dietary fatty acid intake can profoundly affect the reproductive success and may further determine the post-natal development of her offspring. These effects are suggested to be caused by physiological influences of these nutrients already prior to conception, but even by modulating the processes of gestation and parturition [36, 37, 38]. Here we could show that diets high in PUFAs and SFAs differently affect reproductive performances in female guinea pigs. These influences, however, were not related to differences in saliva cortisol and plasma estrogen levels at the time of conception among the dietary groups in the present study. Especially PUFAs can positively affect the endocrine system, particularly the hypothalamic-pituitary-adrenal and -gonadal axes and related cortisol, estrogen, and progesterone secretion rates, whereas EPA and AA are also precursors of prostaglandins. All these hormones interactively modulate reproductive processes, including the estrus cycle, conception, the process of gestation, and even parturition [3, 39]. Although we found no differences in cortisol and estrogen concentrations, differences in the analyzed plasma fatty acids indicate modified availabilities of precursors for prostaglandin synthesis, cell membrane composition, and the energy balance.

The dietary supplementations significantly affected the plasma fatty acid status, determined by the percentage of single fatty acids, and, presumably, the availability of these molecules for metabolic functions. Supplementations with walnut oil (high in PUFAs) significantly decreased the percentage of SFAs and increased n-3 and n-6 PUFAs, resulting in the highest P:S ratios. Supplementations with coconut fat (high in SFAs) significantly increased the percentage of SFAs and decreased n-3 PUFAs, resulting in higher n-6:n-3 ratios in these animals. PUFAs and SFAs are seemingly metabolized in different ways: PUFAs are much faster oxidized and SFAs rather stored in abdominal fat [11, 40]. A higher P:S ratio may therefore contribute to the short term energy supply, which would be required for maintaining daily physiological functions, while a lower P:S ratio may indicate that more fat is stored. Hence, the significantly decreased body mass in SFA females at conception seems to be a contradiction, because a higher body fat accumulation could be assumed to result in a higher body mass. However, previous findings in guinea pigs indicate that a lower body mass can also be related to a higher percentage of body fat, although this was found in fetuses [41]. Energetic needs in SFA females were possibly not covered by direct energy supplies via dietary PUFAs but rather via internal energy reserves and perhaps resulted in a reduced body mass. In relation to a lower n-3 PUFA status, this may have negatively influenced the reproductive output in SFA females, because especially n-3

PUFAs are suggested to promote reproductive functions [38]. However, as supplemented females did not exhibit a higher body mass at conception compared to control females, the ad libitum feeding of guinea pig pellets definitely counteracted a possible energetic imbalance among the dietary groups.

Dietary fatty acids differently affected the reproductive output in the studied animals and resulted in a relatively low litter size and total birth mass in the SFA group, which reduced the gestational effort for mothers in the following, as shown by the lowest body mass increase. The majority of SFA females gave birth to a maximum of three pups, whereas control and PUFA females had litter sizes of up to five pups. Ovulation rates and the number of produced and released ova may determine the litter size in rodents and can be highly affected by dietary PUFAs, with n-3 enhancing and n-6 possibly decreasing the number of ova and pups per litter [4, 42]. The increased plasma n-6:n-3 ratio in SFA females may have caused a lower ova production and decreased litter sizes. A possibly lower energy supply by dietary SFA intakes [11, 40] could have negatively affected the production of ova too, in contrast to higher P:S and lower n-6:n-3 ratios as found in PUFA females.

Unbalanced energy intakes in guinea pigs, including food restriction and overfeeding, can generally impact on reproduction and result in decreased litter sizes and individual birth mass [41, 43]. Both diets high in PUFAs and SFAs can diminish these effects, resulting in an increased offspring body mass [6, 8]. Interestingly, no differences were found in the individual birth mass among the dietary treatments in the present study, although birth mass tended to decrease with litter size. Guinea pig pups born in larger litters usually show a lower body mass and also a higher mortality rate [13]. We did not detect such differences among the dietary groups, but a lower pup survival rate at birth was found in larger litters of control females. The fact that even dead-born pups were fully developed and still covered in their amniotic sacs indicates that they died at birth, probably because mothers did not rupture the amniotic sacs and the pups therefore asphyxiated soon after birth [44]. Mothers can be assumed to be exhausted by giving birth especially to large litters, whereas dietary fatty acids may provide additional energy during the process of parturition, thereby facilitating maternal care for the neonates, or even affect the morphology of the birth canal, resulting in an easier parturition. A higher energy intake and the resulting improved body condition of a mother may also favor the birth of male offspring [45], but we found no differences in the sex ratio at birth or in the ratio of dead and alive born male and female pups. These results are in contrast to a variety of mechanisms describing how diet and available energy may skew the sex ratio [46], which has also been reported for high-fat diets [9]. The present results suggest that guinea pigs generally produce more male offspring, which is apparently not affected by dietary fat intakes. As relatively low mortality rates and sex ratios higher than 0.5 at birth were detected in all dietary groups, it can be concluded that females used in this study were in good body conditions.

An increased survival rate in large litters is not only of major importance regarding the lifetime reproductive success, but also in terms of a mother's investment during gestation, as reflected in the relative body mass increase during gestation caused by the developing fetuses. A very high body mass increase during gestation represents a significant energetic effort in guinea pigs especially during the second half of gestation [47]. As the litter size and total birth mass in SFA mothers were generally lower, at least compared to PUFA females, this obviously resulted in a reduced gestational effort in these animals. PUFA females, in contrast, showed highest gestational efforts and were seemingly able to support the developing pups especially in case of larger litters by an increased energetic investment during gestation, which resulted in a similar individual birth mass among different litter sizes and groups. An increased energy allocation in the developing fetuses could have also eliminated a negative relationship between litter size and individual birth mass, which usually occurs in guinea pigs [13, 14]. The relatively low litter sizes and total litter masses in SFA females, which resulted in the low gestational efforts, perhaps enabled these animals to provide the developing fetuses with adequate energy, which might have not been possible in case of larger litters.

All females revealed a significantly increased body mass after parturition compared to conception, indicating available internal energy reserves for the lactation period after the exhausting parturition. Although female guinea pigs energetically invest more in their offspring during gestation compared to lactation [16], the short lactation period is usually characterized by a regular decrease in the mothers' body mass [26]. SFA females in the present study showed the most pronounced body mass gain from conception to post parturition. This may indicate that these animals were able to accumulate higher internal energy reserves during gestation, perhaps due to the lower number of developing fetuses and a related lower energetic investment. PUFA females, however, seemed to adjust their own body mass to the litter size and total litter mass, as the total litter mass positively affected the mother's body mass after parturition. This was not detected in control and SFA females and also a previous study in guinea pigs did not reveal such an effect [43]. However, a higher gestational effort, litter size, and total litter mass could be assumed to impair a mother's body condition after parturition. As this is apparently not the case in guinea pigs, it can be concluded that these animals are able to accumulate additional energy reserves during gestation, which may be enhanced by feeding on dietary PUFAs. A lower litter size and total litter mass in SFA females may have further been related to less pronounced placental tissues [48], which perhaps also caused a decreased energy allocation in establishing and maintaining these tissues. This would also be supported by a general positive link to the total litter mass, as more and heavier fetuses require more space in the uterus and have to be provided with more oxygen and nutrients, resulting in larger and heavier placentas [41].

CONCLUSIONS

SFAs seem to reduce the litter size in guinea pigs, which would reduce the lifetime reproductive success probably due to inadequate available energy for reproductive functions and for developing a higher number of fetuses. In this way, SFAs may also decrease the gestational effort, which perhaps favors body condition in these animals during and after gestation due to a lower energy allocation required for the developing fetuses. Dietary PUFAs seem to modify energy accumulations in pregnant guinea pigs in relation to the litter size and total litter mass towards the maximum reproductive success. Although both types of fatty acids seem to affect female reproduction at different levels, they may both maximize the survival rate at birth and therefore the different gestational efforts would be justified. These promising results may further help to understand how dietary fatty acids can modulate reproductive processes and the reproductive success, perhaps by primarily affecting the mother's body condition. This demonstrates the significant influence of dietary fatty acids on mammalian reproduction, since not only the developing fetuses but also a mother's reproductive performance may be highly influenced by an adjustment of pre-fertilization dietary PUFA and SFA intakes. However, a limitation of this study is the lack of knowledge which male fertilized the female, because a high rate of multiple paternities could have been possible due to our experimental setup [49]. Paternal effects in relation to dietary fatty acids (e.g. number of mutation rates [50]) should be considered in order to fully interpret the whole impact of these nutrients on reproduction in mammals.

ACKNOWLEDGEMENTS

The authors would like to thank the following staff of the University of Vienna: Alexandra Christian and Verena Brudermann, for the help in animal keeping, and Elisabeth Pschernig, responsible for hormonal analyses. We would also like to thank Michael Nemeth for his help in constructing the experimental setup.

Funding

MN was supported by a PhD completion grant of the University of Vienna during the preparation of the manuscript. The funders played no role in study design, performance of the experiment, data analysis, or writing of the manuscript.

Availability of data and materials

All data generated or analyzed during this study are included in this published article and its Additional file 1.

Authors' contributions

MN designed and performed the experiment, analyzed the data, and wrote the manuscript. EM designed the experiment and edited the manuscript. CS performed the experiment, analyzed the data and edited the manuscript. KHW performed the experiment and edited the manuscript. RQ performed the experiment and edited the manuscript. BW designed the experiment, wrote and edited the manuscript. All authors read and approved the final manuscript.

Competing interests

The authors declare that they have no competing interests.

Consent for publication

Not applicable.

Ethics approval and consent to participate

Experiments were conducted in accordance with EU Directive 2010/63/EU for animal experiments and the Austrian laws for animal experiments and animal keeping. The study has been checked and approved by the internal board on animal ethics and experimentation of the Faculty of Life Sciences of the University of Vienna (# 2014–005) and the Austrian Federal Ministry of Science and Research (BMWF-66.006/0024-II/3b/2013).

Additional files

Additional file 1: Dataset.xls; Dataset supporting the results and conclusions of this article. (XLS 56 kb)

REFERENCES

1. Schneider JE. Energy balance and reproduction. Physiol Behav. 2004;81:289–317.
2. Funston RN. Fat supplementation and reproduction in beef females. J Anim Sci. 2004;82(E-Suppl):E154–61.
3. Wathes DC, Abayasekara DRE, Aitken RJ. Polyunsaturated fatty acids in male and female reproduction. Biol Reprod. 2007;77:190–201.
4. Broughton KS, Bayes J, Culver B. High α-linolenic acid and fish oil ingestion promotes ovulation to the same extent in rats. Nutr Res. 2010;30:731–8.
5. Ambrose DJ, Kastelic JP, Corbett R, Pitney PA, Petit HV, Small JA, et al. Lower pregnancy losses in lactating dairy cows fed a diet enriched in α-linolenic acid. J Dairy Sci. 2006;89:3066–74.
6. Rebollar PG, García-García RM, Arias-Álvarez M, Millán P, Rey AI, Rodríguez M, et al. Reproductive long-term effects, endocrine response and fatty acid profile of rabbit does fed diets supplemented with n-3 fatty acids. Anim Reprod Sci. 2014;146:202–9.
7. Santillán ME, Vincenti LM, Martini AC, Fiol de Cuneo M, Ruiz RD, Mangeaud A, et al. Developmental and neurobehavioral effects of perinatal exposure to diets with different ω-6:ω-3 ratios in mice. Nutrition. 2010;26:423–31.
8. Bilbo SD, Tsang V. Enduring consequences of maternal obesity for brain inflammation and behavior of offspring. FASEB J. 2010;24:2104–15.
9. Alexenko AP, Mao J, Ellersieck MR, Davis AM, Whyte JJ, Rosenfeld CS, et al. The contrasting effects of Ad libitum and restricted feeding of a diet very high in saturated fats on sex ratio and metabolic hormones in mice. Biol Reprod. 2007;77:599–604.
10. Fernandez ML, West KL. Mechanisms by which dietary fatty acids modulate plasma lipids. J Nutr. 2005;135:2075–8.
11. Sanz M, Lopez-Bote CJ, Menoyo D, Bautista JM. Abdominal fat deposition and fatty acid synthesis are lower and β-oxidation is higher in broiler chickens fed diets containing unsaturated rather than saturated fat. J Nutr. 2000;130:3034–7.
12. Künkele J. Effects of litter size on the energetics of reproduction in a highly precocial rodent the guinea pig. J Mammal. 2000;81:691–700.
13. Manjeli Y, Tchoumboue J, Njwe RM, Teguia A. Guinea-pig productivity under traditional management. Trop Anim Health Prod. 1998;30:115–22.
14. Czarnecki R, Adamski M. Factors influencing litter size and birthweight in the newborn long-haired Guinea pigs (Cavia aperea f. porcellus). J Appl Anim Res. 2016;44:71–6.
15. Künkele J, Trillmich F. Are precocial young cheaper? Lactation energetics in the guinea pig. Physiol Zool. 1997;70:589–96.

16. Künkele J. Energetics of gestation relative to lactation in a precocial rodent, the guinea pig (Cavia porcellus). J Zool. 2000;250:533–9.
17. Berry M, Gray C, Wright K, Dyson R, Wright I. Premature guinea pigs: A new paradigm to investigate the late-effects of preterm birth. J Dev Orig Health Dis. 2015;6:143–8.
18. ssniff® Complete feeds for rabbits and guinea pigs. http://www.ssniff.com/documents/04_catalogue_rabbit_guinea_engl_1.pdf. Accessed 7 Dec 2016.
19. Ferraz AC, Delattre AM, Almendra RG, Sonagli M, Borges C, Araujo P, et al. Chronic ω-3 fatty acids supplementation promotes beneficial effects on anxiety, cognitive and depressive-like behaviors in rats subjected to a restraint stress protocol. Behav Brain Res. 2011;219:116–22.
20. Nemeth M, Millesi E, Wagner KH, Wallner B. Sex-Specific Effects of Diets High in Unsaturated Fatty Acids on Spatial Learning and Memory in Guinea Pigs. PLoS One. 2015;10:e0140485.
21. R Core Team. R: A language and environment for statistical computing. Vienna: R Foundation for Statistical Computing; 2015. http://www.R-project.org/.
22. Nemeth M, Millesi E, Wagner KH, Wallner B. Effects of diets high in unsaturated fatty acids on socially induced stress responses in guinea pigs. PLoS One. 2014;9:e116292.
23. Spector AA. Plasma free fatty acid and lipoproteins as sources of polyunsaturated fatty acid for the brain. J Mol Neurosci. 2001;16:159–65.
24. Sachser N, Dürschlag M, Hirzel D. Social relationships and the management of stress. Psychoneuroendocrinology. 1998;23:891–904.
25. Weir BJ. The management and breeding of some more hystricomorph rodents. Lab Anim. 1970;4:83–97.
26. Schöpper H, Palme R, Ruf T, Huber S. Chronic stress in pregnant guinea pigs (Cavia aperea f. porcellus) attenuates long-term stress hormone levels and body weight gain, but not reproductive output. J Comp Physiol B. 2011;181:1089–100.
27. Ishii O. Observations on the sexual cycle of the guinea pig. Biol Bull. 1920;38:237–50.
28. Nemeth M, Pschernig E, Wallner B, Millesi E. Non-invasive cortisol measurements as indicators of physiological stress responses in guinea pigs. Peer J. 2016;4:e1590.
29. Palme R, Möstl E. Biotin-streptavidin enzyme immunoassay for the determination of oestrogens and androgens in boar faeces. In: Görög S, editor. Proceedings of the 5th Symposium on the Analysis of Steroids. Budapest: Akademiai Kiado; 1993. p. 111–7.
30. Palme R, Möstl E. Measurement of cortisol metabolites in faeces of sheep as a parameter of cortisol concentration in blood. Z Säugetierkd. 1997;62:192–7.
31. Wagner KH, Auer E, Elmadfa I. Content of trans fatty acids in margarines, plant oils, fried products and chocolate spreads in Austria. Eur Food Res Technol. 2000;210:237–41.
32. Pinheiro J, Bates D, DebRoy S, Sarkar D, R Core Team. nlme: Linear and Nonlinear Mixed Effects Models. R package version 31–122. 2015. http://CRAN.R-project.org/package=nlme.
33. De Rosario-Martinez H. phia: Post-Hoc Interaction Analysis. R package version 0.2–1. 2015. http://CRAN.R-project.org/package=phia.
34. Pohlert T. The Pairwise Multiple Comparison of Mean Ranks Package (PMCMR). R package. 2014. http://CRAN.R-project.org/package=PMCMR.
35. Fox J. Effect Displays in R for Generalised Linear Models. J Stat Softw. 2003;8:1–27.
36. Sullivan EL, Riper KM, Lockard R, Valleau JC. Maternal high-fat diet programming of the neuroendocrine system and behavior. Horm Behav. 2015;76:153–61.
37. Innis SM. Perinatal biochemistry and physiology of long-chain polyunsaturated fatty acids. J Pediatr. 2003;143:S1–8.
38. Gulliver CE, Friend MA, King BJ, Clayton EH. The role of omega-3 polyunsaturated fatty acids in reproduction of sheep and cattle. Anim Reprod Sci. 2012;131:9–22.

39. Abayasekara DRE, Wathes DC. Effects of altering dietary fatty acid composition on prostaglandin synthesis and fertility. Prostaglandins Leukot Essent Fatty Acids. 1999;61:275–87.
40. Shimomura Y, Tamura T, Suzuki M. Less body fat accumulation in rats fed a safflower oil diet than in rats fed a beef tallow diet. J Nutr. 1990;120:1291–6.
41. Kind KL, Roberts CT, Sohlstrom AI, Katsman A, Clifton PM, Robinson JS, et al. Chronic maternal feed restriction impairs growth but increases adiposity of the fetal guinea pig. Am J Physiol Regul Integr Comp Physiol. 2005;288:R119–26.
42. Trujillo EP, Broughton KS. Ingestion of n-3 polyunsaturated fatty acids and ovulation in rats. J Reprod Fertil. 1995;105:197–203.
43. Michel CL, Bonnet X. Influence of body condition on reproductive output in the guinea pig. J Exp Zool. 2012;317:24–31.
44. Tejml P, Šoch M, Brouček J, Jirotková D, Smutný L, Zábranský L, et al. Factors influencing behaviour of guinea pig females during the birth. Adv Environ Sci Energy Plann. 2015;2015:81–4.
45. Trivers RL, Willard DE. Natural selection of parental ability to vary the sex ratio of offspring. Science. 1973;179:90–2.
46. Rosenfeld CS, Roberts RM. Maternal diet and other factors affecting offspring sex ratio: A review. Biol Reprod. 2004;71:1063–70.
47. Engle WA, Lemons JA. Composition of the fetal and maternal guinea pig throughout gestation. Pediatr Res. 1986;20:1156–60.
48. Elias AA, Ghaly A, Matushewski B, Regnault TRH, Richardson BS. Maternal Nutrient Restriction in Guinea Pigs as an Animal Model for Inducing Fetal Growth Restriction. Reprod Sci. 2016;23:219–27.
49. Asher M, Lippmann T, Epplen JT, Kraus C, Trillmich F, Sachser N. Large males dominate: ecology, social organization, and mating system of wild cavies, the ancestors of the guinea pig. Behav Ecol Sociobiol. 2008;62:1509–21.
50. Kong A, Frigge ML, Masson G, Besenbacher S, Sulem P, Magnusson G, et al. Rate of de novo mutations, father's age, and disease risk. Nature. 2012;488:471–5.

CHAPTER 18

CRISPR/Cas9 Targeting Events Cause Complex Deletions and Insertions at 17 Sites in the Mouse Genome

Ha Youn Shin[1,2,*,], Chaochen Wang[1,*], Hye Kyung Lee[1,3,*], Kyung Hyun Yoo[1,4], Xianke Zeng[1], Tyler Kuhns[1], Chul Min Yang[1], Teresa Mohr[1], Chengyu Liu[5] & Lothar Hennighausen[1,**]**

[1] *Laboratory of Genetics and Physiology, National Institute of Diabetes and Digestive and Kidney Diseases, US National Institutes of Health, Bethesda, Maryland 20892, USA.*

[2]*Department of Biomedical Science and Engineering, Konkuk University, Seoul 05029, Republic of Korea.*

[3]*Department of Cell and Developmental Biology & Dental Research Institute, Seoul National University, Seoul 110-749, Republic of Korea.*

[4]*Department of Life Systems, Sookmyung Women's University, Seoul 140-742, Republic of Korea.*

[5] *Transgenic Core, National Heart, Lung, and Blood Institute, US National Institutes of Health, Bethesda, Maryland 20892, USA.*

ABSTRACT

Although CRISPR/Cas9 genome editing has provided numerous opportunities to interrogate the functional significance of any given genomic site, there is a paucity of data on the extent of molecular scars inflicted on the mouse genome. Here we interrogate the molecular consequences of CRISPR/Cas9-mediated deletions at 17 sites in four loci of the mouse genome. We sequence targeted sites in 632 founder mice and analyse 54 established lines. While the median deletion size using single sgRNAs is 9 bp, we also obtain large deletions of up to 600 bp. Furthermore, we show unreported asymmetric deletions and large insertions of middle repetitive sequences. Simultaneous targeting of distant loci results in the removal of the intervening sequences. Reliable deletion of juxtaposed sites is only achieved through two-step targeting. Our findings also demonstrate that an extended analysis of F1 genotypes is required to obtain conclusive information on the exact molecular consequences of targeting events.

INTRODUCTION

The clustered regularly interspaced short palindromic repeat (CRISPR)-Cas9 system has been developed into an effective genome engineering tool[1,2,3,4,5]. CRISPR/Cas9 gene editing replaced previous gene-targeting technologies[6] and the direct injection of Cas9 and guide RNAs into zygotes provides an efficient and cost effective means to target the mouse genome[7]. The application of the CRISPR/Cas9 system in genome engineering requires two components, the single guide RNA (sgRNA) and the Cas9 nuclease[8]. Cas9 nuclease recognizes the protospacer adjacent motif (PAM) in the targeted region, which is adjacent to sgRNA, and creates double-strand breaks. Double-strand breaks are rapidly repaired by either non-homologous end joining (NHEJ) or homology-directed repair. NHEJ-mediated DNA repair often creates short deletions, occasionally large deletions, insertions and point mutations[7,9,10]. CRISPR/Cas9 has been used successfully to disrupt individual and multiple target genes and knock-in mice were generated to investigate biological functions[7,9,11,12,13,14,15,16,17,18]. Site-specific genomic modification has also provided new opportunities to interrogate the biological significance of transcriptional enhancers[19,20,21,22,23,24].

Although the CRISPR/Cas9 system has received vast attention, only a relatively small number of reports have described its application to introduce mutations in the mouse germline and the extent of molecular consequences has not been systematically explored. Our laboratory has targeted 17 genomic sites in the mouse genome and the respective mutations were analysed in 632 founders. Additional detailed analyses were performed in 54 lines established from specific founders. Specifically, we deleted sequences in enhancers bound by the transcription factor (TF) STAT5 (refs 19, 25, 26) and sequences recognized by CTCF , a protein known to aid in the establishment of functional chromatin loops. We investigated the molecular consequences on targeting sites with single sgRNAs and identified prevalent asymmetric deletions, preferred sites at which deletion occurs and large deletions. We also investigated strategies to delete juxtaposed sites and determined that only a sequential two-step, but not a one-step, targeting approach yielded reliable results. Our analyses permitted an assessment of target specificities, deletion efficiencies and size distributions based on one- and two-step targeting approaches.

RESULTS

Asymmetric deletions using single sgRNAs

CRISPR/Cas9 gene editing has been used successfully to target the mouse genome (Supplementary Table 1). To obtain in-depth knowledge on the extent of molecular consequences at target sites, we analysed deletions introduced at 17 loci in the mouse genome. By directly injecting the Cas9 machinery into mouse zygotes, we targeted five enhancers bound by the cytokine-sensing TF STAT5 in three genomic loci, *Stat5a (A)*20, *Socs2 (B)*26 and *Wap (C)*19 (Figure 1). We also targeted 10 sites bound by CCCTC-binding factor (CTCF), a DNA-binding protein proposed to participate in generating chromatin loops27, in the *Wap (D)* and *Csn (E)* loci; Supplementary Figure 1). Lastly, we targeted one enhancer bound by STAT5 and NFIB, a TF involved in epithelial cell differentiation28,29,30 in the *Csn (F)* locus (Supplementary Figure 1). These genomic sites were targeted individually or in combination. Juxtaposed sites within a given locus were targeted either simultaneously or successively, that is, in a one-step or two-step procedure (Figure 1, Supplementary Notes 1–3 and Supplementary Table 2). Four distinct targeting strategies were pursued (detailed diagrams shown in Figs 2, 3, 4). We targeted individual TF-binding sites with only one corresponding sgRNA (Type 1), targeting individual TF-binding sites with more than one sgRNA (Type 2) and more than one TF-binding site with several sgRNAs (Type 3). In addition to targeting more than one site simultaneously (one-step, Type 3), we also targeted them successively (two-steps, Type 4). To accomplish this, we first targeted specific TF-binding sites and generated homozygous mutant mice, which subsequently served as hosts for targeting additional sites in the same gene locus. Molecular consequences of targeting events at 17 sites were investigated in more than 630 founder mice and 54 established lines, using both polymerase chain reaction (PCR) and DNA sequencing. The number of founders from each targeted site is shown in Supplementary Table 2. Deleted sequences from each individual founder are shown in Supplementary Notes 1–3. Each founder is the result of a distinct deletion induced by a given targeting event.

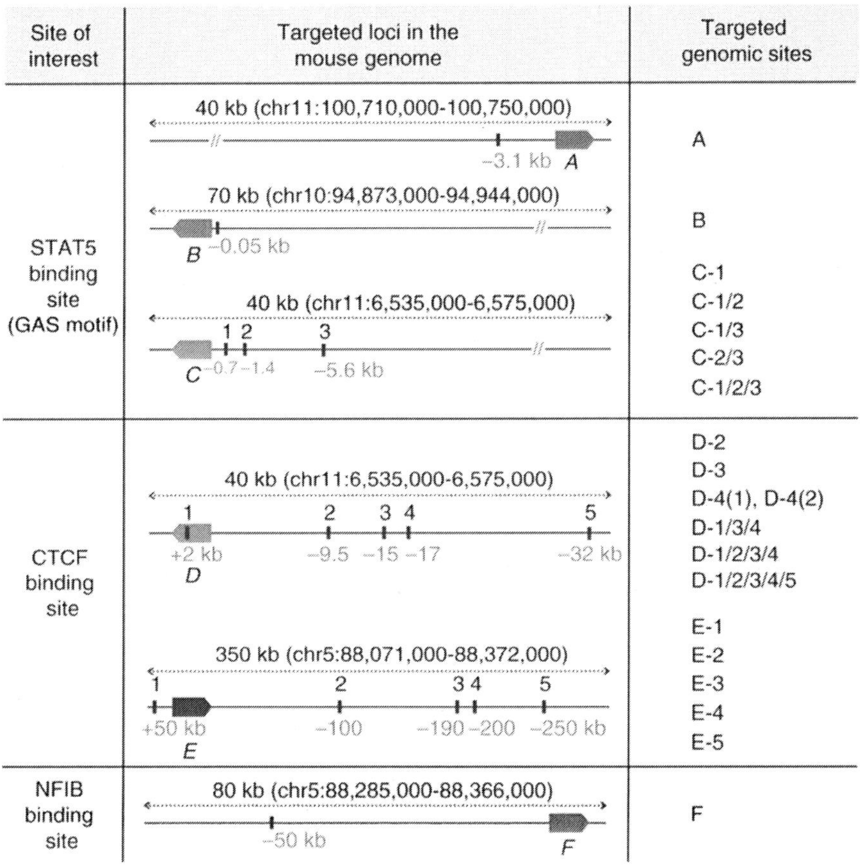

Figure 1: Targeting 17 sites in the mouse genome with CRISPR/Cas9.

STAT5 TF-binding sites (GAS motif), CTCF-binding regions and an NFIB-binding site were targeted for deletion. STAT5-binding sites in three gene loci (A, *Stat5* (ref. 20); B, *Socs2* (ref. 26); C, *Wap*19) were targeted individually (A, B and C-1). The three STAT5-binding sites in the *Wap* super-enhancer19 were deleted in four different combinations (C-1/2, C-1/3, C2/3 and C-1/2/3). Combined deletion of two or more STAT5-binding sites within the same gene locus was accomplished through a successive (two-steps) or simultaneous (one-step) targeting. A total of 11 CTCF-binding sites in two gene loci (D, *Wap*; E, *Csn*) were targeted individually and in different combinations. (F), an NFIB-binding site in the *Csn* locus was targeted (Supplementary Figure 1). Target mutations were identified in 632 founder mice and 54 mutant lines were established. The positions of reference genes in the respective loci are indicated as coloured boxes. A, *Stat5a*; B, *Socs2*; C and D, *Wap*; E, *Csn1s1*; F, *Csn3*.

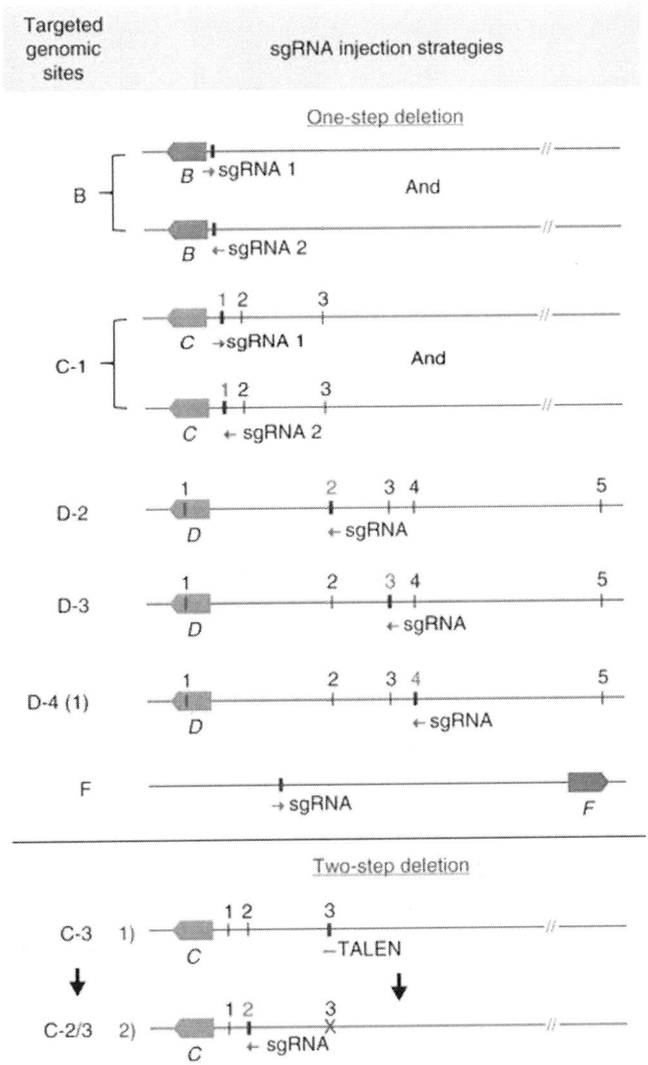

Figure 2: Targeting individual genomic sites with corresponding single sgRNAs. Two genomic sites (B and C-1) were targeted independently with two individual sgRNAs (1 and 2). Four individual genomic sites (D-2, D-3, D-4(1) and F) were targeted with one sgRNA each. The deletion of two juxtaposed genomic sites (C-2/3) was generated in two steps. A single sgRNA-targeting site C-2 was injected into zygotes from mice carrying already a deletion in site C-3. The deletion of site C-3 was generated by TALEN19. B, *Socs2*; C and D, *Wap*; E, *Csn1s1*; F, *Csn3*. The red numbers refer to the sites being targeted in the respective experiments.

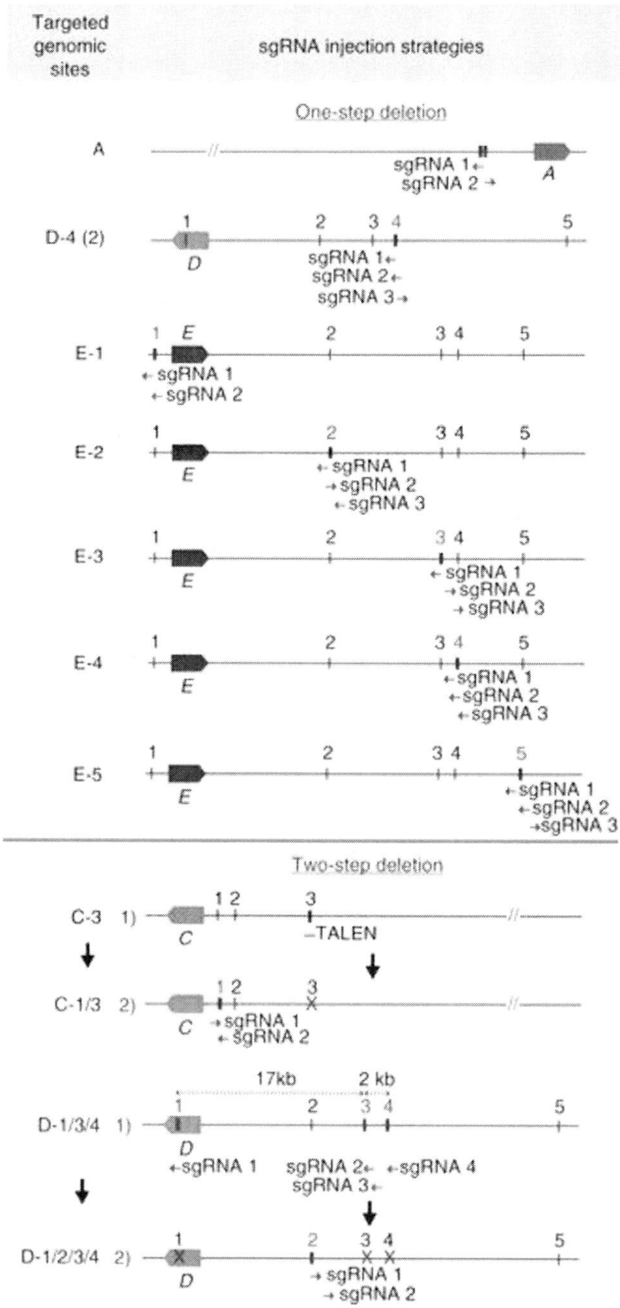

Figure 3: Targeting individual genomic sites with more than one sgRNA.

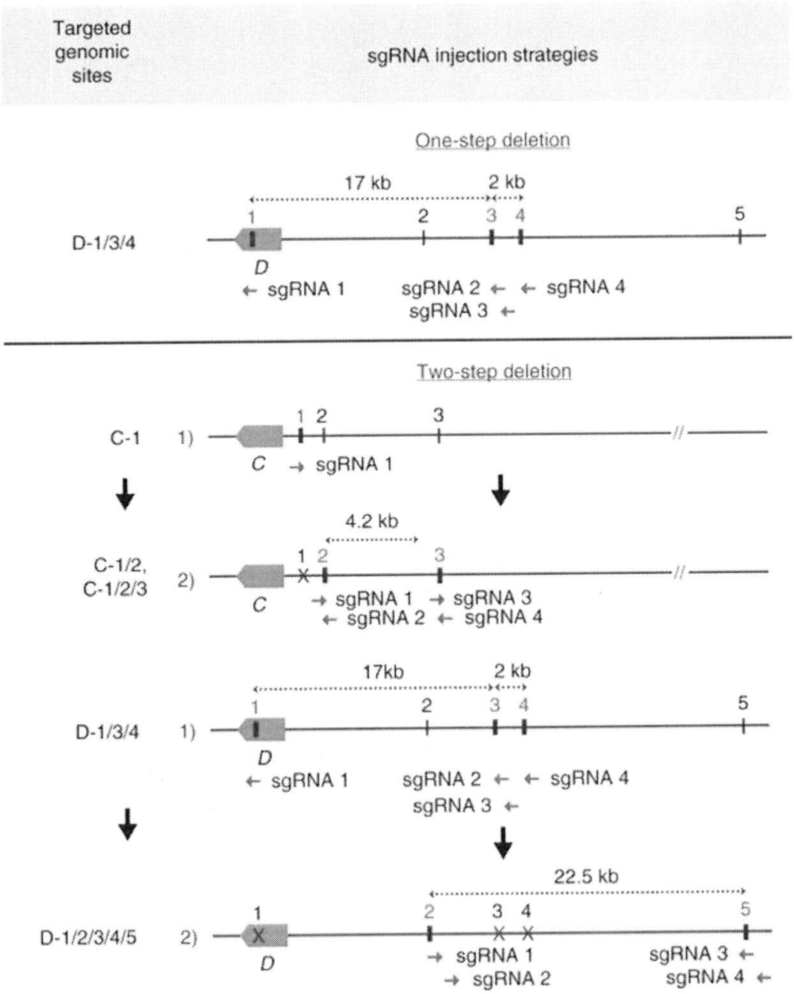

Figure 4: Targeting more than one genomic site with several sgRNAs. Three juxtaposed genomic sites (D-1/3/4) were targeted simultaneously with four sgRNAs. Combined deletion of more than one genomic site (C-1/2, C-1/2/3 and D-1/2/3/4/5) was accomplished in two steps. Four sgRNAs targeting sites C-2 and C-3 were simultaneously injected into zygotes from mice carrying already a deletion in site C-1 (ref. 19). Four sgRNAs targeting sites D-2 and D-5, which are 23 kb apart from each other, were simultaneously injected into mice carrying already deletions in sites C-1/3/4. C and D, *Wap.* The red numbers refer to the sites being targeted in the respective experiments.

Each of seven individual genomic sites (A, D-4(2), E-1, E-2, E-3, E-4 and E-5) was simultaneously targeted with two or three sgRNAs. Combined deletions of more than one genomics site (C-1/3 and D-1/2/3/4) were accomplished in two steps. Two sgRNAs targeting site C-1 were simultaneously injected into zygotes from mice carrying already a deletion in site C-3 (ref. 19). Two sgRNAs targeting site D-2 were simultaneously injected into mice carrying already a deletion in sites C-1/3/4. A, *Stat5*; C and D, *Wap*; E, *Csn1s1*. The red numbers refer to the sites being targeted in the respective experiments.

Cas9 nuclease recognizes the PAM, typically the NGG sequence adjacent to sgRNA in target DNA, and induces a double-strand break between the third and fourth nucleotides from PAM31, but the orientation of deletions has not been reported. We examined the possibility of preferential orientations and distinguished between symmetric and asymmetric deletions (Figure 5a). Only deletions obtained from injections of single sgRNAs were analysed, thereby avoiding effects of multiple variables (Figure 2 and Supplementary Note 1). Deletions upstream of the Cas9-cutting site that were equal or less than 1.5-fold compared to downstream ones were defined as symmetric. Deletions exceeding 1.5-fold at either end were considered asymmetric. More than 80% of the deletions detected in 139 founders obtained from targeting nine different sites were asymmetric and extended in either direction of the Cas9-cutting site (Figure 5b). Notably, more than 70% of the deletions exceeded a two-fold difference. Asymmetric deletions were prevalent in founders from all nine genomic sites (from 50 to 100% of frequency), suggesting that this result was not linked to specific sgRNA sequences (Figure 5c). Symmetric deletions were preferentially observed in small deletions of less than 10 bp (Figure 5d). From all asymmetric deletions, 59% extended towards the 5′ end and 41% towards the 3′ end of the sgRNA (Figure 5e,f). However, this was not statistically significant (*P*=0.6). The 18 previously published studies studies that have targeted individual loci in the mouse genome with single sgRNAs have not specifically addressed the symmetry of deletions10·11·17·18·31·32·33·34·35·36·37·38·39·40·41·42·43·44. Deleted sequences were available from seven studies11·18·31·34·39·43·44 (Supplementary Data 1), but only one18 showed large enough data sets to permit a direct comparison with ours. Kim and colleagues18 targeted two genomic sites based on 84 founders and with a cutoff of two-fold, 82% of the deletions were asymmetric compared to 73% in our study with 139 mice representing nine genomic sites (Supplementary Figure 2a). With a cutoff of 1.5-fold, 89% of the deletions were asymmetric compared to 82% in our study. While the frequency of asymmetric deletions was similar, the maximum deletion size was 585 bp in our study compared to 269 bp and the deletion sizes of top 50% are bigger in our study (Supplementary Figure 2b).

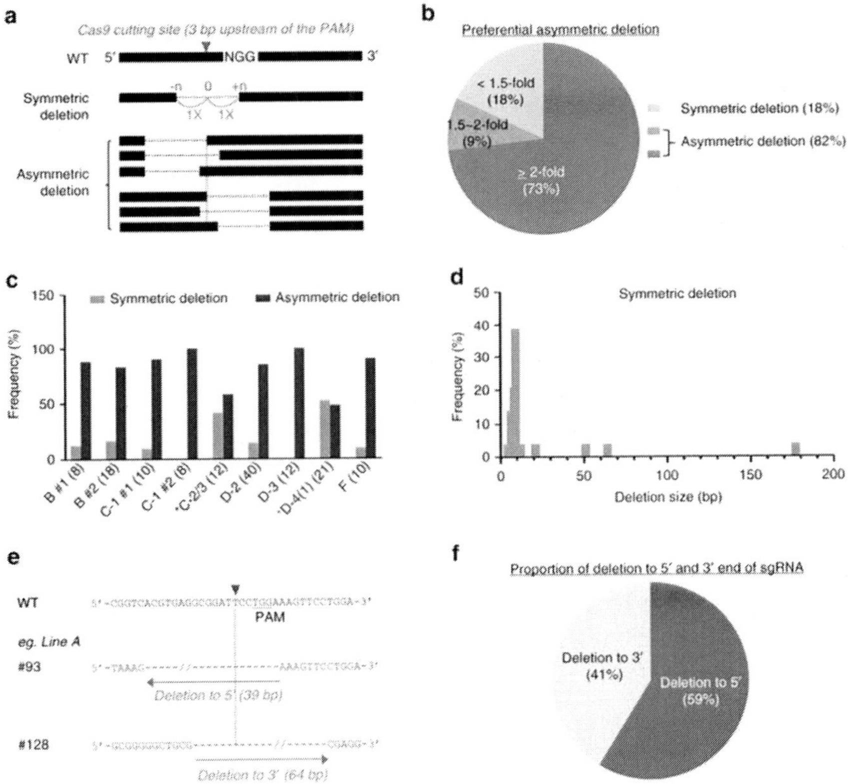

Figure 5: Asymmetric deletions. (**a**) Schematic diagram of symmetric and asymmetric deletions detected in mice targeted with a single sgRNA. Red triangle, Cas9-cutting site three base pairs upstream of the PAM sequence. Symmetric deletions were defined as those with an equal or less than 1.5-fold ratio between the upstream and downstream Cas9-cutting site. In asymmetric deletions, the difference at either site was more than 1.5-fold than at the other site. (**b**) Percentage of symmetric and asymmetric deletions identified in CRISPR/Cas9-targeted mice (n=139). More than 80% of deletions were asymmetric and more than 70% of the deletions exceeded a two-fold difference. Only deletions obtained from a single sgRNA injection were analysed to avoid the effect of multiple variables. Deletions obtained from a single sgRNA injection: deletions targeting TF-binding site B, C-1, C-2/3, D-2, D-3, D-4(1) and F. (**c**) Ratio of symmetric and asymmetric deletions obtained with each sgRNA. B #1, n=8; B #2, n=18; C-1 #1, n=10; C-1 #2, n=8; C-2/3, n=12, D-2, n=40; D-3, n=12; D-4(1), n=21; F, n=10. Asterisk (*), sgRNAs with identical deletions identified in more than one half of the founders. (**d**) Frequency of symmetric deletions obtained with different deletion sizes. (**e**) Representative examples of deletions towards the 5′ end and 3′ end of sgRNA. If the deletion at the upstream Cas9-cutting site was longer (≥1.5-fold) than that at the downstream one, it was defined as a 5′ deletion and vice versa. (**f**) Percentage of deletions towards the 5′ end and 3′ end of sgRNA.

Deletions preferentially occur at repeat sequences

We had noticed that with any given sgRNA 45% of the mutant founders carried apparently identical deletions, although mutations are supposed to be independent from each other. On detailed examination of such prevalent deletions, we determined that they frequently occurred at repeat sequences in targeted regions (Figure 6a). Notably, single or duplicated units of repeat sequences were retained at the deletion site. More than 60% of mutant founders from one specific genomic site (D-4), a CTCF-binding site in the *Wap* locus, carried the exact same deletion with the repeat sequences aligned at one end (Figure 6b, upper panel). Over 30% of mutant founders from another genomic site (C-2/3), two STAT5-binding sites within the *Wap* super-enhancer[19], had the same repeat sequence aligned at both ends (Figure 6b, bottom panel). Notably, 80% of deletions derived from one sgRNA (C-2/3) occurred at repeat sequences (Supplementary Figure 3). Among the entire cohort of founders (56) carrying deletions at repeat sequences, 65% had deletions within a single copy of repeat sequences and 35% of founders had the duplication of repeat sequences (Figure 6c). We exclusively analysed deletions obtained on targeting the genome with single sgRNAs (Figure 2 and Supplementary Note 1). Deletions with repeat sequences aligned at one end were probably due to microhomology-mediated repair as previously reported *in vitro*[45,46,47,48,49,50] and *in vivo*[40,51,52,53,54,55,56](Supplementary Table 1). However, to date the frequency of such deletion patterns had not been examined systematically in a large cohort and deletions with repeat sequences aligned at both ends had never been reported. Although the molecular mechanism that generates deletions with repeat sequences at both ends is unclear, microhomology-mediated end joining may facilitate the deletion with the repeat sequences aligned at one end[57] (Supplementary Figure 4). These results indicate that CRISPR-based deletions do not simply occur randomly, but with preferential patterns.

Large deletions created by single sgRNAs in zygotes

Zhou *et al.*[11] have reported that injections of dual adjacent sgRNAs at a given site not only improved the deletion efficiency, but also increased the deletion size in nine founder mice. To further assess whether large deletions are also obtained on targeting single sites, we investigated the extent of deletions obtained on injection of one or more sgRNAs corresponding to a single genomic site (Figure 7a). Although the median deletion size obtained with single sgRNAs (9 bp) (Figure 2 and Supplementary Note 1) was shorter than that gained with more than one adjacent sgRNAs (84 bp) (Figure 3 and Supplementary Note 2), we also observed large deletions of up to 600 bp with single sgRNAs (Figure 7b).

Notably, in one experiment the majority of founders (83%) exceeded the average deletion size (49 bp) and ~40% harboured deletions over 200 bp (Supplementary Figure 5). The deletion sizes generated by individual sgRNAs or more than one sgRNAs were independent of the guanine-cytosine (GC) content of the sgRNA or the distances between sgRNAs (Supplementary Figure 6).

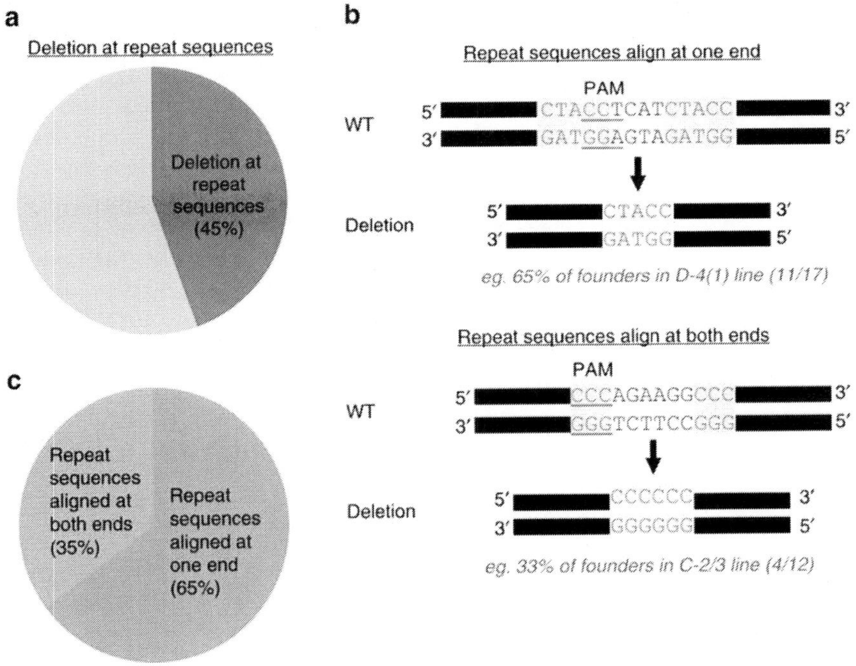

Figure 6: Preferential deletions at repeat sequences. (**a**) Average frequency of deletions found at repeat sequences by single sgRNA injections. Repeat sequences were aligned in ~45% at one end or both ends (total n=122; deletion at repeat sequences, n=56). Only deletions obtained from injections with single sgRNAs were analysed to avoid effects of multiple variables. Deletions obtained from single sgRNA injections: Deletions targeting TF-binding site B, C-1, C-2/3, D-2, D-3, D-4(1) and F. (**b**) Representative examples of repeat sequences aligned at one end (upper panel) and both ends (bottom panel). More than 60% of mutant founder mice targeting the genomic site D-4 exhibited the exact same deletion that retained only a single copy of repeat sequence. More than 30% of mutant founder mice targeting the genomic site C-2/3 showed the exact same deletion and repeat sequences remained at both ends. (**c**) Percentage of repeat sequences aligned at one end and both ends in founder mice carrying deletions at repeat sequences.

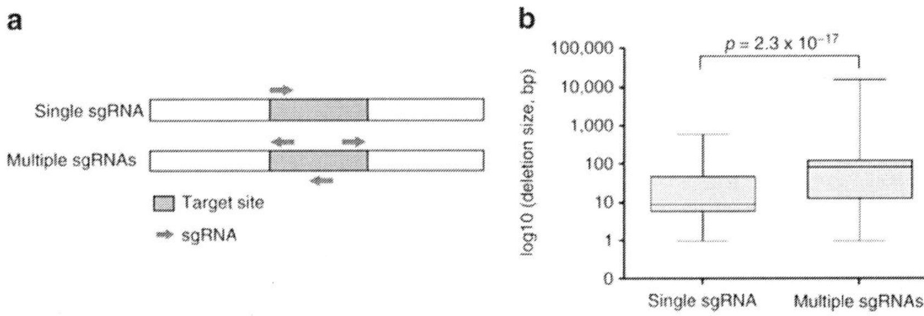

Figure 7: Large deletions obtained with single sgRNAs. (a) Diagram of single and multiple sgRNAs targeting specific sites. Targeted sites are shown in purple and sgRNAs are indicated as cyan arrows. (**b**) Comparison of deletion sizes generated by single sgRNAs and multiple sgRNAs injected into mouse zygotes (total number of founder mice, n=243; single sgRNA injection at a single site, n=122; multiple sgRNA injections at a single site, n=121). Single sgRNA injection at a single site, deletions obtained by targeting TF-binding site B, C-1, C-2/3, D-2, D-3, D-4(1) and F; multiple sgRNA injection at a single site, deletions obtained by targeting TF-binding site A, C-1/3, D-4(2), D-1/2/3/4, E-1, E-2, E-3, E-4 and E-5. The median deletion size obtained with single sgRNAs (9 bp) was smaller than that with multiple sgRNAs (84 bp). The deletion size generated with a single sgRNA was up to 600 bp. Median, middle bar inside the box; IQR, 50% of the data; whiskers, 1.5 times the IQR.

Sequential versus simultaneous deletion of adjacent sequences

Large deletions have been reported on simultaneously targeting more than one adjacent sites[7,9,13,14] (Supplementary Table 1). However, it is not clear if the deletion of juxtaposed sites can be achieved efficiently through the co-injection of the respective sgRNAs or whether a sequential deletion approach would be more robust. We addressed these questions and compared the deletion patterns obtained from mutant mice generated by simultaneous (one-step) or sequential (two-step) injection of sgRNAs covering seven sites (Figure 8a, Figure 4 and Supplementary Note 3). While the range of short deletions (<400 bp) obtained with both strategies was not significantly different (average deletion size of ~8 bp per site and ~39 bp per site), large deletions (>400 bp) of up to 24 kb were only obtained by co-targeting loci (Figure 8b). Strikingly, among 45 founders obtained on simultaneously injecting sgRNAs for two sites more than 50% of the deletions were classified as large (Figure 8c). Importantly, we initially failed to identify these large deletions due to the PCR screening strategy, which typically amplifies short fragments (~400 bp) spanning individual sites. These large deletions were only detected using serial PCR primers spanning the entire target region.

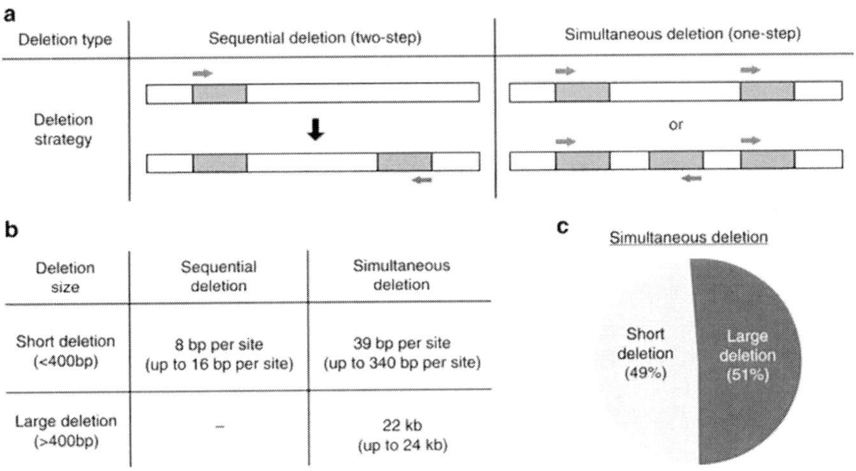

Figure 8: Deletion sizes obtained on sequentially (two-steps) and simultaneously (one-step) targeting the mouse genome.

(**a**) Schematic diagram of deleting several sites in a single gene locus using sequential or simultaneous sgRNA targeting. The targeted site is shown in purple and sgRNAs are indicated as cyan arrows. (**b**) Comparison of deletion sizes obtained from sequential and simultaneous sgRNA-targeting approaches. Sequential sgRNA injections, deletions obtained by targeting TF-binding site C-2/3; simultaneous sgRNA injections, deletions obtained by targeting TF-binding sites C-1/2, C-1/2/3, D-1/3/4 and D-1/2/3/4/5. Deletions smaller than 400 bp and identified by typical PCR genotyping method were called 'short deletion'. Those over 400 bp were considered 'large deletions'. Results are shown as the mean (total number of founder mice, n=65; sequential deletion, n=20; simultaneous deletion, n=45). (**c**) The percentage of short and large deletions obtained from mutant mice generated by simultaneous injection with more than one sgRNA.

Through in-depth sequence analysis, we identified two distinct deletion patterns, the 'stitched large deletion' and the 'continuous large deletion' (Supplementary Figure 7). In one experiment, we observed a combination of short and large deletions (>2 kb) in a 7 kb region, which resulted in the stitched large deletions (Supplementary Figure 7a). In another experiment, we also observed continuous large deletions over 20 kb in size that removed the entire sequence between sgRNAs. Strikingly, all nine founders from one particular experiment harboured these large deletions (Supplementary Figure 7b and Supplementary Note 3). Based on the definition of microhomology-based deletions45, we did not observe any microhomology-based large deletions with any given sgRNA injection method. Collectively, our results indicate that although the simultaneous targeting can rapidly generate deletions of multiple sites, it frequently creates large deletions, possibly removing potential regulatory elements. Although time consuming, two-step targeting appears to be the more reliable approach to precisely delete individual sites within a given locus.

INSERTIONS

In addition to deletions, we also observed insertions (Supplementary Notes 1–3 and Supplementary Table 3). The frequency of insertions was 4% on targeting individual genomic sites with single sgRNAs, 10% on targeting individual genomic sites with more than one sgRNAs and 6% on simultaneously targeting more than one genomic site with several sgRNAs (Supplementary Table 3). We have also observed two different types of insertions, insertion combined with deletions (Type A) and insertion only (Type B) (Supplementary Notes 1–3). Although most insertions consisted of only a few nucleotides, we also observed a large insertion of 800 nucleotides comprised of repetitive sequences (Supplementary Note 3a).

Avoiding genotyping pitfalls linked to large deletions

Standard PCR genotyping methods are usually employed to screen for desired mutations in founder mice (F0). Initially, we designed PCR primers to examine short (400–500 bp) genomic regions surrounding targeted sites. However, the presence of large deletions in one allele with the other allele being wild-type can cause misleading PCR results as only the wild-type allele would be amplified. These mice would be incorrectly categorized as wild-type based on their apparent genotype, which failed to detect the mutant allele (Figure 9a). Similarly, mice carrying deletions on both alleles could be misidentified. A regular PCR strategy would detect the small deletion on one allele but would miss a larger one on the second allele that extends the location of one or both primers. These mice would misleadingly appear to be homozygous and the 'hidden deletion' would go unnoticed. Indeed, most of the 'homozygous' founders identified in our study were not genuine homozygous mutants but rather compound heterozygotes (Supplementary Note 3). Only the use of PCR spanning the entire targeted loci of up to 30 kb revealed the biallelic complexity of CRISPR/Cas-induced deletions. The presence of biallelic deletions of different sizes at multiple target sites was even more difficult to decipher (Figure 9b). In our hands, 7 out of 30 founders were initially incorrectly categorized due to the large deletion (Supplementary Table 4). Lastly, when the large deletions were generated in intervening regions between two target sites, they were frequently missed if only the target sites were sequenced (Figure 9c). Such large deletions could compromise the validity of biological studies and to avoid such problems, alternative genotyping methods, such as genomic qPCR58 or even whole-genome sequencing, might be necessary.

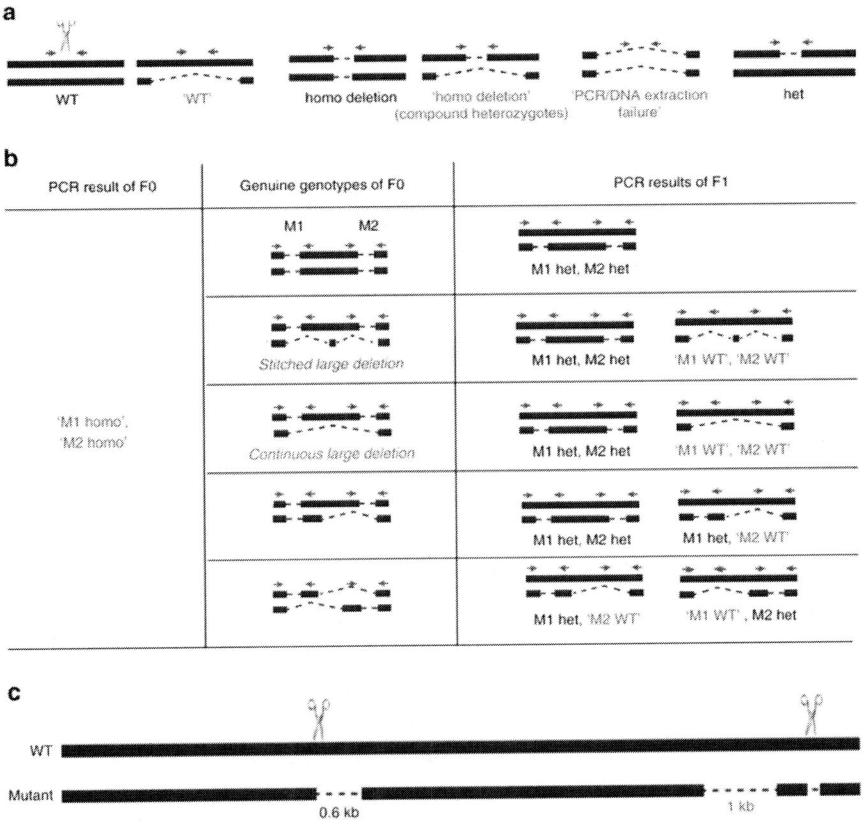

Figure 9: Screening for compound heterozygote deletions that are disguised as homozygote deletions. (**a**) Diagrams depict misleading genotyping results caused by large deletions at one of the two alleles. Misleading results are shown in red with quotation marks. All possible genotypes generated from single site mutagenesis are shown. (**b**) Diagrams depict examples of potential misleading genotyping results caused by stitched large deletions and continuous large deletions. Blue and red

arrows indicate PCR primers. Misleading results are shown in red with quotation marks. WT, wild-type; het, heterozygous; homo, homozygous; M1: mutated site 1; M2: mutated site 2. (**c**) Schematic demonstration of a mutant derived from simultaneous injection of two sgRNAs, which carries a 1 kb deletion between two targeting sites in addition to the desired short deletions at the target sites.

To decode complex genotypes, especially large deletions obtained with two or more sgRNAs targeting juxtaposed sites, we used serial PCRs spanning sequences within loci as well as outside primers spanning entire loci (Supplementary Figure 7c). Using this strategy, we easily identified deletions of more than 22 kb. In summary, the simultaneous sgRNA injections resulted in complex deletions and the generation of F1 mice is required to decode their exact genomic architecture. Moreover, simultaneous targeting of sites separated by up to 23 kb results in the deletion of the entire region. Thus, to restrict deletions to the desired sites we propose a sequential, two-step, targeting approach.

DISCUSSION

Although the CRISPR/Cas9 technology has been rapidly adopted as the premier genome editing tool for a range of organisms, much remains to be learned about the molecular consequences obtained on targeting the mouse genome, the collateral damage and the scars left behind. Based on more than 630 founder mice and 54 established lines covering 17 genomic sites, we have now acquired a more detailed understanding of deletion patterns obtained on injection of single or multiple sgRNAs into mouse zygotes. Preferential deletion patterns and large deletions were frequently obtained on targeting the mouse genome with single sgRNAs. Our studies also highlight that attempts to individually delete juxtaposed sites in a given locus through the co-injection of several sgRNAs almost exclusively results in large deletions spanning the entire sequence between the outside sgRNAs. This problem can be avoided by sequentially targeting sites, a time consuming, yet reliable, two-step process.

It has been suggested that CRISPR/Cas9-mediated DNA cleavage is randomly repaired by the NHEJ pathway59. However, we have observed hitherto unreported deletion patterns, which provided additional insight into the targeting accuracy. It has been reported that the seed sequence next to the PAM site are crucial for determining the target specificity60 and Cas9 usually cleaves between the third and fourth nucleotides from PAM31. Our study demonstrated that the majority of CRISPR/Cas9-mediated deletions using single sgRNAs occur asymmetrically, independent of the targeted locus. We also analysed data from a study by Kim's laboratory18 and found a similar preference of asymmetric deletions. These findings will help in predicting potential deletion sites and they may further aid sgRNA design to achieve targeted deletions at a high efficiency. Based on our findings that deletions frequently occur at repeat sequences a rationale design for sites with a high frequency of reproducible deletions should be possible.

When co-cutting a chromosome with two sgRNAs, the excised DNA fragment is occasionally inverted and re-integrated. Boroviak *et al.*61have elegantly demonstrated that such inversions happen frequently when two sgRNAs flanking large pieces of DNA (0.155–1.15 Mb) were co-injected into mouse zygotes. We have not specifically investigated the presence of inversions in mice generated by co-injecting two nearby sgRNAs. However, DNA sequencing analyses of PCR fragments spanning both cutting sites have not revealed any inversion of the intervening sequences. Possibly, the short DNA pieces excised in our experiments were quickly degraded by exonucleases and therefore they do not have a chance to be re-integrated into the chromosomes as the large pieces do. In one experiment, we deliberately deleted a 22 kb fragment by co-injecting sgRNAs, and all nine founder mice were apparently homozygous. Only analyses on the F1 generation determined that these mice were compound heterozygous, with the two alleles carrying slightly different deletions.

We have observed large deletions of up to 600 bp induced by single sgRNAs. We also obtained deletions of up to 24 kb induced by multiple sgRNAs targeting more than one juxtaposed site, which is in agreement with other studies62. Large deletions are easily overlooked in conventional PCR screening strategies and the observation that the two alleles can routinely harbour deletions of distinctly different sizes further complicates their identification. Based on our analyses, we suggest that true molecular changes can only be identified in the F1 generation. Moreover, mosaic mutations offer additional challenges that need to be sorted out in the F1 generation. To rapidly obtain homozygous mutants, apparent identical homozygous founders can be bred to each other. However, the possible misinterpretation of founder genotypes will yield undesirable compound heterozygosity. Thus, it would be prudent to rely only on genotypes from the F1 generation obtained by breeding founders with wild-type mice. Large deletions at target sites and even at some distance require additional analyses, possibly whole-genome sequencing, such as those reported to screen the off-target sites63.

METHODS

Mice

CRISPR/Cas9-targeted founder mice were obtained from the transgenic core of the National Heart, Lung, and Blood Institute. Eight-week-old C57BL/6 mice were purchased from Charles River and bred with founder mice to segregate the mosaicism. All animal procedures were followed by the National Institutes of Health, National Institute of Diabetes, Digestive and Kidney Diseases guidelines for the care and use of laboratory animals.

Design of CRISPR sgRNAs and microinjection into mouse zygotes

The CRISPR sgRNAs were designed based on the nearest PAM of the target sequence and their off-target scores were evaluated by the online tool at crispr.mit.edu[19,20]. Each sgRNAs were cloned into the pDR274 plasmid vector (Addgene #42250), and *in vitro* transcribed using the MEGAshortscript T7 kit (Life Technologies). Cas9 mRNA was *in vitro* synthesized from the MLM3613 plasmid vector (Addgene #42251) using the mMESSAGE mMACHINE T7 kit (Life Technologies). Cas9 mRNA (100 ng μl^{-1}) and sgRNAs (50 ng μl^{-1}) were microinjected into the cytoplasm of fertilized eggs of superovulated B6CBAF1/J female mice (JAX) and implanted into oviducts of pseudopregnant fosters (Swiss Webster, Taconic Farm). To target individual genomic sites, one or more than one sgRNA (up to three sgRNAs) were designed and injected independently or simultaneously. To target more than one genomic site, several sgRNAs (up to a total of four sgRNAs, two sgRNAs for each site) were injected simultaneously or sequentially. When we injected two sgRNAs, we normally used 50 ng μl^{-1} for each sgRNA. When more than two sgRNAs are injected, we reduced the concentration proportionally, so that the total sgRNA concentration does not exceed 100 ng μl^{-1} (that is, 25 ng μl^{-1} for each sgRNA was used when four sgRNAs are co-injected). Some sgRNAs caused lethality. When that happens, we normally diluted both Cas9 mRNA and sgRNA (first by 4-folds and then by 10-folds), which often improves litter sizes. The detailed targeting strategies for individual genomic sites were shown in Supplementary Notes 1, 2 and 3. The number of sgRNAs injected into the individual genomic sites: (1) Targeting individual genomic sites with single sgRNAs (Site B and C-1, two independent sgRNAs; site C-2/3, D-2, D-3, D-4(1), F, one sgRNA), (2) Targeting individual genomic sites with more than one sgRNA (Site A, C-1/3, D-1/2/3/4, E-1, two sgRNAs simultaneously; site

D-4(2), E-2, E-3, E-4, E-5, three sgRNAs simultaneously) and (3) Targeting more than one genomic site with several sgRNAs (Site C-1/2, C-1/2/3, D-1/3/4, D1/2/3/4/5, total four sgRNAs simultaneously).

Generation of CRISPR mouse lines and genotyping

Founder mice were bred with wild-type mice to obtain heterozygous F1 mice. F1 mice with an identical genotype were interbred to generate F2 homozygous mice. All mice were genotyped by PCR amplification of genomic DNA isolated from the tip of the tail, followed by Sanger sequencing. Large deletions were identified by serial PCR genotyping using primers that were designed to amplify ~400 bp encompassing the target sequence or long-range PCR.

Statistical analyses

All samples used for statistical analyses were randomly selected, and blinding was not applied. Before any statistical analysis, a Shapiro–Wilk normality test was applied to validate the normal distribution of the data. Significance for box plots was determined using Wilcoxon signed-rank test. A binomial test was employed to examine the statistical significance of prevalent orientation of deletions.

Data availability

The authors declare that the data supporting the findings of this study are available within the article and its Supplementary Information files or from the corresponding author on reasonable request.

Additional information

How to cite this article: Shin, H. Y. *et al.* CRISPR/Cas9 targeting events cause complex deletions and insertions at 17 sites in the mouse genome. *Nat. Commun.* **8**, 15464

Publisher's note: Springer Nature remains neutral with regard to jurisdictional claims in published maps and institutional affiliations.

ACKNOWLEDGEMENTS

This research was funded by the IPR of the NIDDK/NIH and by a grant of the Korean Health Technology R&D Project, Ministry of Health & Welfare, Republic of Korea (HI15C1184) to H.K.L.

REFERENCES

1. Jinek, M. et al. RNA-programmed genome editing in human cells. Elife 2, e00471 (2013).
2. Cong, L. et al. Multiplex genome engineering using CRISPR/Cas systems. Science 339, 819–823 (2013).
3. Mali, P. et al. RNA-guided human genome engineering via Cas9. Science 339, 823–826 (2013).
4. Lander, E. S. The heroes of CRISPR. Cell 164, 18–28 (2016).
5. Komor, A. C., Badran, A. H. & Liu, D. R. CRISPR-based technologies for the manipulation of eukaryotic genomes. Cell 168, 20–36 (2017).
6. Thomas, K. R. & Capecchi, M. R. Site-directed mutagenesis by gene targeting in mouse embryo-derived stem cells. Cell 51, 503–512 (1987).
7. Wang, H. et al. One-step generation of mice carrying mutations in multiple genes by CRISPR/Cas-mediated genome engineering. Cell 153, 910–918 (2013).
8. Jinek, M. et al. A programmable dual-RNA-guided DNA endonuclease in adaptive bacterial immunity. Science 337, 816–821 (2012).
9. Wang, L. et al. Large genomic fragment deletion and functional gene cassette knock-in via Cas9 protein mediated genome editing in one-cell rodent embryos. Sci. Rep. 5, 17517 (2015).
10. Nakao, H. et al. A possible aid in targeted insertion of large DNA elements by CRISPR/Cas in mouse zygotes. Genesis 54, 65–77 (2016).
11. Zhou, J. et al. Dual sgRNAs facilitate CRISPR/Cas9-mediated mouse genome targeting. FEBS J. 281, 1717–1725 (2014).
12. Yang, H. et al. One-step generation of mice carrying reporter and conditional alleles by CRISPR/Cas-mediated genome engineering. Cell 154, 1370–1379 (2013).
13. Fujii, W., Kawasaki, K., Sugiura, K. & Naito, K. Efficient generation of large-scale genome-modified mice using gRNA and CAS9 endonuclease. Nucleic Acids Res. 41, e187 (2013).
14. Hara, S. et al. Microinjection-based generation of mutant mice with a double mutation and a 0.5 Mb deletion in their genome by the CRISPR/Cas9 system. J. Reprod. Dev. 62, 531–536 (2016).
15. Seruggia, D., Fernandez, A., Cantero, M., Pelczar, P. & Montoliu, L. Functional validation of mouse tyrosinase non-coding regulatory DNA elements by CRISPR-Cas9-mediated mutagenesis. Nucleic Acids Res. 43, 4855–4867 (2015).
16. Yen, S. T. et al. Somatic mosaicism and allele complexity induced by CRISPR/Cas9 RNA injections in mouse zygotes. Dev. Biol. 393, 3–9 (2014).
17. Han, Y., Slivano, O. J., Christie, C. K., Cheng, A. W. & Miano, J. M. CRISPR-Cas9 genome editing of a single regulatory element nearly abolishes target gene expression in mice--brief report. Arterioscler. Thromb. Vasc. Biol. 35, 312–315 (2015).
18. Sung, Y. H. et al. Highly efficient gene knockout in mice and zebrafish with RNA-guided endonucleases. Genome Res. 24, 125–131 (2014).
19. Shin, H. Y. et al. Hierarchy within the mammary STAT5-driven Wap super-enhancer. Nat. Genet. 48, 904–911 (2016).

20. Metser, G. et al. An autoregulatory enhancer controls mammary-specific STAT5 functions. Nucleic Acids Res. 44, 1052–1063 (2016).
21. Huang, J. et al. Dynamic control of enhancer repertoires drives lineage and stage-specific transcription during hematopoiesis. Dev. Cell 36, 9–23 (2016).
22. Korkmaz, G. et al. Functional genetic screens for enhancer elements in the human genome using CRISPR-Cas9. Nat. Biotechnol. 34, 192–198 (2016).
23. Diao, Y. et al. A new class of temporarily phenotypic enhancers identified by CRISPR/Cas9-mediated genetic screening. Genome Res. 26, 397–405 (2016).
24. Li, Y. et al. CRISPR reveals a distal super-enhancer required for Sox2 expression in mouse embryonic stem cells. PLoS ONE 9, e114485 (2014).
25. Willi, M., Yoo, K. H., Wang, C., Trajanoski, Z. & Hennighausen, L. Differential cytokine sensitivities of STAT5-dependent enhancers rely on Stat5 autoregulation. Nucleic Acids Res. 44, 10277–10291 (2016).
26. Zeng, X., Willi, M., Shin, H. Y., Hennighausen, L. & Wang, C. Lineage-specific and non-specific cytokine-sensing genes respond differentially to the master regulator STAT5. Cell Rep. 17, 3333–3346 (2016).
27. Ghirlando, R. & Felsenfeld, G. CTCF: making the right connections. Genes Dev. 30, 881–891 (2016).
28. Grunder, A. et al. Nuclear factor I-B (Nfib) deficient mice have severe lung hypoplasia. Mech. Dev. 112, 69–77 (2002).
29. Chang, C. Y. et al. NFIB is a governor of epithelial-melanocyte stem cell behaviour in a shared niche. Nature 495, 98–102 (2013).
30. Robinson, G. W. et al. Coregulation of genetic programs by the transcription factors NFIB and STAT5. Mol. Endocrinol. 28, 758–767 (2014).
31. Shen, B. et al. Generation of gene-modified mice via Cas9/RNA-mediated gene targeting. Cell Res. 23, 720–723 (2013).
32. Young, S. A. et al. CRISPR/Cas9-mediated rapid generation of multiple mouse lines identified ccdc63 as essential for spermiogenesis. Int. J. Mol. Sci. 16, 24732–24750 (2015).
33. Mashiko, D. et al. Feasibility for a large scale mouse mutagenesis by injecting CRISPR/Cas plasmid into zygotes. Dev. Growth Differ. 56, 122–129 (2014).
34. Horii, T. et al. Validation of microinjection methods for generating knockout mice by CRISPR/Cas-mediated genome engineering. Sci. Rep. 4, 4513 (2014).
35. Qin, W. et al. Efficient CRISPR/Cas9-mediated genome editing in mice by zygote electroporation of nuclease. Genetics 200, 423–430 (2015).
36. Hu, J. et al. Endothelin signaling activates Mef2c expression in the neural crest through a MEF2C-dependent positive-feedback transcriptional pathway. Development 142, 2775–2780 (2015).
37. Menoret, S. et al. Homology-directed repair in rodent zygotes using Cas9 and TALEN engineered proteins. Sci. Rep. 5, 14410 (2015).
38. Tsuchiya, Y. et al. Disruption of MeCP2 attenuates circadian rhythm in CRISPR/Cas9-based Rett syndrome model mouse. Genes Cells 20, 992–1005 (2015).
39. Shinmyo, Y. et al. CRISPR/Cas9-mediated gene knockout in the mouse brain using in utero electroporation. Sci. Rep. 6, 20611 (2016).
40. Ott de Bruin, L. et al. Rapid generation of novel models of RAG1 deficiency by CRISPR/Cas9-induced mutagenesis in murine zygotes. Oncotarget 7, 12962–12974 (2016).
41. Miyasaka, Y. et al. Heterozygous mutation of Ush1g/Sans in mice causes early-onset progressive hearing loss, which is recovered by reconstituting the strain-specific mutation in Cdh23. Hum. Mol. Genet. 25, 2045–2059 (2016).
42. Guan, Y. et al. CRISPR/Cas9-mediated somatic correction of a novel coagulator factor IX gene mutation ameliorates hemophilia in mouse. EMBO Mol. Med. 8, 477–488 (2016).
43. An, L. et al. Efficient generation of FVII gene knockout mice using CRISPR/Cas9 nuclease and truncated guided RNAs. Sci. Rep. 6, 25199 (2016).

44. Marusugi, K. et al. Functional validation of tensin2 SH2-PTB domain by CRISPR/Cas9-mediated genome editing. J. Vet. Med. Sci. 78, 1413–1420 (2016).
45. Bae, S., Kweon, J., Kim, H. S. & Kim, J. S. Microhomology-based choice of Cas9 nuclease target sites. Nat. Methods 11, 705–706 (2014).
46. Qu, X. et al. Zinc finger nuclease: a new approach for excising HIV-1 proviral DNA from infected human T cells. Mol. Biol. Rep. 41, 5819–5827 (2014).
47. Ghezraoui, H. et al. Chromosomal translocations in human cells are generated by canonical nonhomologous end-joining. Mol. Cell 55, 829–842 (2014).
48. Li, J. et al. Efficient inversions and duplications of mammalian regulatory DNA elements and gene clusters by CRISPR/Cas9. J. Mol. Cell Biol. 7, 284–298 (2015).
49. Lu, G. et al. Ligase I and ligase III mediate the DNA double-strand break ligation in alternative end-joining. Proc. Natl Acad. Sci. USA 113, 1256–1260 (2016).
50. Chuai, G. et al. Deciphering relationship between microhomology and in-frame mutation occurrence in human CRISPR-based gene knockout. Mol. Ther. Nucleic Acids 5, e323 (2016).
51. Gratz, S. J. et al. Genome engineering of Drosophila with the CRISPR RNA-guided Cas9 nuclease. Genetics 194, 1029–1035 (2013).
52. Ge, D. T., Tipping, C., Brodsky, M. H. & Zamore, P. D. Rapid screening for CRISPR-directed editing of the drosophila genome using white coconversion. G3 (Bethesda) 6, 3197–3206 (2016).
53. He, M. D. et al. Efficient ligase 3-dependent microhomology-mediated end joining repair of DNA double-strand breaks in zebrafish embryos. Mutat. Res. 780, 86–96 (2015).
54. Peng, D., Kurup, S. P., Yao, P. Y., Minning, T. A. & Tarleton, R. L. CRISPR-Cas9-mediated single-gene and gene family disruption in Trypanosoma cruzi. MBio 6, e02097–14 (2014).
55. Singer, M. et al. Zinc finger nuclease-based double-strand breaks attenuate malaria parasites and reveal rare microhomology-mediated end joining. Genome Biol. 16, 249 (2015).
56. Zhang, W. W. & Matlashewski, G. CRISPR-Cas9-mediated genome editing in Leishmania donovani. MBio 6, e00861 (2015).
57. McVey, M. & Lee, S. E. MMEJ repair of double-strand breaks (director's cut): deleted sequences and alternative endings. Trends Genet. 24, 529–538 (2008).
58. Spielmann, M. et al. Exome sequencing and CRISPR/Cas genome editing identify mutations of ZAK as a cause of limb defects in humans and mice. Genome Res. 26, 183–191 (2016).
59. Maruyama, T. et al. Increasing the efficiency of precise genome editing with CRISPR-Cas9 by inhibition of nonhomologous end joining. Nat. Biotechnol. 33, 538–542 (2015).
60. Semenova, E. et al. Interference by clustered regularly interspaced short palindromic repeat (CRISPR) RNA is governed by a seed sequence. Proc. Natl Acad. Sci. USA 108, 10098–10103 (2011).
61. Boroviak, K., Doe, B., Banerjee, R., Yang, F. & Bradley, A. Chromosome engineering in zygotes with CRISPR/Cas9. Genesis 54, 78–85 (2016).
62. Zhang, L. et al. Large genomic fragment deletions and insertions in mouse using CRISPR/Cas9. PLoS ONE 10, e0120396 (2015).
63. Bolukbasi, M. F., Gupta, A. & Wolfe, S. A. Creating and evaluating accurate CRISPR-Cas9 scalpels for genomic surgery. Nat. Methods 13, 41–50 (2016).

CHAPTER 19

In vitro Developmental Ability of Ovine Oocytes Following Intracytoplasmic Injection with Freeze-dried Spermatozoa

Maite Olaciregui, Victoria Luño, Paula Domingo , Noelia González & Lydia Gil

Animal Pathology Department of Veterinary Faculty, Universidad de Zaragoza, Miguel Servet 177, 50013, Zaragoza, Spain.

ABSTRACT

Freeze-drying (FD) is a new and alternative method to preserve spermatozoa in refrigeration or at room temperature. Suitable protection is required to maintain the sperm DNA integrity during the whole process and storage. The aim of this study was to examine the effect of rosmarinic acid and storage temperature on the DNA integrity of freeze-dried ram sperm. In addition, we evaluated the *in vitro* developmental ability to the blastocyst stage of oocytes injected with freeze-dried sperm. Ram sperm was freeze-dried in basic medium and in this medium supplemented with 105 μM rosmarinic acid. The vials were stored for 1 year at 4 °C and at room temperature. Frozen sperm was used as control. After rehydration, sperm DNA damage was evaluated, observing that the percentage of spermatozoa with DNA damage decreased significantly in the presence of rosmarinic acid, without differences between the two storage temperatures. Moreover, no differences were observed between the freeze-dried group and the frozen-thawed group in terms of blastocyst formation rate. We proved for the first time that ovine spermatozoa can be lyophilized effectively, stored at room temperature for long term, reconstituted and further injected into oocytes with initial embryo development.

INTRODUCTION

Freeze-drying sperm preservation is a new and attractive method by which spermatozoa can be long-term stored both in a refrigerator and at room temperature. That means that no liquid nitrogen is required, which considerably reduces the costs of doses storage and shipping. Nowadays, this method has been applied to different domestic species, such as rabbit[1], horse[2,3], cat[4], dog[5,6], boar[7] and bull[8]. However, we did not find any study in which this technique had already been applied on ram sperm.

In spite of the advantages of this method, the main limitation is still the damage caused on the sperm. In particular, the spermatozoa lose their motility after freeze-drying and intracytoplasmic sperm injection (ICSI) is needed to fertilize oocytes[9]. Besides, sperm DNA might be damaged by the action of DNAases or oxidative stress during the freezing and freeze-drying processes. Kusakabe *et al.*[10] suggested that the release of endogenous nucleases from the plasma membrane of damaged spermatozoa after sperm freeze-drying is the most likely reason to produce structural chromosome aberrations. They found that chromosome integrity of mouse spermatozoa can be maintained during freeze-drying when the spermatozoa are suspended in a simple Tris-HCl-buffered solution containing 50 mM EGTA and 50 mM NaCl.

On the other hand, oxidative stress might represent another mechanism of damage, leading to spermatozoa DNA fragmentation during freeze-drying and rehydration procedures[10,11]. Exposure of DNA to ROS (reactive oxygen species) induces a wide variety of chemical changes that include cross-linking, base modification and DNA strand breakage[12]. It has been already reported that the ram is one of the species with the fastest sperm DNA degradation[13]. The beneficial effect of antioxidant therapy on the oxidative stress in mammalian spermatozoa has widely been studied[14,15,16,17,18,19]. It has been well characterized that the addition of antioxidants to freezing extenders decreases the detrimental effects of ROS[20]. Rosmarinic acid (RA) is a natural antioxidant that protects cells from oxidative stress *in vitro*[21]. It has been proved that the incorporation of RA in freezing extenders improved sperm quality after cryopreservation[20]. However, to our knowledge, no studies have been performed to evaluate the effect of antioxidants on freeze-dried ram sperm.

Furthermore, the DNA of the freeze-dried sperm can also be damaged if the adequate protection is not provided during the storage[22]. Preservation at a low temperature is important for a stable long-term storage of freeze-dried spermatozoa, being 4 °C the most suitable temperature for this purpose[23, 24]. Nevertheless, storage at room temperature would be ideal and would allow a simple transport of freeze-dried samples around the world[25]. Kusakabe *et al.*[10, 11] already determined that freeze-dried samples can be transported for long distances at ambient temperatures without any consequence regarding chromosomal damage or loss of developmental potential. However, it is still required to further develop new strategies in order to be able to transport and to preserve indefinitely sperm samples with no impact on sperm quality and excluding liquid nitrogen or dry ice.

The aim of this study was to examine the effect of rosmarinic acid and the influence of the storage temperature on the DNA integrity of freeze-dried ram sperm. In addition, we evaluated the *in vitro* developmental ability to the blastocyst stage of injected ovine oocytes with freeze-dried ram sperm.

RESULTS

Effect of rosmarinic acid and storage temperature on DNA integrity

Halomax reflected the DNA fragmentation in each sperm head; the bigger size of halos indicated the more accelerated level of DNA fragmentation (Figure 1). The results obtained by using the sperm chromatin dispersion test to analyze the DNA integrity of the freeze-dried ram sperm are showed in Figure 2. The sperm samples evaluated after thawing (control) showed higher percentage of sperm with intact DNA (97.4%) than sperm freeze-dried with EGTA without rosmarinic acid supplementation. Regardless of the storage temperature, when RA was added to freeze-drying medium the percentage of spermatozoa with DNA damage decrease significantly ($p = 0.003$). The spermatozoa lyophilized with EGTAR had 2.9% chromatin damaged cells (large and spotty halo) whereas sperm lyophilized with EGTA had 4.4% chromatin damaged cells. For both storage temperatures, 4 °C ($p = 0.004$) and 22 °C ($p = 0.032$), the DNA damage levels in EGTA group were higher than those in EGTAR and control groups

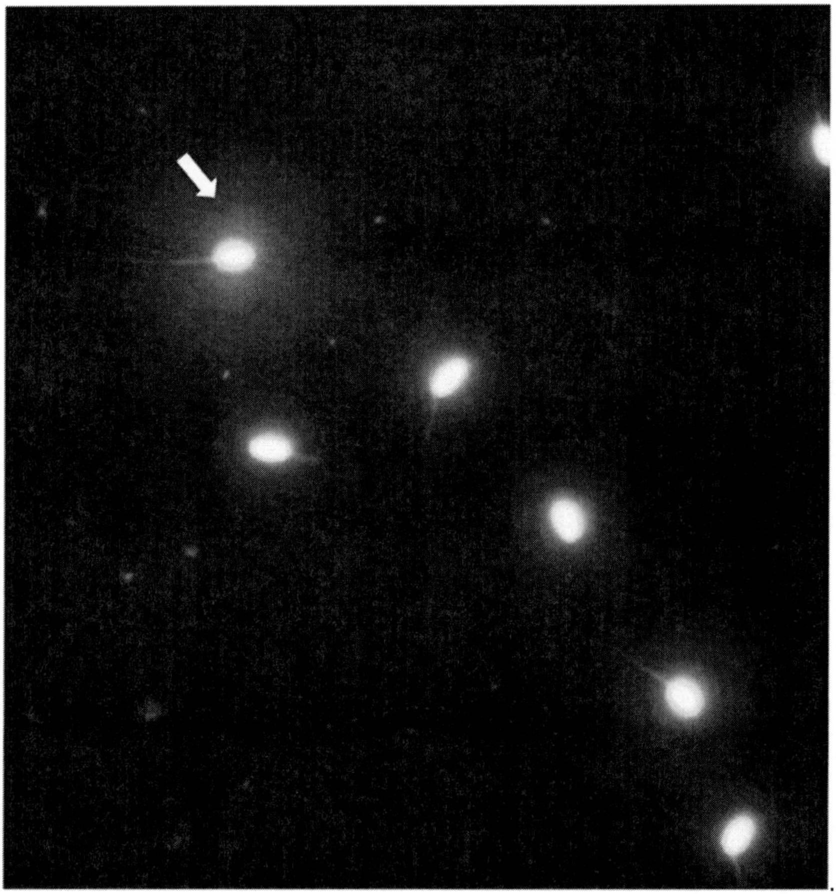

Figure 1. SCD-processed freeze-dried sperm sample with Syber-Green fluorescence: sperm nuclei with fragmented DNA exhibit a large and spotty halo of chromatin dispersion (arrow). Sperm nuclei that exhibited small and compact halos of chromatin dispersion corresponded to spermatozoa with unfragmented DNA.

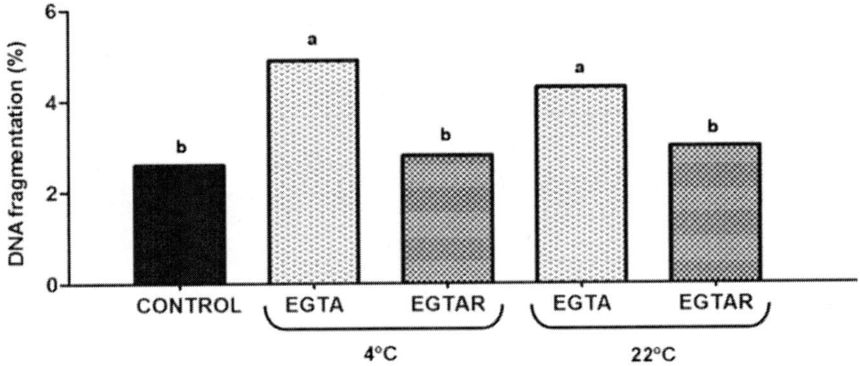

Figure 2. Effects of rosmarinic acid and storage temperature on DNA integrity of freeze-dried ram sperm. Sperm DNA fragmentation was assessed by Sperm cromatin dispersion test (SCD). Five replicated trials were carried out for each group and a minimum of 300 spermatozoa were counted per semen sample. Different letters indicate significant differences at the $P < 0.05$ level, among the four treatments studied.

When the temperature of storage was analyzed, there were no significant differences between the two temperatures studied ($p = 0.269$). Regardless of the freeze-drying medium used, the percentage of sperm with intact DNA no differed when they were storage at 4 °C (96.2%) or at room temperature (96.5%).

Freeze-dried sperm in both treatments and stored at room temperature were used for injection into IVM oocytes for examination of subsequent *in vitro* developmental competence.

Embryonic development obtained after ICSI with frozen-thawed or lyophilized spermatozoa

The nuclear status analyzed after injection and cultured for 16 hours are shown in Table 1. Zygotes were categorized as normally fertilized if one female and one male pronuclei (2 PN) were formed Figure 3; whereas oocytes with two polar bodies (PBs) and one PN were considered activated without sperm participation. There were no significant differences in term of fertilization rate among the different freeze-dried groups studied, and among freeze-dried sperm groups and frozen thawed control group.

Table 1: Female and male pronucleus formation in *in vitro* matured oocytes after injection with frozen-thawed and freeze-dried sperm.

Group	N° Oocytes examined	% of oocyte with 2 PB_S and 2 PN_S	% of oocytes with 2 PB_S and 1 PN
Frozen-thawed	30	64.52	12.30
EGTA	30	57.10	9.63
EGTAR	30	61.44	10.26

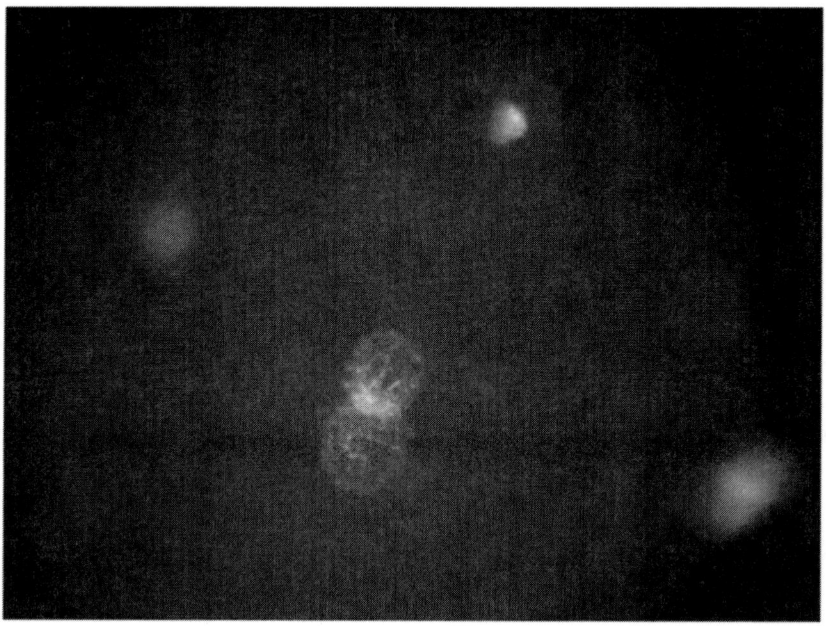

Figure 3. Pronuclear stage evaluated 16 hours after injection with freeze-dried sperm and stained with Hoechst 33342: two polar bodies and two pronucleus.

Results of embryonic development from the two groups of freeze-dried sperm (EGTA and EGTAR) and frozen-thawed sperm are shown in Table 2. There were no significant differences among all groups studied on the cleavage rate evaluated 48 h post ICSI. The rate of development to the blastocyst stage was similar for the frozen-thawed control group and freeze-dried group, but superior to parthenogenetic control (sham injection). In addition, no differences were observed between the two freeze-dried sperm groups related to the blastocyst formation rate. Representative embryos resulting from ICSI are shown in Figure 4.

Table 2: *In vitro* development of ovine oocytes after injection with fresh or freeze-dried sperm.

Group	N° Oocytes examined	Cleavage (%)	Embryo development (%)
Frozen-thawed	49	34.7	24.5
EGTA	101	36.6	25.6
EGTAR	87	36.8	24.5
Sham injection	53	30.2	9.4

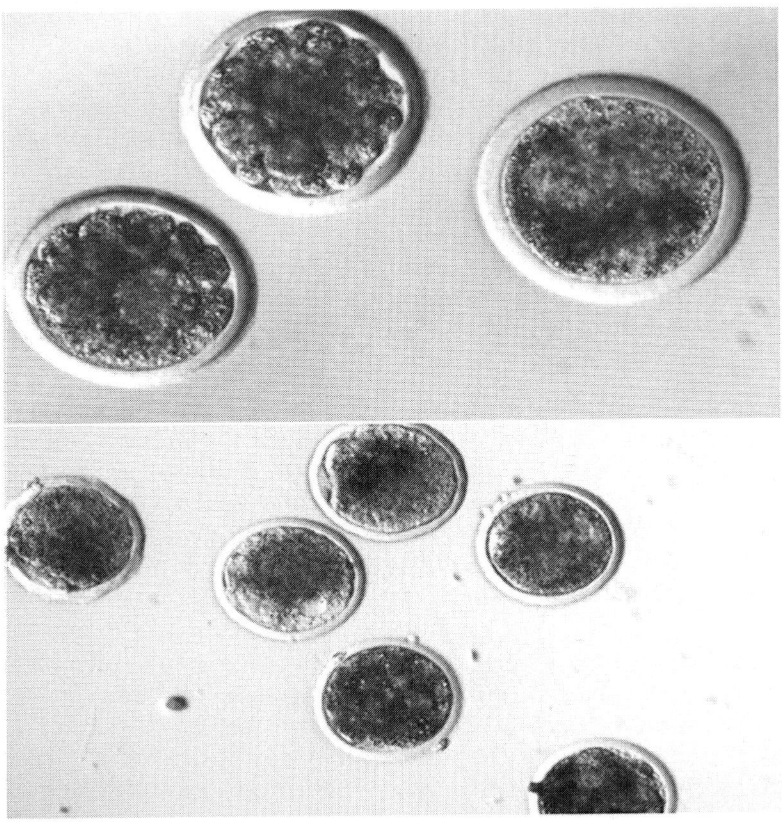

Figure 4. Ovine embryos developed from oocytes 168 h after injection with freeze-dried sperm.

DISCUSSION

The freeze-drying technique has recently been applied as a novel method to maintain animal sperm samples. It is expected to be a future tool in animal reproduction technology in order to improve biobanking and storage of genetic diversity26.

The present study has demonstrated for the first time that highly effective ovine fertilization and blastocyst development were pursued by using freeze-dried ram spermatozoa and ICSI. A significant proportion of oocytes reached the blastocyst stage when they were injected with freeze-dried spermatozoa. Previous studies using boar sperm suggested that activation and development of oocytes impaired significantly when they were injected with freeze-dried spermatozoa, in comparison with those produced with fresh spermatozoa9, 27. However, in this study a similar proportion of blastocysts was obtained when the oocytes were injected with either frozen-thawed or with freeze-dried ram sperm. Kwon *et al.*9 reported that substances from sperm heads responsible for oocyte activation are inhibited or inactivated during the freeze-drying processes. In the present study EGTA was added to the freeze-dying extender and no differences were observed between control group (frozen-thawed) and freeze-dried groups in terms of blastocyst formation rate. These results can be explained by the fact that chelating agent inactivates sperm DNase and protects sperm DNA from against disruption during the freeze-drying steps and the subsequent storage10. Therefore, chromosomal damage and further negative impact on embryo development can be avoided by using EGTA on freeze-dried spermatozoa.

Another potential source of damage with negative consequences on the quality of embryonic development is the oxidative damage of sperm DNA12. Oxidative stress is created when the cellular generation of ROS overwhelms the antioxidant defenses27, specially detected during freezing, freeze-drying and rehydration processes. During the cryopreservation, reactive oxygen species promote DNA fragmentation28; this is the reason by which the addition of antioxidants to sperm extenders became one of the most suitable alternatives to improve the viability of cryopreserved spermatozoa in different mammalian species such as bull29, boar30 and ram31. It was already suggested that protection of sperm DNA is one of the most important factors for the correct development of oocytes after fertilization; although spermatozoa with fragmented DNA are able to fertilize oocytes, the rate and quality of pregnancy rates decreased significantly32, 33. Several studies16, 17, 18, 19, 34 have shown the beneficial effect of antioxidant therapy on the oxidative stress of mammalian spermatozoa. The addition of antioxidants has important benefits on the quality and fertility of ram semen after liquid storage at different temperatures35. To date, there have not been studies focus on determine the antioxidant effect on freeze dried sperm. We have demonstrated for the first time that the presence of antioxidant in the EGTA solution, which was used to freeze-drying, efficiently preserved the DNA integrity of ovine sperm. Therefore, results of this investigation demonstrated that rosmarinic acid has a protective effect against sperm DNA fragmentation during ram sperm freeze-drying process. Rosmarinic acid is a natural antioxidant with therapeutic effects against oxidative stress on cell *in vitro*36, 37. In previous studies20, rosmarinic acid showed a high-efficiency free radical-scavenging effect, and it is an important supplement for boar semen cryopreservation. They found that rosmarinic acid supplementation improved the post-thaw quality of boar spermatozoa and the ability to fertilize oocytes *in vitro*. In the present study, freeze-dried semen samples in a medium supplemented with 105 μM rosmarinic acid showed the best results in terms of sperm DNA integrity after rehydration. However, fertilization rate and embryonic development did not differ between oocytes injected with sperm freeze-dried in the presence of rosmarinic acid and those samples lacking this compound. It is well known that some DNA repair factors exist in the oocyte38, 39, and fertilized oocytes have the capacity to solve DNA damaging processes40. It is true that certain levels of sperm DNA fragmentation have no impact on the fertilization rate and do not compromise blastocyst yield after ICSI41. However, when the DNA damage is recognized as irreparable, embryos are excluded by cell cycle arrests or activation of apoptotic pathways42. Moreover, *in vitro* culture of oocytes and embryos may lead to dysregulation of many genes43,44,45 resulting in low cellular viability and long-term embryo viability because of a worse competence for DNA damage repair. This repair capacity mainly depends on the length of sperm DNA fragmentation and the cytoplasmic quality of the oocyte. Single-strand breaks (SSBs) could be quickly repaired by oocytes after fertilization, but double-strand breaks (DSBs) could be responsible for chromosome aberrations and loss of genetic materials. Thus, the repair of DSBs in oocytes is more difficult than that of SSBs46.

In this study, rosmarinic acid did not increase cleavage rates *in vitro*, although rosmarinic acid-treated sperm exhibited a lower degree of DNA fragmentation. It has been indicated that fertilization with sperm exposed to a DNA damaging agent alters the expression of DNA repair genes at the one-cell stage rat preimplantation-embryo47. It is actually possible that the increase in DNA damage on the sperm group without rosmarinic acid leads to an increase in the expression of DNA repair genes in those oocytes injected with spermatozoa from that group.

Men *et al.*27 showed that trehalose performed a protective effect against sperm DNA fragmentation on freeze-dried boar sperm, but fertilization and embryonic development were not improved. That study included the analysis of the expression profile of some DNA repair genes in porcine oocytes after intra-cytoplasmic sperm injection[7] and they concluded that expression of DNA repair genes in fertilized oocytes after ICSI using fresh or freeze-dried sperm, in the presence or in the absence of trehalose, was not different. However, the results of our present study indicated the possibility that the ability of oocytes for DNA repair might neutralize the differences in the level of sperm DNA damage between the two experimental groups. It has been reported that sperm DNA damage levels are directly related to embryonic cleavage rates, so that the activation of additional DNA repair pathways forces delays in cell division[48,49,50] resulting in a poor quality blastocyst[51].

In the present study, we have generated novel data regarding the influence of the storage temperature on freeze-dried sperm samples. We proved that freeze-dried ram spermatozoa can maintain their DNA integrity for long-term storage at room temperature. The first report regarding storage temperature on freeze-dried mouse sperm showed that the fertilizing ability of sperm was impaired when samples were stored at 25 °C[52]. In addition, freeze-dried rat spermatozoa stored at 4 °C for 1 year exhibited lower frequent chromosomal abnormalities than those stored at 25 °C[53]. Several reports indicated that preservation at a low temperature is important for long-term stable preservation of freeze-dried spermatozoa, being 4 °C the most optimal temperature for the preservation of freeze-dried spermatozoa[54,55]. One of the main goals of freeze-drying studies is the long term storage and simple shipment of spermatozoa at ambient temperature. However, for freeze-dried spermatozoa subjected to the present procedure, long-term storage at 25 °C has not been achieved yet in any mammalian specie. Kwon *et al.*[9] observed the development to the blastocyst stage only in those oocytes injected with freeze-dried boar sperm stored at 4 °C and not in those injected with freeze-dried sperm stored at 25 °C. On the other hand, normal offspring obtained from oocytes fertilized with freeze-dried mouse spermatozoa that were air-transported at ambient temperatures from Japan to the USA was obtained[56,57]. In the present study freeze-dried semen samples were stored at 25 °C for one year, leading to the obtention of normal fertilization and blastocyst formation after that sperm was injected in oocytes.

In conclusion, our data confirm for the first time that ovine spermatozoa can be effectively lyophilized, long-term stored at room temperature and successful for embryo development after injected into oocytes. We also demonstrated that the addition of rosmarinic acid to the medium improves sperm DNA integrity after freeze-drying procedures.

METHODS

Chemicals and media

Unless otherwise stated, all chemicals were from Sigma-Aldrich Co (Alcobendas, Madrid, Spain). The freeze-drying media tested were: EGTA: 10 mM Tris-HCl buffer, 50 mM NaCl and 50 mM EGTA; and EGTAR: EGTA supplemented with 105 μM rosmrinic acid20. The final pH of both solutions was adjusted to 8.0–8.2.

The medium used for the collection and washing of cumulus oocyte complexes (COCs) was Dulbeco's phosphate buffered saline (DPBS) composed of 136.89 mM NaCl, 2.68 mM KCl, 8.1 mM Na_2HPO_4 and 1.46 mM $CaCl_2 2H_2O$, and supplemented with heparin (50 IU/ml).

The medium used for oocyte maturation was TCM199 supplemented with sheep serum (SS 10% v/v), FSH (0.01 IU/ml), LH (0.02 IU/ml), sodium pyruvate (0.2 mM), estradiol-17b (1 μg/ml), glutamine (2 mM), cysteamine (0.02 mg/ml) and penicillin-streptomycin (0.05 mg/ml). The medium osmolarity was adjusted to 275 mOsm. ICSI was performed in 10 μl droplets of SOF (Synthetic Oviduct Fluid) supplemented with 0.4% BSA and 20 mM of Hepes (BHSOF) under mineral oil. The medium used for *in vitro* culture of sperm-injected oocytes was SOF supplemented with 2% (v/v) BME-essential amino acids, 1% (v/v) MEM-nonessential amino acids, 5% (v/v) Fetal Bovine Serum, 1 mM glutamine, 8 mg/mL BSA and 40 μg/ml gentamicin.

Sperm collection and freeze-drying procedures

The study was performed following approval by the Veterinary Ethical Committee of University of Zaragoza. The care and use of animals were performed according to the Spanish Policy for Animal Protection RD1201/05, which meets the European Union Directive 86/609 on the protection of animals used for experimental and other scientific purposes.

Sperm was collected from three mature Rasa Aragonesa rams, in the breeding season (October–December) in a twice a week seminal collection regime using an artificial vagina.

Semen quality was previously tested in our laboratory, selecting samples in which all sperm parameters were in the normal ranges for ram semen (motility >85%; total morphology abnormalities <15%).

In each replica, ejaculates from three rams were pooled and divided into three equal fractions including control (cryopreserved) and two test groups (freeze-dried). Fraction to freeze was suspended in Andromed® extender to reach a final semen concentration of 100 × $10^6$and cooled to 4 °C in 2 h. Then semen samples were packed in 0.25 ml plastic straws and placed 4 cm above the level of liquid nitrogen for 10 min. Finally straws were plunged into liquid nitrogen to store at −196 °C. The fractions to freeze-dry were suspended in the freeze-drying media EGTA or EGTAR to a concentration of 150 × 10^6/mL and kept at 37 °C for 10 min. 150 µl of sperm suspension from each group was placed into individual cryo tubes (Labcon North America, USA) and plunged into liquid nitrogen for 5 min. Then samples were transferred to a programmable freeze-dryer (LyoBeta 25, Telstar, USA) previously cooled to −50 °C. Samples from the two groups were submitted to a freeze-dying process under the same conditions: a primary drying at a pressure of 0.053 mbar and temperature of −68 °C, and a secondary drying at a pressure of 0.018 mbar and a temperature of 20 °C. Finally, the vials were stored at two different temperatures: 4 °C and at room temperature, for 1 year until used for DNA evaluation and ICSI.

Sperm preparation and DNA fragmentation analysis

Frozen samples were thawed in a water bath at 37 °C during 21 s for DNA fragmentation analysis. Freeze-dried sperm samples were rehydrated by adding 150 µl Milli-Q water. The sperm suspension was centrifuged for 2 min at 600 g and the supernatant was removed. The sperm pellet was resuspended in 500 µl phosphate buffered saline (PBS) for DNA fragmentation analysis.

Sperm from different treatments were evaluated by the Sperm Chromatin Dispersion test (SCD) specifically designed for ram spermatozoa (Ovis-Halomax®). The DNA fragmentation analysis in all groups was performed following the manufacturer's instructions. In brief, the lysis solution was placed at room temperature (22 °C). Then, an eppendorf tube containing agarose was placed in a water bath at 95 °C–100 °C for five minutes, and then transferred in a water bath at 37 °C for five minutes. Meanwhile, 25 µl of each diluted sperm sample was added to an empty eppendorf tube, and 50 ml of liquefied agarose was then transferred into the tube and gently mixed. The temperature of the tubes was maintained at 37 °C. Then, a drop of 2 µl of the cell suspension was placed onto marked wells and each drop was covered with a 24 × 24 mm glass coverslip. The slides were held in a horizontal position throughout the entire process. The slides were placed on a cold surface precooled at 4 °C in a fridge to solidify the agarose. After 5 min the slides were taken out of the fridge and the coverslips were gently removed. Then, the slides were fully immersed horizontally in 10 ml of lysis solution for five minutes. Subsequently, the preparation was introduced into a bath of distilled water for 5 min and then dehydrated by immersion in 2 successive baths of ethanol at 70% and 100% for 2 min each. Finally, the slides were allowed to air-dry before staining. All the slides were stained using a commercial kit for green fluorescence staining (Fluogreen, Halotech DNA SL, Spain). 2 μl of green fluorochrome and mountain medium (1:1; vol/vol) was placed into the well of slide for fluorescent staining of sperm chromatin. The samples were evaluated using fluorescent microscopy (Olympus BX-40, Olympus U-RFL-T, Tokyo, Japan) at magnification 400X and a minimum of 300 spermatozoa were counted per semen sample. Five replicated trials were carried out for each group. Sperm showing a small and compacted halo around a compacted nuclear core contained intact DNA and sperm that displayed a large and spotty halo around the nuclear core corresponded to those sperm with fragmented DNA.

Oocyte collection and *in vitro* maturation

Oocytes were collected from ovaries of adult sheep at a local slaughterhouse and transported to the laboratory at 35 °C in PBS medium within 1 h. After washing three times the ovaries with saline solution, follicles 2–6 mm in diameter were aspirated. COCs were selected and placed into 50 μl droplets of maturation medium for 24 h at 38.5 °C in a humidified 5% CO_2 atmosphere under mineral oil. After maturation, oocytes were freed from cumulus cells by gentle aspiration with a pipette in maturation medium supplemented with 0.1% (w/v) hyaluronidase. Denuded oocytes with the first polar body (PB) were harvested under a stereomicroscope and used for ICSI.

Intracytoplasmic sperm injection

After rehydration or thawed, sperm samples were suspended in BHSOF medium and kept at room temperature for no longer than 2 hours during ICSI. Microinjection was performed at 37 °C with the micro injector (Integra TI). The cumulus-free oocytes were placed in 10 μl drops of BHSOF medium on the cover of a plastic dish. 2 μl of sperm suspension was transferred to a drop of 5 μl of 7% polyvinylpyrrolidone, which was prepared close to the drops containing the oocytes. All drops were covered with paraffin oil. A single sperm in the suspension was aspirated from its tail into the injection pipette, and the pipette was moved to the drop containing the oocyte. The sperm was injected into the ooplasm. The first polar body was either in the 6 or 12 o'clock position, and the injection pipette was in the 3 o'clock position. During the injection, cytoplasm was aspirated to approve that the oolema was broken. This process was repeated until all the oocytes had been injected. The parthenogenetic control was performed by sham injections in a similar manner. And as control group, oocytes were injected with frozen-thawed sperm. Injected oocytes were washed three times with BHSOF and recovered in this medium before activation (5% CO_2 in air at 38.5 °C). Within 1 h after injection, the oocytes in all groups were activated by adding 5 µM of Ca ionophore for 5 min. After activation, the sperm-injected oocytes were washed three times and *in-vitro* cultured for 6 days (5% CO_2 in air at 38.5 °C). 16 h after ICSI putative zygotes were fixed for 30 min (0.5% glutaraldehyde in DPBS), stained for 15 min (1% Hoechst 33342 in DPBS), washed in DPBS and examined under an epifluorescence microscope to evaluate the nuclear stage. The criteria used for defining fertilization versus parthenogenetic activation was the presence of two polar bodies and both female and male pronuclei. The cleavage rate and embryo development (compact morulae and blastocyst) were determined at 48 and 144 hours respectively, using an inverted microscope (Nikon Eclipse TE2000-S).

Statistical analysis

The statistical analysis was performed using SPSS, version 17.0 for Windows. Data concerning DNA fragmentation, pronuclear formation and embryo development were expressed in percentages and analyzed using chi-squared test. Differences were considered significant if $P < 0.05$.

Additional Information

Publisher's note: Springer Nature remains neutral with regard to jurisdictional claims in published maps and institutional affiliations.

ACKNOWLEDGEMENTS

The authors would like to thank Gonzalo Romeo, chief of freeze-drying service of the Animal Experimentation Service of Veterinary Faculty of

Zaragoza, and Mercazaragoza (Zaragoza, Spain) for providing the ovaries. This study was supported by Government of Aragon Research Groups (Fondo Social Europeo, DGA) and IA_2.

REFERENCES

1. Liu, J. L. et al. Freeze-dried sperm fertilization leads to full-term development in rabbits. Biol. Reprod. 70, 1776–1781 (2004).
2. Choi, Y. H., Varner, D. D., Love, C. C., Hartman, D. L. & Hinrichs, K. Production of live foals via intracytoplasmic injection of lyophilized sperm and sperm extract in the horse. Reproduction 142, 529–538 (2011).
3. Olaciregui, M., Luño, V., Marti, J. I., Aramayona, J. & Gil, L. Freeze-dried stallion spermatozoa: evaluation of two chelating agents and comparative analysis of three sperm DNA damage assays. Andrología 48(9), 900–906 (2016).
4. Ringleb, J., Waurich, R., Wibbel, G., Streich, W. J. & Jewgenow, K. Prolonged storage of epididymal spermatozoa does not affect their capacity to fertilise in vitromatured domestic cat (Felis catus) oocytes when using ICSI. Reprod. Fertil. Dev. 23, 818–825 (2011).
5. Watanabe, H., Asano, T. Y., Abe, Y., Fukui, Y. & Suzuki, H. Pronuclear formation of freeze-dried canine spermatozoa microinjected into mouse oocytes. J. Assist. Reprod. Genet. 26, 531–536 (2009).
6. Olaciregui, M., Luño, V., Gonzalez, N., de Blas, I. & Gil, L. Freeze-dried dog sperm: Dynamics of DNA integrity. Cryobiology 71, 286–290 (2015).
7. Men, N. et al. Expressions of DNA repair genes in porcine oocytes before and after fertilization by ICSI using freeze-dried sperm. Anim Sci J. 87(11), 1325–1333 (2016).
8. Hara, H. et al. Adverse effect of cake collapse on the functional integrity of freeze-dried bull spermatozoa. Cryobiology 68(3), 354–360 (2014).
9. Kwon, I. K., Park, K. E. & Niwa, K. Activation, pronuclear formation, and development in vitroof pig oocytes following intracytoplasmic injection of freeze-dried spermatozoa. Biol Reprod 71, 1430–1436 (2004).
10. Kusakabe, H., Szczygiel, M. A., Whittingham, D. G. & Yanagimachi, R. Maintenance of genetic integrity in frozen and freeze-dried mouse spermatozoa. Proc Natl Acad Sci USA 98, 13501–13506 (2001).
11. Kusakabe, H., Yanagimachi, R. & Kamiguchi, Y. Mouse and human spermatozoa can be freeze-dried without damaging their chromosomes. Hum. Reprod. 23(2), 233–239 (2008).
12. Twigg, J. P., Irvine, D. S. & Aitken, R. J. Oxidative damage to DNA in human spermatozoa does not preclude pronucleus formation at intracytoplasmic sperm injection. Hum. Reprod 13, 1864–1871 (1998).
13. López-Fernández, C. et al. Dynamics of sperm DNA fragmentation in domestic animals III: Ram. Theriogenology 70, 898–908 (2008).
14. Malo, C. et al. Antioxidant supplementation improves boar sperm characteristics and fertility after cryopreservation: comparison between cysteine and rosemary (Rosmarinus officinalis). Cryobiology 61, 142–147 (2010).
15. Agarwal, A., Aponte-Mellado, A., Premkumar, B. J., Shaman, A. & Gupta, S. The effects of oxidative stress on female reproduction: a review. Reprod. Biol. Endocrinol. 29, 10–49 (2012).
16. Zhang, W., Yi, K., Chen, C., Hou, X. & Zhou, X. Application of antioxidants and centrifugation for cryopreservation of boar spermatozoa. Anim Reprod Sci 132(3–4), 123–128 (2012).

17. Zanganeh, Z. et al. Does rosemary aqueous extract improve buck semen cryopreservation? Small Rumin Res. 114, 120–125 (2013).
18. Zalata, A. et al. In vitrostudy of cypermethrin on human spermatozoa and the possible protective role of vitamins C and E. Andrologia 46(10), 1141–1147 (2014).
19. Sariozkan, S. et al. Antioxidative effects of cysteamine, hyaluronan and fetuin on post-thaw semen quality, DNA integrity and oxidative stress parameters in the Brown Swiss bull. Andrologia 47(2), 138–147 (2015).
20. Luño, V., Gil, L., Olaciregui, M., González, N., Jerez., R. & de Blas, I. Rosmarinic acid improves function and in vitrofertilising ability of boar sperm after cryopreservation. Cryobiology 69, 157–162 (2014).
21. Oteiza, P. I., Erlejman, A. G., Verstraeten, S. V., Keen, C. L. & Fraga, C. G. Flavonoid–membrane interactions: a protective role of flavonoids at the membrane surface? Clin Dev. Immunol. 12, 19–25 (2005).
22. Martins, C. F., Bao, S. N., Dodea, M. N., Correa, G. A. & Rumpf, R. Effects of freeze-drying on cytology, ultrastructure, DNA fragmentation, and fertilizing ability of bovine sperm. Theriogenology 67, 1307–1315 (2007).
23. Kaneko, T. & Serikawa, T. Long-term preservation of freeze-dried mouse spermatozoa. Cryobiology 64, 211–214 (2012).
24. Kaneko., T. & Serikawa, T. Successful long-term preservation of rat sperm by freeze-drying. PLoS ONE 7(4), e35043,
25. Kaneko, T. & Nakagata, N. Relation between storage temperature and fertilizing ability of freeze-dried mouse spermatozoa. Comp. Med. 55, 140–144 (2005).
26. Kaneko, T. Sperm freeze-drying and micro-insemination for biobanking and maintenance of genetic diversity in mammals. Reprod. Fertil. Dev. 2016 10.1071/RD15386.
27. Men, N. T. et al. Effect of trehalose on DNA integrity of freeze-dried boar sperm, fertilization, and embryo development after intracytoplasmic sperm injection. Theriogenology 80, 1033–1044 (2013).
28. Thomson, L. K. et al. Cryopreservation-induced human sperm DNA damage is predominantly mediated by oxidative stress rather than apoptosis. Hum. Reprod 24(9), 2061–2070 (2009).
29. Foote, R. H., Brockett, C. C. & Kaproth, M. T. Motility and fertility of bull sperm in whole milk extender containing antioxidants. Anim Reprod Sci 71, 13–23 (2002).
30. Funahashi, H. & Sano, T. Select antioxidants improve the function of extended boar semen stored at 10 8C. Theriogenology 63, 605–616 (2005).
31. Maxwell, W. M. C. & Stojanov, T. Liquid storage of ram semen in the absence or presence of some antioxidants. Reprod. Fertil. Dev. 8, 1013–1020 (1996).
32. Henkel, R. et al. DNA fragmentation of spermatozoa and assisted reproduction technology. Reprod. Biomed. 7, 477–484 (2003).
33. Henkel, R. et al. Influence of deoxyribonucleic acid damage on fertilization and pregnancy. Fertil. Steril. 81, 965–972 (2004).
34. Malo, C., Gil, L., Cano, R., Martínez, F. & Galé, I. Antioxidant effect of rosemary (Rosmarinus officinalis) on boar epididymal spermatozoa during cryopreservation. Theriogenology 75, 1735–1741 (2011).
35. Mata-Campuzanoa, M. et al. Refrigerated storage of ram sperm in presence of Trolox and GSH antioxidants: Effect of temperature, extender andstorage time. Anim. Reprod.Sci. 151, 137–147 (2014).
36. Kodama, H., Yamaguchi, R., Fukuda, J., Kasai, H. & Tanaka, T. Increased oxidative deoxyribonucleic acid damage in the spermatozoa of infertile male patients. Fertil. Steril. 68, 519–524 (1997).
37. Oteiza, P. I., Erlejman, A. G., Verstraeten, S. V., Keen, C. L. & Fraga, C. G. Flavonoid–membrane interactions: a protective role of flavonoids at the membrane surface? Clin Dev. Immunol. 12, 19–25 (2005).
38. Pedersen, R. A. & Brandriff, B. Radiation- and drug-induced DNA repair in mammalian oocytes and embryos. Basic Life Sci. 15, 389–410 (1980).

39. Roest, H. P. The ubiquitin-conjugating DNA repair enzyme HR6A is a maternal factor essential for early embryonic development in mice. Mol Cell Biol. 24, 5485–5495 (2004).
40. Zini, A. et al. Potential adverse effect of sperm DNA damage on embryo quality after ICSI. Humanit Reprod 20, 3476–3480 (2005).
41. Wossidlo., M. et al. Dynamic link of DNA demethylation, DNA strand breaks and repair in mouse zygotes. Embo J. 29, 1877–1888 (2010).
42. Jaroudi, S. et al. Expression profiling of DNA repair genes in human oocytes and blastocysts using microarrays. Hum. Reprod. 24, 2649–2655 (2009).
43. Zheng, P., Schramm, R. D. & Latham, K. E. Developmental regulation and in vitroculture effects on expression of DNA repair and cell cycle checkpoint control genes in rhesus monkey oocytes and embryos. Biol. Reprod. 72, 1359–1369 (2005).
44. Jones, G. M. et al. Gene expression profiling of human oocytes following in vivoor in vitromaturation. Hum.Reprod. 23, 1138–1144 (2008).
45. Salhab, M. et al. In vitromaturation of oocytes alters gene expression and signaling pathways in bovine cumulus cells. Mol. Reprod. Develop. 80, 166–182 (2013).
46. Enciso, M., Sarasa, J., Agarwal, A., Fernández, J. L. & Gosálvez, J. A two-tailed Comet assay for assessing DNA damage in spermatozoa. Reprod. Biomed. Online 18, 609–616 (2009).
47. Harrouk, W., Codrington, A., Vinson, R., Robaire, B. & Hales, B. F. Paternal exposure to cyclophosphamide induces DNA damage and alters the expression of DNA repair genes in the rat preimplantation embryo. Mutat. Res. 461, 229–241 (2000).
48. Su, T. T., Walker, J. & Stumpff, J. Activating the DNA damage checkpoint in a developmental context. Curr Biol 10, 119–126 (2000).
49. Wells, D. et al. Expression of genes regulating chromosome segregation, the cell cycle and apoptosis during human preimplantation development. Hum Reprod. 20, 1339–1348 (2005).
50. Bazrgar, M. et al. DNA repair signalling pathway genes are overexpressed in poor-quality pre-implantation human embryos with complex aneuploidy. Eur J Obstet Gynecol Reprod Biol 175, 152–156 (2014).
51. Ivec, M., Kovacic, B. & Vlaisavljevic, V. Prediction of human blastocyst development from morulas with delayed and/or incomplete compaction. Fertil Steril. 96, 1473–1478 (2011).
52. Wakayama, T. & Yanagimachi, R. Development of normal mice from oocytes injected with freeze-dried spermatozoa. Nat. Biotechnol. 16, 639–641 (1998).
53. Hochi, S., Watanabe, K., Kato, M. & Hirabayashi, M. Live rats resulting from microinjection of oocytes with spermatozoa freeze-dried and stored for one year. Mol. Reprod. Dev. 70, 1776–1781 (2008).
54. Kaneko, T. & Nakagata, N. Relation between storage temperature and fertilizing ability of freeze-dried mouse spermatozoa. Comp. Med. 55, 140–144 (2005).
55. Kawase, Y., Hani, T., Kamada, N., Jishage, K. & Suzuki, H. Effect of pressure at primary drying of freeze-drying mouse sperm reproduction ability and preservation potential. Reproduction 133, 841–846 (2007).
56. Kaneko, T. Preservation of mammalian sperm by freeze-drying. Cryobiology 69(3), 510 (2014).
57. Kaneko, T. Simple sperm preservation by freeze-drying for conserving animal strains. Methods Mol Biol. 239, 317–329,

Index